普通高等教育"九五"国家教委重点教材

北京高等教育精品教材
BEIJING GAODENG JIAOYU JINGPIN JIAOCAI

教育部科技进步一等奖
国家科技进步三等奖

累计发行200万册

清华大学 计算机系列教材

沈美明　温冬婵　编著

IBM-PC汇编语言程序设计
（第2版）

清华大学出版社
北京

内 容 简 介

本书主要阐述 80x86 汇编语言程序设计方法和技术。全书共分四部分：第 1 章和第 2 章为基础知识部分；第 3 章和第 4 章为编程工具部分，主要内容为 80x86 的指令系统与寻址方式，以及包括伪操作在内的汇编语言程序格式；第 5 章～第 9 章说明编程方法，内容包括循环、分支、子程序等基本程序结构，程序设计的基本方法和技术，多模块连接技术，宏汇编技术，以中断为主的输入输出程序设计方法，以及 BIOS 和 DOS 系统功能调用；第 10 章和第 11 章为实际应用部分，说明图形显示、发声和磁盘文件存取技术。全书提供了大量程序实例，每章后均附有习题。

本书是在 1991 年第 1 版的基础上，融会了 10 年来教学与科研的新成果改编成的。其第 1 版曾先后获得过原电子工业部工科电子类专业优秀教材一等奖、教育部科技进步一等奖、国家科技进步三等奖等；受到广大读者的欢迎，累计发行达 130 多万册。

本书适用于高等院校以及大、中专院校作为"汇编语言程序设计"课程的教材。本书也适于初学者使用，只要具有一种高级语言程序设计基础的读者，都可通过学习本书掌握汇编语言程序设计技术。

本书封面贴有清华大学出版社防伪标签，无标签者不得销售。
版权所有，侵权必究。举报：010-62782989，beiqinquan@tup.tsinghua.edu.cn。

图书在版编目（CIP）数据

IBM PC 汇编语言程序设计/沈美明，温冬婵编著. —2 版. —北京：清华大学出版社，2001（2024.12重印）
（清华大学计算机系列教材）
ISBN 978-7-302-04664-6

Ⅰ. Ⅰ… Ⅱ. ①沈… ②温… Ⅲ. 汇编语言－程序设计－高等学校－教材 Ⅳ. ①TP313
中国版本图书馆 CIP 数据核字（2007）第 046511 号

责任编辑：白立军
责任印制：宋　林

出版发行：清华大学出版社
网　　址：https://www.tup.com.cn，https://www.wqxuetang.com
地　　址：北京清华大学学研大厦 A 座　　　　邮　编：100084
社 总 机：010-83470000　　　　　　　　　　　邮　购：010-62786544
投稿与读者服务：010-62776969，c-service@tup.tsinghua.edu.cn
质量反馈：010-62772015，zhiliang@tup.tsinghua.edu.cn

印 装 者：三河市铭诚印务有限公司
经　　销：全国新华书店
开　　本：185mm×260mm　　　印　张：31　　　字　数：711 千字
印　　次：2024 年 12 月第 49 次印刷
定　　价：79.00 元

产品编号：004664-07/TP

作者简历

沈美明　1959年毕业于清华大学自动控制系计算机专业，留校任教至今。现为清华大学计算机系教授、博士生导师。多年来一直从事"汇编语言程序设计"课程的教学工作以及并行与分布计算机系统、并行程序开发环境等方面的科研工作。先后主持多项国家"863"计划、国防科技预研和自然科学基金项目等。近年来和合作者一起在国内外刊物和学术会议上发表论文80余篇。曾获国家科技进步三等奖，并多次获部委级科技进步奖。

温冬婵　1970年毕业于清华大学自动控制系，现为清华大学计算机系副主任、教授。多年来一直承担"汇编语言程序设计"课程的教学工作。科研方向为并行与分布计算机系统、并行编译技术等。参加了多项国家"863"计划、国防预研和自然科学基金项目等，曾多次获得国家部委科技进步奖，所编写的教材获得部委一等奖和国家科技进步三等奖。

近年来，与他人合著学术专著一部，编著和翻译著作9本，在国内外刊物和学术会议上发表论文50余篇。

序

清华大学计算机系列教材已经出版发行了近30种,包括计算机专业的基础数学、专业技术基础和专业等课程的教材,覆盖了计算机专业大学本科和研究生的主要教学内容。这是一批至今发行数量很大并赢得广大读者赞誉的书籍,是近年来出版的大学计算机教材中影响比较大的一批精品。

本系列教材的作者都是我熟悉的教授与同事,他们长期在第一线担任相关课程的教学工作,是一批很受大学生和研究生欢迎的任课教师。编写高质量的大学(研究生)计算机教材,不仅需要作者具备丰富的教学经验和科研实践,还需要对相关领域科技发展前沿的正确把握和了解。正因为本系列教材的作者们具备了这些条件,才有了这批高质量优秀教材的出版。可以说,教材是他们长期辛勤工作的结晶。本系列教材出版发行以来,从其发行的数量、读者的反映、已经获得的许多国家级与省部级的奖励、以及在各个高等院校教学中所发挥的作用上,都可以看出本系列教材所产生的社会影响与效益。

计算机科技发展异常迅速、内容更新很快。作为教材,一方面要反映本领域基础性、普遍性的知识,保持内容的相对稳定性;另一方面,又需要跟踪科技的发展,及时地调整和更新内容。本系列教材都能按照自身的需要及时地做到了这一点,如《计算机组成与结构》一书十年中共出版了三版,其他如《数据结构》等也都已出版了第二版,使教材既保持了稳定性,又达到了先进性的要求。本系列教材内容丰富、体系结构严谨、概念清晰、易学易懂,符合学生的认识规律,适合于教学与自学,深受广大读者的欢迎。系列教材中多数配有丰富的习题集和实验,有的还配备多媒体电子教案,便于学生理论联系实际地学习相关课程。

随着我国进一步的开放,我们需要扩大国际交流,加强学习国外的先进经验。在大学教材建设上,我们也应该注意学习和引进国外的先进教材。但是,计算机系列教材的出版发行实践以及它所取得的效果告诉我们,在当前形势下,编写符合国情的具有自主版权的高质量教材仍具有重大意义和价值。它与前者不仅不矛盾,而且是相辅相成的。本系列教材的出版还表明,针对某个学科培养的要求,在教育部等上级部门的指导下,有计划地组织任课教师编写系列教材,还能促进对该学科科学、合理的教学体系和内容的研究。

我希望今后有更多、更好的我国优秀教材出版。

清华大学计算机系教授,中科院院士
张钹

再 版 前 言

汇编语言是计算机能够提供给用户使用的最快而又最有效的语言,也是能够利用计算机所有硬件特性并能直接控制硬件的惟一语言。因而,对程序的空间和时间要求很高的场合,汇编语言的应用是必不可少的。至于很多需要直接控制硬件的应用场合,则更是非用汇编语言不可了。

《IBM-PC 汇编语言程序设计》(第 2 版)是高等院校计算机科学与技术专业"汇编语言程序设计"必修课所用教材。它的第 1 版(1991 年版)曾被评为 1992 年第四届全国科技类优秀畅销书;获得 1996 年电子工业部第三届工科电子类专业优秀教材一等奖;1999年教育部科技进步一等奖;以及 1999 年国家科技进步三等奖。

在本书的第一版中,我们选用了以 8086 为 CPU 的 PC 机作为基础机型来组织教学。第 2 版是在第 1 版的基础上增加了有关当今计算机技术发展的新内容,其中包括 8086 后继机型(80x86)所提供的指令及寻址方式;汇编程序 MASM 新版本所提供的伪操作及高级汇编语言技术;保护模式的编程基础等,以便满足广大读者使用高档微机的需要。由于第 2 版内容的进一步充实,因此书的篇幅也有明显增加。为照顾不同需求,本书对第 1 版中某些部分作了删减,以便突出重点,达到课程的基本要求。本书可供高等院校及大、中专院校作为"汇编语言程序设计"课程的教材使用。但本书也适合于初学者使用,只要有一种高级语言程序设计基础,都可以通过学习本书掌握汇编语言程序设计技术。

全书由"基础理论、编程工具、编程方法和实际应用"四部分(共 11 章)组成。第 1 章和第 2 章为基础理论部分。包括数制、码制等基础知识,计算机组成及基本原理。第 3 章和第 4 章介绍编程工具。包括指令系统、寻址方式、伪操作和汇编语言格式。第 5 章~第9 章说明编程方法。包括循环、分支、子程序等基本程序结构,宏汇编技术,中断等输入输出程序设计方法,BIOS 和 DOS 系统功能调用方法,以及多个模块的连接技术。第 10 章和第 11 章为实际应用。包括图形显示、发声和磁盘文件存取技术。这四个组成部分构成一个完整的系统。书中提供了大量程序例题,每章之后均附有若干习题,便于读者复习及检查学习效果。同时为了能适应各种类型院校的不同要求,各章之间相互配合而又自成体系,易于为不同类型院校按其要求适当加以裁剪,所以本教材的适用面比较宽。

本书为清华大学计算机科学与技术系"汇编语言程序设计"课程的教材。该课程课内80 学时,其中讲课 48 学时,上机实践 32 学时,课内外学时比例为 1∶1.5。讲课内容为第1 章~第 9 章,第 10 章和第 11 章则结合实验由学生自学并上机。选用本教材的各校均可根据教学计划规定的学时灵活安排。为便于查阅,本书把指令系统集中在第 3 章,因此所占篇幅较大。在讲课过程中,为使学生尽可能早些上机,开始编程训练,可把有关指令分散到其后各章讲述。例如,把转移类指令放在循环与分支程序设计一章,把转子与返回指令放在子程序结构一章,把中断指令放在输入输出程序设计一章,等等。课程的上机安排可参考与本书配套的《IBM-PC 汇编语言程序设计实验教程》一书,根据课程上机时数及学生的水平,选用相应的实验。

本书的第 1 章～第 7 章由沈美明编写,第 8 章～第 11 章由温冬婵编写。书中如有错误或不当之处,欢迎读者不吝批评指正。

编著者

2001 年 8 月

目 录

再版前言 …………………………………………………………………………… I

第 1 章 基础知识 ………………………………………………………………… 1
1.1 进位记数制与不同基数的数之间的转换 …………………………………… 1
1.1.1 二进制数 ………………………………………………………………… 1
1.1.2 二进制数和十进制数之间的转换 ……………………………………… 2
1.1.3 十六进制数及其与二进制、十进制数之间的转换 …………………… 4
1.2 二进制数和十六进制数运算 …………………………………………………… 6
1.2.1 二进制数运算 …………………………………………………………… 6
1.2.2 十六进制数运算 ………………………………………………………… 6
1.3 计算机中数和字符的表示 ……………………………………………………… 7
1.3.1 数的补码表示 …………………………………………………………… 7
1.3.2 补码的加法和减法 ……………………………………………………… 9
1.3.3 无符号整数 ……………………………………………………………… 11
1.3.4 字符表示法 ……………………………………………………………… 11
1.4 几种基本的逻辑运算 …………………………………………………………… 12
1.4.1 "与"运算(AND) ……………………………………………………… 12
1.4.2 "或"运算(OR) ………………………………………………………… 13
1.4.3 "非"运算(NOT) ……………………………………………………… 13
1.4.4 "异或"运算(XOR Exclusive-OR) ………………………………… 13
习题 …………………………………………………………………………………… 14

第 2 章 80x86 计算机组织 ……………………………………………………… 15
2.1 80x86 微处理器 ………………………………………………………………… 15
2.2 基于微处理器的计算机系统构成 ……………………………………………… 17
2.2.1 硬件 ……………………………………………………………………… 17
2.2.2 软件 ……………………………………………………………………… 18
2.3 中央处理机 ……………………………………………………………………… 19
2.3.1 中央处理机 CPU 的组成 ……………………………………………… 19
2.3.2 80x86 寄存器组 ………………………………………………………… 20
2.4 存储器 …………………………………………………………………………… 24
2.4.1 存储单元的地址和内容 ………………………………………………… 24
2.4.2 实模式存储器寻址 ……………………………………………………… 26
2.4.3 保护模式存储器寻址 …………………………………………………… 30

2.5 外部设备 ··· 31
  习题 ·· 33

第 3 章 80x86 的指令系统和寻址方式 ································ 35
  3.1 80x86 的寻址方式 ·· 36
    3.1.1 与数据有关的寻址方式 ···································· 36
    3.1.2 与转移地址有关的寻址方式 ································ 44
  3.2 程序占有的空间和执行时间 ·· 46
  3.3 80x86 的指令系统 ·· 47
    3.3.1 数据传送指令 ·· 47
    3.3.2 算术指令 ·· 58
    3.3.3 逻辑指令 ·· 68
    3.3.4 串处理指令 ·· 75
    3.3.5 控制转移指令 ·· 85
    3.3.6 处理机控制与杂项操作指令 ································ 104
  习题 ·· 107

第 4 章 汇编语言程序格式 ·· 117
  4.1 汇编程序功能 ··· 117
  4.2 伪操作 ·· 118
    4.2.1 处理器选择伪操作 ·· 118
    4.2.2 段定义伪操作 ·· 118
    4.2.3 程序开始和结束伪操作 ···································· 126
    4.2.4 数据定义及存储器分配伪操作 ······························ 127
    4.2.5 表达式赋值伪操作 EQU ···································· 133
    4.2.6 地址计数器与对准伪操作 ·································· 134
    4.2.7 基数控制伪操作 ·· 136
  4.3 汇编语言程序格式 ·· 137
    4.3.1 名字项 ·· 137
    4.3.2 操作项 ·· 138
    4.3.3 操作数项 ·· 138
    4.3.4 注释项 ·· 144
  4.4 汇编语言程序的上机过程 ·· 146
    4.4.1 建立汇编语言的工作环境 ·································· 146
    4.4.2 建立 ASM 文件 ·· 146
    4.4.3 用 MASM 程序产生 OBJ 文件 ································ 147
    4.4.4 用 LINK 程序产生 EXE 文件 ································ 152
    4.4.5 程序的执行 ·· 153
    4.4.6 COM 文件 ·· 153

习题 155

第 5 章 循环与分支程序设计 160

5.1 循环程序设计 160
- 5.1.1 循环程序的结构形式 160
- 5.1.2 循环程序设计方法 161
- 5.1.3 多重循环程序设计 172

5.2 分支程序设计 176
- 5.2.1 分支程序的结构形式 176
- 5.2.2 分支程序设计方法 176
- 5.2.3 跳跃表法 180

5.3 如何在实模式下发挥 80386 及其后继机型的优势 183
- 5.3.1 充分利用高档机的 32 位字长特性 184
- 5.3.2 通用寄存器可作为指针寄存器 187
- 5.3.3 与比例因子有关的寻址方式 188
- 5.3.4 各种机型提供的新指令 191

习题 193

第 6 章 子程序结构 196

6.1 子程序的设计方法 196
- 6.1.1 过程定义伪操作 196
- 6.1.2 子程序的调用和返回 198
- 6.1.3 保存与恢复寄存器 198
- 6.1.4 子程序的参数传送 199
- 6.1.5 增强功能的过程定义伪操作 217

6.2 子程序的嵌套 224

6.3 子程序举例 225

习题 240

第 7 章 高级汇编语言技术 246

7.1 宏汇编 246
- 7.1.1 宏定义、宏调用和宏展开 246
- 7.1.2 宏定义中的参数 249
- 7.1.3 LOCAL 伪操作 252
- 7.1.4 在宏定义内使用宏 253
- 7.1.5 列表伪操作 255
- 7.1.6 宏库的建立与调用 258
- 7.1.7 PURGE 伪操作 261

7.2 重复汇编 261
- 7.2.1 重复伪操作 262

7.2.2 不定重复伪操作 …… 264
7.3 条件汇编 …… 265
7.3.1 条件伪操作 IF 的使用举例 …… 266
7.3.2 条件伪操作 IF1 的使用举例 …… 268
7.3.3 条件伪操作 IFNDEF 的使用举例 …… 270
7.3.4 条件伪操作 IFB 的使用举例 …… 274
7.3.5 条件伪操作 IFIDN 的使用举例 …… 276
习题 …… 278

第 8 章 输入输出程序设计 …… 282
8.1 I/O 设备的数据传送方式 …… 282
8.1.1 CPU 与外设 …… 282
8.1.2 直接存储器存取(DMA)方式 …… 282
8.2 程序直接控制 I/O 方式 …… 283
8.2.1 I/O 端口 …… 283
8.2.2 I/O 指令 …… 284
8.2.3 I/O 程序举例 …… 285
8.3 中断传送方式 …… 289
8.3.1 8086 的中断分类 …… 290
8.3.2 中断向量表 …… 293
8.3.3 中断过程 …… 296
8.3.4 中断优先级和中断嵌套 …… 297
8.3.5 中断处理程序 …… 299
习题 …… 313

第 9 章 BIOS 和 DOS 中断 …… 315
9.1 键盘 I/O …… 316
9.1.1 字符码与扫描码 …… 317
9.1.2 BIOS 键盘中断 …… 318
9.1.3 DOS 键盘功能调用 …… 319
9.2 显示器 I/O …… 324
9.2.1 字符属性 …… 324
9.2.2 BIOS 显示中断 …… 327
9.2.3 DOS 显示功能调用 …… 335
9.3 打印机 I/O …… 336
9.3.1 DOS 打印功能 …… 337
9.3.2 打印机的控制字符 …… 338
9.3.3 BIOS 打印功能 …… 342
9.4 串行通信口 I/O …… 345

  9.4.1 串行通信接口 ……………………………………………… 346
  9.4.2 串行口功能调用 ……………………………………………… 348
 习题 ……………………………………………………………………………… 353

第 10 章 图形与发声系统的程序设计 …………………………………… 355
 10.1 显示方式 …………………………………………………………… 355
  10.1.1 显示分辨率 …………………………………………………… 355
  10.1.2 BIOS 设置显示方式 ………………………………………… 356
 10.2 视频显示存储器 ……………………………………………………… 359
  10.2.1 图形存储器映像 ……………………………………………… 359
  10.2.2 数据到颜色的转换 …………………………………………… 361
  10.2.3 直接视频显示 ………………………………………………… 363
 10.3 EGA/VGA 图形程序设计 …………………………………………… 367
  10.3.1 读写像素 ……………………………………………………… 367
  10.3.2 图形方式下的文本显示 ……………………………………… 373
  10.3.3 彩色绘图程序 ………………………………………………… 376
  10.3.4 动画显示技术 ………………………………………………… 381
 10.4 通用发声程序 ………………………………………………………… 383
  10.4.1 可编程时间间隔定时器 8253/54 …………………………… 384
  10.4.2 扬声器驱动方式 ……………………………………………… 387
  10.4.3 通用发声程序 ………………………………………………… 388
  10.4.4 80x86 PC 的时间延迟 ……………………………………… 390
 10.5 乐曲程序 …………………………………………………………… 392
  10.5.1 音调与频率和时间的关系 …………………………………… 392
  10.5.2 演奏乐曲的程序 ……………………………………………… 393
  10.5.3 键盘控制发声程序 …………………………………………… 396
 习题 ……………………………………………………………………………… 397

第 11 章 磁盘文件存取技术 ………………………………………………… 400
 11.1 磁盘的记录方式 ……………………………………………………… 400
  11.1.1 磁盘记录信息的地址 ………………………………………… 400
  11.1.2 磁盘系统区和数据区 ………………………………………… 402
  11.1.3 磁盘目录及文件分配表 ……………………………………… 402
 11.2 文件代号式磁盘存取 ………………………………………………… 404
  11.2.1 路径名和 ASCIZ 串 ………………………………………… 405
  11.2.2 文件代号和错误返回代码 …………………………………… 406
  11.2.3 文件属性 ……………………………………………………… 407
  11.2.4 写磁盘文件 …………………………………………………… 408
  11.2.5 读磁盘文件 …………………………………………………… 413

  11.2.6　移动读写指针 …………………………………… 417
 11.3　字符设备的文件代号式 I/O ………………………………… 423
 11.4　BIOS 磁盘存取功能 …………………………………………… 427
  11.4.1　BIOS 磁盘操作 ……………………………………… 427
  11.4.2　状态字节 ……………………………………………… 429
  11.4.3　BIOS 磁盘操作举例 ………………………………… 430
 习题 …………………………………………………………………… 433

附录 ……………………………………………………………………… 435
 附录1　80x86 指令系统一览 …………………………………… 435
 附录2　伪操作与操作符 ………………………………………… 454
 附录3　中断向量地址一览 ……………………………………… 469
 附录4　DOS 系统功能调用(INT 21H) ………………………… 471
 附录5　BIOS 功能调用 …………………………………………… 477

参考文献 ………………………………………………………………… 482

第1章 基础知识

1.1 进位记数制与不同基数的数之间的转换

1.1.1 二进制数

进位记数制是一种计数的方法,习惯上最常用的是十进制记数法。一个任意的十进制可以表示为:

$$a_n a_{n-1} \cdots a_0 \cdot b_1 b_2 \cdots b_m$$

其含义是

$$a_n \cdot 10^n + a_{n-1} \cdot 10^{n-1} + \cdots + a_i \cdot 10^i + \cdots + a_0 \cdot 10^0 + b_1 \cdot 10^{-1} + b_2 \cdot 10^{-2} + \cdots + b_j \cdot 10^{-j} + \cdots + b_m \cdot 10^{-m}$$

其中 $a_i(i=0,1,\cdots,n), b_j(j=1,2,\cdots,m)$ 是 $0,1,2,3,4,5,6,7,8,9$ 十个数码中的一个。

十进制数的基数为 10,即其数码的个数为 10,且遵循逢十进一的规则。上式中相应于每位数字的 10^k 称为该位数字的权,所以每位数字乘以其权所得到的乘积之和即为所表示数的值。例如:

$$12345.67 = 1\times10^4 + 2\times10^3 + 3\times10^2 + 4\times10^1 + 5\times10^0 + 6\times10^{-1} + 7\times10^{-2}$$

十进制数是人们最熟悉、最常用的一种数制,但它不是惟一的数制。例如,计时用的时、分、秒就是按 60 进制计数的。基数为 r 的 r 进制数的值可表示为:

$$a_n \cdot r^n + a_{n-1} \cdot r^{n-1} + \cdots + a_0 \cdot r^0 + b_1 \cdot r^{-1} + b_2 \cdot r^{-2} + \cdots + b_m \cdot r^{-m}$$

其中 a_i, b_j 可以是 $0,1,\cdots,r-1$ 中的任一个数码,r^k 则为各位数相应的权。

计算机中为便于存储及计算的物理实现,采用了二进制数。二进制数的基数为 2,只有 0,1 两个数码,并遵循逢 2 进 1 的规则,它的各位权是以 2^k 表示的,因此二进制数 $a_n a_{n-1} \cdots a_0 \cdot b_1 b_2 \cdots b_m$ 的值是:

$$a_n \cdot 2^n + a_{n-1} \cdot 2^{n-1} + \cdots + a_0 \cdot 2^0 + b_1 \cdot 2^{-1} + b_2 \cdot 2^{-2} + \cdots + b_m \cdot 2^{-m}$$

其中 a_i, b_j 为 0,1 两个数码中的一个。例如:

$$101101_2 = 1\times2^5 + 1\times2^3 + 1\times2^2 + 1\times2^0 = 45_{10}$$

其中数的下角标表示该数的基数 r,即二进制的 101101 与十进制的 45 等值。

n 位二进制数可以表示 2^n 个数。例如 3 位二进制数可以表示 8 个数,它们是:

二进制数	000	001	010	011	100	101	110	111
相应的十进制数	0	1	2	3	4	5	6	7

而 4 位二进制数则表示十进制的 0~15 共 16 个数,如下所示:

二进制数	0000	0001	0010	0011	0100	0101	0110	0111
相应的十进制数	0	1	2	3	4	5	6	7
二进制数	1000	1001	1010	1011	1100	1101	1110	1111
相应的十进制数	8	9	10	11	12	13	14	15

为便于人们阅读及书写，经常使用八进制数或十六进制数来表示二进制数。它们的基数和数码如表 1.1 所示。

表 1.1 几种常用的进位计数制的基数和数码

进位计数制	基 数	数 码
十六进制数	16	0，1，2，3，4，5，6，7，8，9，A，B，C，D，E，F
十进制数	10	0，1，2，3，4，5，6，7，8，9
八进制数	8	0，1，2，3，4，5，6，7
二进制数	2	0，1

按同样的方法，读者可以很容易地掌握八进制和十六进制数的表示方法。可以看出，23_{10} 可以表示为 17_{16}、27_8 及 10111_2，1.375_{10} 可以表示为 1.6_{16}、1.3_8 及 1.011_2 等。在计算机里，通常用数字后面跟一个英文字母来表示该数的数制。十进制数一般用 D(decimal)、二进制数用 B(binary)、八进制数用 O(octal)、十六进制数用 H(hexadecimal) 来表示。例如：117D，1110101B，0075H，…。当然也可以用这些字母的小写形式，本书的后面就采用了这种表示方法。

1.1.2 二进制数和十进制数之间的转换

1. 二进制数转换为十进制数

各位二进制数码乘以与其对应的权之和即为该二进制数相对应的十进制数。例如：
$$1011100.10111B = 2^6 + 2^4 + 2^3 + 2^2 + 2^{-1} + 2^{-3} + 2^{-4} + 2^{-5} = 92.71875D$$

2. 十进制数转换为二进制数

十进制数转换为二进制数的方法很多，这里只说明比较简单的降幂法及除法两种。

(1) 降幂法

首先写出要转换的十进制数，其次写出所有小于此数的各位二进制权值，然后用要转换的十进制数减去与它最相近的二进制权值，如够减则减去并在相应位记以 1；如不够减则在相应位记以 0 并跳过此位；如此不断反复，直到该数为 0 为止。

例 1.1 N=117D，小于 N 的二进制权为：

        64 32 16 8 4 2 1

对应的二进制数是 1 1 1 0 1 0 1

计算过程如下：

$$117-2^6=117-64=53 \quad (a_6=1)$$
$$53-2^5=53-32=21 \quad (a_5=1)$$
$$21-2^4=21-16=5 \quad (a_4=1)$$
$$(a_3=0)$$
$$5-2^2=5-4=1 \quad (a_2=1)$$
$$(a_1=0)$$
$$1-2^0=1-1=0 \quad (a_0=1)$$

所以 N=117D=1110101B。

例 1.2 N=0.8125D,小于此数的二进制权为:

$$0.5 \quad 0.25 \quad 0.125 \quad 0.0625$$

对应的二进制数是 $\quad 1 \quad\quad 1 \quad\quad 0 \quad\quad 1$

计算过程如下:

$$0.8125-2^{-1}=0.8125-0.5=0.3125 \quad (b_1=1)$$
$$0.3125-2^{-2}=0.3125-0.25=0.0625 \quad (b_2=1)$$
$$(b_3=0)$$
$$0.0625-2^{-4}=0.0625-0.0625=0 \quad (b_4=1)$$

所以 N = 0.8125D = 0.1101B。

(2) 除法

把要转换的十进制数的整数部分不断除以 2,并记下余数,直到商为 0 为止。

例 1.3 N=117D

$$117/2=58 \quad (a_0=1)$$
$$58/2=29 \quad (a_1=0)$$
$$29/2=14 \quad (a_2=1)$$
$$14/2=7 \quad (a_3=0)$$
$$7/2=3 \quad (a_4=1)$$
$$3/2=1 \quad (a_5=1)$$
$$1/2=0 \quad (a_6=1)$$

所以 N =117D =1110101B。

对于被转换的十进制数的小数部分则应不断乘以 2,并记下其整数部分,直到结果的小数部分为 0 为止。

例 1.4 N=0.8125D

$$0.8125\times 2=1.625 \quad (b_1=1)$$

$$0.625 \times 2 = 1.25 \quad (b_2 = 1)$$
$$0.25 \times 2 = 0.5 \quad (b_3 = 0)$$
$$0.5 \times 2 = 1.0 \quad (b_4 = 1)$$

所以 N=0.8125D=0.1101B。

1.1.3 十六进制数及其与二进制、十进制数之间的转换

我们知道,在计算机内部,数的运算和存储都是采用二进制的。但是,二进制数对于人们阅读、书写及记忆都是很不方便的。十进制数虽然是人们最熟悉的一种进位计数制,但它与二进制数之间并无直接的对应关系。为了便于人们对二进制数的描述,应该选择一种易于与二进制数相互转换的数制。显然,使用 2^n 作为基数的数制是能适合人们的这种要求的,常用的有八进制数和十六进制数,这里主要介绍十六进制数。

1. 十六进制数的表示

计算机中存储信息的基本单位为一个二进制位(bit),它可以用来表示 0 和 1 两个数码。此外,由于计算机中常用的字符是采用由 8 位二进制数组成的一个字节(byte)来表示的,因此字节也成为计算机中存储信息的单位。计算机的字长一般都选为字节的整数倍,如 16 位、32 位、64 位等。一个字节由 8 位组成,它可以用两个四位组(又称半字节)来表示,所以用十六进制数来表示二进制数是比较方便的。

十六进制数的基数是 16,共有 16 个数码,它们是 0,1,2,3,4,5,6,7,8,9,A,B,C,D,E,F。其中 A 表示十进制的 10,余类推。它们与二进制和十进制数的关系如下:

二进制数	十进制数	十六进制数
0000	0	0
0001	1	1
0010	2	2
0011	3	3
0100	4	4
0101	5	5
0110	6	6
0111	7	7
1000	8	8
1001	9	9
1010	10	A
1011	11	B
1100	12	C
1101	13	D
1110	14	E
1111	15	F

对应于十六进制数中各位的权是 16^k。

2. 十六进制数和二进制数之间的转换

由于十六进制数的基数是 2 的幂，所以这两种数制之间的转换是十分容易的。一个二进制数，只要把它从低位到高位每 4 位组成一组，直接用十六进制数来表示就可以了。

例 1.5   0011   0101   1011   1111
                 3   5   B   F

亦即 0011010110111111B＝35BFH

反之，把十六进制数中的每一位用 4 位二进制数表示，就形成相应的二进制数了。

例 1.6   A   1   9   C
             1010   0001   1001   1100

亦即 A19CH＝1010000110011100B

3. 十六进制数和十进制数之间的转换

各位十六进制数与其对应权值的乘积之和即为与此十六进制数相对应的十进制数。

例 1.7 $N = BF3CH$
            $= 11 \times 16^3 + 15 \times 16^2 + 3 \times 16^1 + 12 \times 16^0$
            $= 11 \times 4096 + 15 \times 256 + 3 \times 16 + 12 \times 1$
            $= 48956D$

十进制数转换为十六进制数也可使用降幂法和除法。

(1) 降幂法

首先写出要转换的十进制数，其次写出小于该数的十六进制权值，然后找出该数中包含多少个最接近它的权值的倍数，这一倍数即对应位的值，用原数减去此倍数与相应位权值的乘积得到一个差值，再用此差值去找低一位的权值的倍数，如此反复直到差值为 0 为止。

例 1.8   N＝48956D   小于 N 的十六进制权值为

                    4096   256   16   1
对应的十六进制数   B   F   3   C

计算过程如下：
$$48956 - 11 \times 4096 = 3900$$
$$3900 - 15 \times 256 = 60$$
$$60 - 3 \times 16 = 12$$
$$12 - 12 \times 1 = 0$$

所以 $N = 48956D = (11)(15)(3)(12)$
                    $= BF3CH$

(2) 除法

把要转换的十进制数的整数部分不断除以 16，并记下余数，直到商为 0 为止。

例 1.9 $N = 48956D$

          $48956/16 = 3059$ ($a_0 = 12$)
          $3059/16 = 191$ ($a_1 = 3$)

$$191/16 = 11 \quad (a_2 = 15)$$
$$11/16 = 0 \quad (a_3 = 11)$$

所以 N=48956D=BF3CH。

对于要转换的十进制数的小数部分，则应不断地乘以 16，并记下其整数部分，直到结果的小数部分为 0 为止。由于其方法与二、十进制数的转换方法是相同的，这里不再举例说明。显然，为把一个十进制数转换为二进制数，可以先把该数转换为十六进制数，然后再转换为二进制数，这样可以减少计算次数；反之，要把一个二进制数转换为十进制数，也可采用同样的办法。

1.2 二进制数和十六进制数运算

1.2.1 二进制数运算

加法规则：   乘法规则：
 0+0= 0   0×0= 0
 0+1= 1   0×1= 0
 1+0= 1   1×0= 0
 1+1= 0（进位 1）   1×1= 1

1.2.2 十六进制数运算

十六进制数的运算可以采用先把该十六进制数转换为十进制数，经过计算后再把结果转换为十六进制数的方法，但这样做比较繁琐。其实，只要按照逢十六进一的规则，直接用十六进制数来计算也是很方便的。

十六进制加法：

当两个一位数之和 S 小于 16 时，与十进制数同样处理；如果两个一位数之和大于或等于 16 时，则应该用 S−16 及进位 1 来取代 S。

例 1.10

```
   05C3H
 + 3D25H
   -----
   42E8H
```

十六进制数的减法也与十进制数类似，够减时可直接相减，不够减时服从向高位借 1 为 16 的规则。

例 1.11

```
   3D25H
 − 05C3H
   -----
   3762H
```

十六进制数的乘法可以用十进制数的乘法规则来计算，但结果必须用十六进制数来表示。

例 1.12

$$
\begin{array}{r}
05C3H \\
\times\ 00ABH \\
\hline
3F61 \\
+\ \ 399E\ \ \ \\
\hline
3D941H
\end{array}
$$

十六进制数的除法可以根据其乘法和减法规则处理,需要时读者可自行处理,这里不再赘述。

1.3 计算机中数和字符的表示

1.3.1 数的补码表示

计算机中的数是用二进制来表示的,数的符号也是用二进制表示的。在机器中,把一个数连同其符号在内数值化表示的数称为机器数。一般用最高有效位来表示数的符号,正数用 0 表示,负数用 1 表示。机器数可以用不同的码制来表示,常用的有原码、补码和反码表示法。由于多数机器的整数采用补码表示法,80x86 机也是这样。这里只介绍补码表示法。

补码表示法中正数采用符号-绝对值表示,即数的最高有效位为 0 表示符号为正,数的其余部分则表示数的绝对值。例如,假设机器字长为 8 位,则 $[+1]_{补} = 00000001$,$[+127]_{补} = 01111111$,$[+0]_{补} = 00000000$。

当用补码表示法来表示负数时则要麻烦一些。负数 X 用 $2^n - |X|$ 来表示,其中 n 为机器的字长。当 $n = 8$ 时,$[-1]_{补} = 2^8 - 1 = 11111111$,而 $[-127]_{补} = 2^8 - 127 = 10000001$,显然,最高有效位为 1 表示该数的符号为负。应该注意,$[-0]_{补} = 2^8 = 00000000$,所以在补码表示法中 0 只有一种表示,即 00000000。对于 10000000 这个数,在补码表示法中被定义为 -128。这样,8 位补码能表示数的范围为 $-128 \sim +127$。

下面介绍一种比较简单的办法来写出一个负数的补码表示:先写出与该负数相对应的正数的补码表示(用符号-绝对值法),然后将其按位求反(即 0 变 1,1 变 0),最后在末位(最低位)加 1,就可以得到该负数的补码表示了。

例 1.13 机器字长为 16 位,写出 N = $-117D$ 的补码表示。

+117D 可表示为	0000	0000	0111	0101
按位求反后为	1111	1111	1000	1010
末位加 1 后	1111	1111	1000	1011
用十六进制数表示为	F	F	8	B

即 $[-117D]_{补}$ = FF8BH

例 1.14 如机器字长为 8 位,则 $-46D$ 的补码表示为:

+46D 的补码表示为	0010	1110
按位求反后为	1101	0001
末位加 1 后	1101	0010
用十六进制数表示为	D	2

即 $[-46]_\text{补} = \text{D2H}$

至此,读者应该已经学会了一个数的补码表示法。在这里,顺便说明一下,用补码表示数时的符号扩展问题。所谓符号扩展是指一个数从位数较少扩展到位数较多(如从8位扩展到16位,或从16位扩展到32位)时应该注意的问题。对于用补码表示的数,正数的符号扩展应该在前面补0,而负数的符号扩展则应该在前面补1。例如,我们已经知道如机器字长为8位,则$[+46]_\text{补}=00101110$,$[-46]_\text{补}=11010010$;如果把它们从8位扩展到16位,则$[+46]_\text{补}=0000000000101110=002\text{EH}$,$[-46]_\text{补}=1111111111010010=$FFD2H。

下面,再来讨论一下 n 位补码表示数的范围问题。8 位二进制数可以表示 $2^8=256$ 个数,当它们是补码表示的带符号数时,它们的表数范围是 $-128 \leqslant N \leqslant +127$。一般说来,$n$ 位补码表示的数的表数范围是:

$$-2^{n-1} \leqslant N \leqslant 2^{n-1}-1$$

所以 $n=16$ 时的表数范围是:$-32768 \leqslant N \leqslant +32767$

表 1.2 表示当 $n=8$ 和 $n=16$ 位时二进制补码数的表数范围。

表 1.2 n 位二进制补码数的表数范围

十进制数	二进制数	十六进制数	十进制数	十六进制数
	n=8		n=16	
+127	01111111	7F	+32767	7FFF
+126	01111110	7E	+32766	7FFE
⋮	⋮	⋮	⋮	⋮
+2	00000010	02	+2	0002
+1	00000001	01	+1	0001
0	00000000	00	0	0000
−1	11111111	FF	−1	FFFF
−2	11111110	FE	−2	FFFE
⋮	⋮	⋮	⋮	⋮
−126	10000010	82	−32766	8002
−127	10000001	81	−32767	8001
−128	10000000	80	−32768	8000

在机器里,为了扩大表数范围,可以用二个机器字(高位字和低位字)来表示一个机器数,这种数称为双字长数或双精度数,其格式如图 1.1 所示。其中高位字的最高有效位为符号位,高位字的低 15 位和整个低位字的 16 位联合组成 31 位数来表示数值,因而低位

字的最高有效位没有符号意义,只有数值意义。双字长数的表数范围可扩大到:
$$-2^{31} \leqslant N \leqslant 2^{31}-1$$
$2^{31} \cong 2.15 \times 10^9$,可见表数范围扩大了许多。图 1.1 中每个字上的 0,15 是位编号,每个机器字都从低位开始给每一位以编号,所以从右至左依次编号为 $0,1,\cdots,15$。

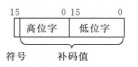

图 1.1 双字长数的表示

80386 及其后继机型的机器字长已扩展为 32 位。为了统一起见,80x86 系统仍称32 位字为双字,这样又有了 8 个字节 64 位的 4 字,其表数范围读者可自行推断。

1.3.2 补码的加法和减法

我们知道,对一个正数的补码表示按位求反后再在末位加1,可以得到与此正数相应的负数的补码表示。我们把这种对一个二进制数按位求反后在末位加 1 的运算称为求补运算,可以证明补码表示的数具有以下特性:

$$[X]_{补} \overset{求补}{\Rightarrow} [-X]_{补} \overset{求补}{\Rightarrow} [X]_{补}$$

在这里,只用例子来说明。由例 1.13 可见:

$$[117]_{补} = 0075H$$
$$[-117]_{补} = FF8BH$$

现对$[-117]_{补}$作求补运算:

$[-117]_{补}$ 为	1111	1111	1000	1011
按位求反后得	0000	0000	0111	0100
末位加 1 后得	0000	0000	0111	0101

此数正是 $[+117]_{补}=0075H$。

这一特性在补码的加、减法运算中很有用。

补码的加法规则是:

$$[X+Y]_{补}=[X]_{补}+[Y]_{补}$$

补码的减法规则是:

$$[X-Y]_{补}=[X]_{补}+[-Y]_{补}$$

其中的$[-Y]_{补}$只要对$[Y]_{补}$求补就可得到。对于这两个规则我们只用例子来说明。读者可以从下面的例子中认识到由于用补码表示数,使计算机中的加、减法运算十分简便,它不必判断数的正负,只要符号位参加运算,便能自动地得到正确的结果。假设机器字长为 8 位,用下列例子来说明补码的加法运算。

例 1.15

```
       十进制              二进制
        25                00011001
     +  32              + 00100000
     ─────              ──────────
        57                00111001
```

例 1.16

```
        32              00100000
   +  (-25)           +11100111
   ─────────          ─────────
         7             00000111
                           ↓
                           1
```

例 1.17

```
        25              00011001
   +  (-32)           +11100000
   ─────────          ─────────
        -7             11111001
```

例 1.18

```
       -25              11100111
   +  (-32)           +11100000
   ─────────          ─────────
       -57             11000111
                           ↓
                           1
```

可以看出,上述例 1.15～例 1.18 的计算结果都是正确的,在例 1.16 和例 1.18 中,从最高有效位向高位的进位由于机器字长的限制而自动丢失,但这并不会影响运算结果的正确性。同时,机器为了某种需要(将在第 3 章中说明)将把这一进位值保留在标志寄存器的进位位 C 中。

下面,再用例 1.19～例 1.22 说明补码的减法运算,机器字长仍假定为 8 位。

例 1.19

```
十进制                                               二进制
    25        00011001    计算机中用对减数求补的      00011001
  - 32       -00100000    方法把减法转化为加法      +11100000
  ─────     ─────────                              ─────────
    -7                                              11111001
```

例 1.20

```
    32        00100000                              00100000
  -(-25)     -11100111                             +00011001
  ─────     ─────────                              ─────────
    57                                              00111001
```

例 1.21

```
    -25       11100111                              11100111
  -(+32)     -00100000                             +11100000
  ─────     ─────────                              ─────────
    -57                                             11000111
                                                       ↓
                                                       1
```

例 1.22

```
    -25       11100111                              11100111
  -(-32)     -11100000                             +00100000
  ─────     ─────────                              ─────────
     7                                              00000111
                                                       ↓
                                                       1
```

可以看出，在机器里，补码减法是用对减数求补后把减法转换为加法进行的，它同样能自动地得到正确的结果。其中例 1.21 和例 1.22 中的由最高有效位向高位的进位同样自动丢失而不会影响运算的结果。

1.3.3 无符号整数

在某些情况下，要处理的数全是正数，此时再保留符号位就没有意义了。我们可以把最高有效位也作为数值处理，这样的数称为无符号整数。8 位无符号数的表数范围是 $0 \leqslant N \leqslant 255$，16 位无符号数的表数范围是 $0 \leqslant N \leqslant 65535$，32 位无符号数的表数范围是 $0 \leqslant N \leqslant 2^{32}-1$。

在计算机中最常用的无符号整数是表示地址的数。此外，如双精度数的低位字也是无符号整数等。在某些情况下，带符号的数(在机器中用补码表示)与无符号数的处理是有差别的，读者在处理数时，应注意它们的区别。

1.3.4 字符表示法

计算机中处理的信息并不全是数，有时需要处理字符或字符串，例如从键盘输入的信息或打印输出的信息都是以字符方式输入输出的。因此，计算机必须能表示字符。字符包括：

字母：A、B、…、Z,a、b、…、z；
数字：0、1、…、9；
专用字符：+、-、*、/、↑、SP(space 空格)…；
非打印字符：BEL(bell 响铃)、LF(line feed 换行)、CR(carriage return 回车)、……

这些字符在机器里必须用二进制数来表示。80x86 机采用目前最常用的美国信息交换标准代码 ASCII(American Standard Code for Information Interchange)来表示。这种代码用一个字节(8 位二进制码)来表示一个字符，其中低 7 位为字符的 ASCII 值，最高位一般用作校验位。表 1.3 列出了用十六进制数表示的部分常用字符的 ASCII 值。

表 1.3 常用字符的 7 位 ASCII 值(用十六进制数表示)

字符	ASCII	字符	ASCII	字符	ASCII	字符	ASCII
NUL	00	4	34	M	4D	f	66
BEL	07	5	35	N	4E	g	67
LF	0A	6	36	O	4F	h	68
FF	0C	7	37	P	50	i	69
CR	0D	8	38	Q	51	j	6A
SP	20	9	39	R	52	k	6B
!	21	:	3A	S	53	l	6C

续表

字符	ASCII	字符	ASCII	字符	ASCII	字符	ASCII
"	22	;	3B	T	54	m	6D
#	23	<	3C	U	55	n	6E
$	24	=	3D	V	56	o	6F
%	25	>	3E	W	57	p	70
&	26	?	3F	X	58	q	71
'	27	@	40	Y	59	r	72
(	28	A	41	Z	5A	s	73
)	29	B	42	[	5B	t	74
*	2A	C	43	\	5C	u	75
+	2B	D	44	]	5D	v	76
,	2C	E	45	↑	5E	w	77
—	2D	F	46	←	5F	x	78
.	2E	G	47	'	60	y	79
/	2F	H	48	a	61	z	7A
0	30	I	49	b	62	{	7B
1	31	J	4A	c	63	|	7C
2	32	K	4B	d	64	}	7D
3	33	L	4C	e	65	~	7E

1.4 几种基本的逻辑运算

1.4.1 "与"运算(AND)

"与"运算又称逻辑乘,可用符号·或∧来表示。如有 A、B 两个逻辑变量(每个变量只能有 0 或 1 两种取值),可能有的取值情况只有 4 种,在各种取值的条件下得到的"与"运算结果如表 1.4 所示;即只有当 A、B 两个变量的取值均为 1 时,它们的"与"运算结果才是 1。

表 1.4

A	B	A∧B
0	0	0
0	1	0
1	0	0
1	1	1

1.4.2 "或"运算(OR)

"或"运算又称逻辑加,可用符号 + 或 ∨ 来表示。"或"运算规则可用表 1.5 所示:即 A、B 两个变量中只要有一个变量取值为 1,则它们"或"运算的结果就是 1。

表 1.5

A	B	A∨B
0	0	0
0	1	1
1	0	1
1	1	1

1.4.3 "非"运算(NOT)

如变量为 A,则它的"非"运算的结果用 $\overline{A}$ 来表示。"非"运算规则如表 1.6 所示:

表 1.6

A	$\overline{A}$
0	1
1	0

1.4.4 "异或"运算(XOR Exclusive-OR)

"异或"运算可以用符号 ∀ 来表示。其运算规则如表 1.7 所示:即当两个变量的取值相异时,它们的"异或"运算结果为 1。

表 1.7

A	B	A∀B
0	0	0
0	1	1
1	0	1
1	1	0

所有的逻辑运算都是按位操作的,下面举例说明:

例 1.23 如果两个变量的取值为 X = 00FFH 和 Y = 5555H,求 Z_1 = X ∧ Y;Z_2 = X ∨ Y;Z_3 = $\overline{X}$;Z_4 = X ∀ Y 的值。

$$X = 0000\ 0000\ 1111\ 1111$$
$$Y = 0101\ 0101\ 0101\ 0101$$
$$Z_1 = 0000\ 0000\ 0101\ 0101 = 0055\text{H}$$
$$Z_2 = 0101\ 0101\ 1111\ 1111 = 55\text{FFH}$$
$$Z_3 = 1111\ 1111\ 0000\ 0000 = \text{FF00H}$$
$$Z_4 = 0101\ 0101\ 1010\ 1010 = 55\text{AAH}$$

习　题

1.1 用降幂法和除法将下列十进制数转换为二进制数和十六进制数：
(1) 369　　(2) 10000　　(3) 4095　　(4) 32767

1.2 将下列二进制数转换为十六进制数和十进制数：
(1) 101101　　(2) 10000000　　(3) 1111111111111111　　(4) 11111111

1.3 将下列十六进制数转换为二进制数和十进制数：
(1) FA　　(2) 5B　　(3) FFFE　　(4) 1234

1.4 完成下列十六进制数的运算，并转换为十进制数进行校核。
(1) 3A＋B7　　(2) 1234＋AF　　(3) ABCD－FE　　(4) 7AB×6F

1.5 下列各数均为十进制数，请用8位二进制补码计算下列各题，并用十六进制数表示其运算结果。
(1) (－85)＋76　　(2) 85＋(－76)　　(3) 85－76
(4) 85－(－76)　　(5) (－85)－76　　(6) －85－(－76)

1.6 下列各数为十六进制表示的8位二进制数，请说明当它们分别被看作是用补码表示的带符号数或无符号数时，它们所表示的十进制数是什么？
(1) D8　　(2) FF

1.7 下列各数均为用十六进制表示的8位二进制数。请说明当它们分别被看作是用补码表示的数或字符的ASCII码时，它们所表示的十进制数及字符是什么？
(1) 4F　　(2) 2B　　(3) 73　　(4) 59

1.8 请写出下列字符串的ASCII码值。
For example,
This is a number 3692.

第 2 章　80x86 计算机组织

80x86 是美国 Intel 公司生产的微处理器系列。该公司成立于 1968 年,1969 年就设计了 4 位的 4004 芯片,1973 年开发出 8 位的 8080 芯片,1978 年正式推出 16 位的 8086 微处理器芯片,由此开始了 Intel 公司的 80x86 微处理器系列的生产历史。本章首先介绍 80x86 微处理器概况,然后简要说明基于微处理器的计算机系统构成,最后将根据汇编语言编程的需要分节介绍存储器、中央处理机和外部设备。

2.1　80x86 微处理器

计算机主要由运算器、控制器、存储器和输入输出设备构成。20 世纪 70 年代初期,由于大规模集成电路技术的发展,已经开始把运算器和控制器集成在一个芯片上,构成中央处理机(central processing unit,CPU),80x86 就是这样一组微处理器系列。

很多计算机厂商把微处理器芯片作为中央处理机,再配上存储器、输入输出设备和系统软件等构成微计算机系统。如由 80386 微处理器芯片构成的微机称为 386 微机;由 80486 微处理器芯片构成的微机称为 486 微机等。

表 2.1　80x86 微处理器概况

型　号	发布年份	字长(位)	晶体管数(万个)	主频(MHz)	数据总线宽度(位)	外部总线宽度(位)	地址总线宽度(位)	寻址空间(B)	高速缓存
8086	1978	16	2.9	4.77	16	16	20	1M	无
8088	1979	16	2.9	4.77	16	8	20	1M	无
80286	1982	16	13.4	6～20	16	16	24	16M	无
80386	1986	32	27.5	12.5～33	32	32	32	4G	有
80486	1989	32	120～160	25～100	32	32	32	4G	8KB
Pentium（586）	1993	32	310～330	60～166	64	64	32	4G	8KB 数据 8KB 指令
Pentium Pro(P6)	1995	32	550+1550	150～200	64	64	36	64G	8KB 数据 8KB 指令 256KB 二级高速缓存
Pentium II	1997	32	750	233～333	64	64	36	64G	32KB 512KB 二级高速缓存,有独立封装和独立总线

表 2.1 给出 Intel 公司生产的 80x86 微处理器系列的一些主要技术数据。从表 2.1 中可以看出这一芯片系列的发展概况。

为了能更清楚地说明问题，下面解释一些名词术语。

晶体管数是指芯片中所包含的晶体管数，它说明器件的集成度；主频是指芯片所用的主时钟频率，它直接影响计算机的运行速度，由于处理器体系结构的差别，同样的主频可能产生不同的计算速度，但主频仍然是反映计算机速度的一个重要指标；数据总线负责计算机中数据在各组成部分之间的传送，数据总线宽度是指在芯片内部数据传送的宽度，外部数据总线宽度则是指芯片内和芯片外交换数据的宽度；地址总线宽度是指专用于传送地址的总线宽度，根据这一数值可确定处理机可以访问的存储器的最大范围（寻址空间），如 20 位地址总线可访问 $2^{20}=1048576$ 个存储单元。在计算机中，为方便起见，在讨论存储器容量时，以 $2^{10}=1024$ 为基本单位，称为 1K，1024K 就称为 1M。所以 20 位地址总线可访问 1M 个存储单元，24 位地址总线可访问 16M 个存储单元，$2^{30}=1024$M 称为 1G，32 位地址总线可访问 4G 个存储单元。在计算机里，8 个二进制位组成一个字节（byte），一般存储器以字节为存储信息的基本单位，用符号 B 来表示。这样，上述存储容量又可称为 1MB、16MB 和 4GB 等。

我们知道，要在计算机上计算一个题目，首先必须用计算机语言把所要计算的题目编制成程序，然后把这一程序（由指令序列组成）连同所要使用的数据一起存入计算机的存储器中。在计算机算题时，要把程序和所用的数据从存储器中取到运算器中进行计算。因此，在计算过程中，为保证运算器能快速运行，存储器也必须源源不断地提供计算所需的指令和数据，并且有与其相应的足够快的速度。为适应这种速度要求，建立了层次结构的存储器组织，其中间层次通常称为主存储器；比其速度更高、但容量较小的一层称为高速缓冲存储器（cache）；比其速度慢、但容量很大的一层称为外存储器，如磁带、磁盘、光盘等。高速缓冲存储器（以下简称高速缓存）对提高计算机的计算速度起很重要的作用，早期是做在芯片之外，随着半导体集成电路技术的发展，从 80386 开始在芯片中已做入少量的高速缓存。在 80486 的芯片中，集成了 8KB 高速缓存；Pentium 中有 8KB 的指令高速缓存和 8KB 的数据高速缓存；Pentium Pro 中除两个 8KB 的高速缓存外，还增加了 256KB 的二级高速缓存，晶体管数中增加的 1 550 万个晶体管就是用于二级高速缓存的；Pentium II 中有 32KB 的高速缓存，它的二级高速缓存容量为 512KB，且具有独立的封装和独立的总线。以上这些措施都是为了提高计算机的速度而采取的。

实际上，在微处理器发展过程中，有很多体系结构方面的措施是无法从表 2.1 的简单描述中看得出来的，这也不属本书所要阐明的问题。但是从总体上已可看出微处理器芯片的发展速度是非常快的，其中重要一环是计算机的运算速度，下面给出的参考数据可以大致地说明这一指标的发展情况。8086 执行一条指令的时间约为 400ns，80286 为 250ns，80486 为 25ns，Pentium 的速度为 80486 的 2 倍，Pentium Pro 为 80486 的 3 倍，Pentium II 比 Pentium Pro 快 10%～25%，所以提高计算机的工作速度可以说是微处理器芯片发展的核心问题。从 80486 起，把协处理器集成到芯片中的目的也是为了提高浮点处理速度。另外，字长的增加有利于提高计算机解题的精度。值得一提的是，从 80286 开始，在机器的工作方式上，除 8086 提供的实模式外，还增加了保护模式的工作方式。在保护模式下,机器可提供虚拟存储的管理和多任务的管理机制。虚拟存储的实现,使计算机可以运行程序空间大于主存储器空间的用户程序。多任务管理的实现,允许多个用户可以同时在机器上工作。从 80386 起,除支持实模式和保护模式外,又增加了一种虚 86 的工作模式。在这种工作模式下,一台机器可同时模拟多个 8086 处理器的工作。所有这些措施都是为了提高微计算机的可用性而开发的,使得微计算机的应用领域更加广泛,促进了微机技术本身的飞速发展。

2.2 基于微处理器的计算机系统构成

计算机系统包括硬件和软件两部分。硬件包括电路、插件板、机柜等;软件则是为了运行、管理和维护计算机而编制的各种程序的总和。

2.2.1 硬件

典型计算机的结构,如图 2.1 所示。其中包括由微处理器芯片构成的中央处理机(CPU)、存储器(memory)和输入输出(I/O)子系统三个主要组成部分,用系统总线把它们连接在一起。

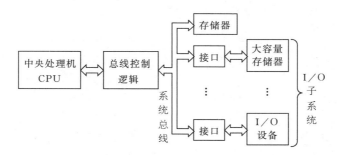

图 2.1 计算机结构

存储器是计算机的记忆部件。人们编写的程序(由指令序列组成)就存放在这里。它也可以存放程序中所用的数据、信息及中间结果。

中央处理机包括运算器和控制器两部分。运算器执行所有的算术和逻辑运算指令;控制器则负责全机的控制工作,它负责把指令逐条从存储器中取出,经译码分析后向全机发出取数、执行、存数等控制命令,以保证正确完成程序所要求的功能。

I/O 子系统一般包括 I/O 设备及大容量存储器两类外部设备。I/O 设备是指负责与计算机的外部世界通信用的输入、输出设备,如显示终端、键盘输入、打印输出等多种类型的外部设备。大容量存储器则是指可存储大量信息的外部存储器,如磁盘、磁带、光盘等。机器内部的存储器则称为内存储器,简称内存,由于内存的容量有限,所以计算机用外存储器作为内存的后援设备,它的容量可以比内存大很多,但存取信息的速度要比内存慢得多。所以,除必要的系统程序外,一般程序(包括数据)是存放在外存中的。只有当运行时,才把它从外存传送到内存的某个区域,再由中央处理机控制执行。

系统总线把 CPU、存储器和 I/O 设备连接起来,用来传送各部分之间的信息。系统总线包括数据线、地址线和控制线三组。数据线传送信息,地址线指出信息的来源和目的地,控制线则规定总线的动作,如方向等。系统总线的工作由总线控制逻辑负责指挥。

2.2.2 软件

计算机软件是计算机系统的重要组成部分,它可以分成系统软件和用户软件两大类。系统软件是由计算机生产厂家提供给用户的一组程序,这些程序是用户使用机器时,为产生、准备和执行用户程序所必需的。用户软件则是用户自行编制的各种程序。图 2.2 表示了计算机软件的层次,这里将简要介绍系统软件的组成。

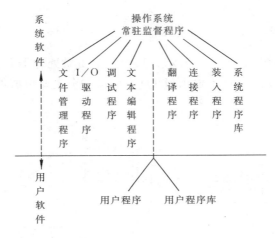

图 2.2 计算机软件层次图

系统软件的核心称为操作系统(operating system)。操作系统是系统程序的集合,它的主要作用是对系统的硬、软件资源进行合理的管理,为用户创造方便、有效和可靠的计算机工作环境。

操作系统的主要部分是常驻监督程序(monitor),只要一开机它就存在于内存中,它可以从用户接收命令,并使操作系统执行相应的动作。

I/O 驱动程序(I/O driver)用来对 I/O 设备进行控制或管理。当系统程序或用户程序需要使用 I/O 设备时,就调用 I/O 驱动程序来对设备发出命令,完成 CPU 和 I/O 设备之间的信息传送。

文件管理程序(file management)用来处理存储在外存储器中的大量信息,它可以和外存储器的设备驱动程序相连接,对存储在其中的信息以文件(file)的形式进行存取、复制及其他管理操作。

文本编辑程序(text editor)用来建立、输入或修改文本,并使它存入内存储器或大容量存储器中。文本是指由字母、数字、符号等组成的信息,它可以是一个用汇编语言或高级语言编写的程序,也可以是一组数据或一份报告。

翻译程序(translator):我们已经知道,计算机是通过逐条地执行组成程序的指令来完成人们所给予的任务的,所以指令就是计算机所能识别并能直接加以执行的语句,当然它是由二进制代码组成的。这种语言称为机器语言,它对于人们显然是很不方便的。既然计算机能识别的惟一语言是机器语言,而用这种语言编写程序又很不方便,所以在计算机语言的发展过程中就出现了汇编语言和高级语言。汇编语言是一种符号语言,它和机器语言几乎一一对应,但在书写时却使用由字符串组成的助记符。例如,加法在汇编语言中一般是用助记符 ADD 表示的,而机器语言则用二进制代码来表示。显然,相对于机器语言来说,汇编语言是易于为人们所理解的,但计算机却不能直接识别汇编语言。汇编程序就是用来把由用户编制的汇编语言程序翻译成机器语言程序的一种系统程序。微机的汇编程序有多种版本,如 MASM、TASM 等。MASM 为 Microsoft 公司开发的汇编程序,TASM(Turbo Assembler)则为 Borland 公司开发的汇编程序,它们都具有较强的功能和宏汇编能力。

高级语言脱离开机器指令用人们更加易于理解的方式来编写程序,当然它们也要翻译成机器语言才能在机器上执行。高级语言的翻译程序有两种方式:一种是先把高级语言程序翻译成机器语言(或先翻译成汇编语言,然后再由汇编程序再次翻译成机器语言)程序,然后再在机器上执行,这种翻译程序称为编译程序(compiler),多数高级语言如 PASCAL、FORTRAN 等都采用这种方式。另一种是直接把高级语言程序在机器上运行,一边解释一边执行,这种翻译程序称为解释程序(interpreter),如 BASIC 就经常采用这种方式。

系统程序中的翻译程序包括汇编程序、解释程序和编译程序。

连接程序(linker)用来把要执行的程序与库文件或其他已经翻译好的子程序(能完成一种独立功能的程序模块)连接在一起,形成机器能执行的程序。

装入程序(loader)用来把程序从外存储器传送到内存储器,以便机器执行。例如,计算机开机后就需要立即启动装入程序把常驻监督程序装入存储器,使机器运转起来。又如,用户程序经翻译和连接后,由连接程序直接调用装入程序,把可执行的用户程序装入内存以便执行。

调试程序(debug)是系统提供给用户的能监督和控制用户程序的一种工具。它可以装入、修改、显示或逐条执行一个程序。微机上的汇编语言程序可以通过 DEBUG 来调试,完成建立、修改和执行等工作。

系统程序库(system library)和用户程序库(user library),各种标准程序、子程序和一些文件的集合称为程序库,它可以被系统程序或用户程序调用。操作系统还允许用户建立程序库,以提高不同类型用户的工作效率。

2.3 中央处理机

2.3.1 中央处理机 CPU 的组成

CPU 的任务是执行存放在存储器里的指令序列。为此,除要完成算术逻辑操作外,还需要担负 CPU 和存储器以及 I/O 之间的数据传送任务。早期的 CPU 芯片只包括运算器和控制器两大部分。从 80386 开始,为使存储器速度能更好地与运算器的速度相匹配,已在芯片中引入高速缓冲存储器。其后生产的芯片随着半导体器件集成度的提高,片内高速缓冲存储器的容量也逐步扩大,但这部分器件就其功能而言还是属于存储器的,本节要说明的是 CPU 芯片中除高速缓冲存储器之外的组成,它们主要由以下三部分组成:

(1) 算术逻辑部件(arithmetic logic unit,ALU)用来进行算术和逻辑运算。

(2) 控制逻辑负责对全机的控制工作,包括从存储器取出指令,对指令进行译码分析,从存储器取得操作数,发出执行指令的所有命令,把结果存入存储器,以及对总线及 I/O 的传送控制等。

(3) 工作寄存器在计算机中起着重要的作用,每一个寄存器相当于运算器中的一个存储单元,但它的存取速度比存储器要快得多。它用来存放计算过程中所需要的或所得到的各种信息,包括操作数地址、操作数及运算的中间结果等。下面专门介绍这些寄存器。

2.3.2 80x86 寄存器组

寄存器可以分为程序可见的寄存器和程序不可见的寄存器两大类。所谓程序可见的寄存器是指在汇编语言程序设计中用到的寄存器,它们可以由指令来指定。而程序不可见的寄存器则是指一般应用程序设计中不用而由系统所用的寄存器。本节将介绍 80x86 中程序可见的那部分寄存器,而程序不可见的寄存器就不加以说明了。

程序可见寄存器可以分为通用寄存器、专用寄存器和段寄存器 3 类。图 2.3 表示了 80x86 的程序可见寄存器组。下面分别加以说明。

1. 通用寄存器

图 2.3 中除阴影区以外的寄存器是 8086/8088 和 80286 所具有的寄存器,它们都是 16 位寄存器。其中 AX、BX、CX、DX 可称为数据寄存器,用来暂时存放计算过程中所用到的操作数、结果或其他信息。它们都可以以字(16 位)的形式访问,或者也可以以字节(8 位)的形式访问。例如,对 AX 可以分别访问高位字节 AH 或低位字节 AL。这 4 个寄存器都是通用寄存器,但它们又可以用于各自的专用目的。

AX(accumulator)作为累加器用,所以它是算术运算的主要寄存器。在乘、除等指令中指定用来存放操作数。另外,所有的 I/O 指令都使用这一寄存器与外部设备传送信息。

BX(base)可以作为通用寄存器使用。此外在计算存储器地址时,它经常用作基址寄存器。

CX(count)可以作为通用寄存器使用。此外常用来保存计数值,如在移位指令、循环(loop)和串处理指令中用作隐含的计数器。

DX(data)可以作为通用寄存器使用。一般在作双字长运算时把 DX 和 AX 组合在一起存放一个双字长数,DX 用来存放高位字。此外,对某些 I/O 操作,DX 可用来存放 I/O 的端口地址。

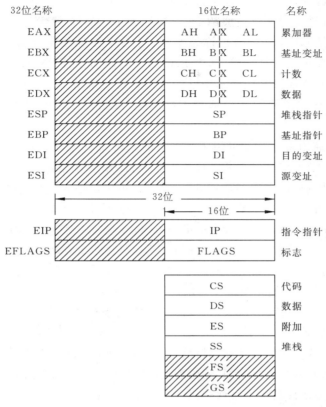

注：1. 对于8086、8088或80286，阴影区域是不可用的。
　　2. FS和GS寄存器无专用名称。
图 2.3　80x86的程序可见寄存器组

SP、BP、SI、DI 四个 16 位寄存器可以像数据寄存器一样在运算过程中存放操作数，但它们只能以字(16 位)为单位使用。此外，它们更经常的用途是在存储器寻址时，提供偏移地址。因此，它们可称为指针或变址寄存器。在这些寄存器中，SP(stack pointer)称为堆栈指针寄存器，BP(base pointer) 称为基址指针寄存器，它可以与堆栈段寄存器 SS 联用来确定堆栈段中的某一存储单元的地址。SP 用来指示段顶的偏移地址，BP 可作为堆栈区中的一个基地址以便访问堆栈中的信息。SI(source index)源变址寄存器和 DI (destination index)目的变址寄存器一般与数据段寄存器 DS 联用，用来确定数据段中某一存储单元的地址。这两个变址寄存器有自动增量和自动减量的功能，所以用于变址是很方便的。在串处理指令中，SI 和 DI 作为隐含的源变址和目的变址寄存器，此时 SI 和 DS 联用，DI 和附加段寄存器 ES 联用，分别达到在数据段和附加段中寻址的目的。有关段地址和偏移地址以及堆栈段的概念，将在 2.4.2 节作详细说明。

对于 80386 及其后继机型则是图 2.3 中所示的完整的寄存器，它们是 32 位的通用寄存器，包括 EAX、EBX、ECX、EDX、ESP、EBP、EDI 和 ESI。在这些机型中，它们可以用来保存不同宽度的数据,如可以用 EAX 保存 32 位数据,用 AX 保存 16 位数据,用 AH 或 AL 保存 8 位数据。在计算机中,8 位二进制数可组成一个字节,8086/8088 和 80286 的字长为 16 位,因此把 2 个字节组成的 16 位数称为字。这样,80386 及其后继的 32 位机就把 32 位数据称为双字,64 位数据称为 4 字。上述 8 个通用寄存器可以以双字的形式或对其低 16 位以字的形式被访问,其中 EAX,EBX,ECX 和 EDX 的低 16 位还可以以字节的形式被访问,这在图 2.3 的表示中已经可以看得很清楚了。当这些寄存器以字或字节形式被访问时,不被访问的其他部分不受影响,如访问 AX 时,EAX 的高 16 位不受影响。

此外,这 8 个通用寄存器还可用于其他目的。在 8086/8088 和 80286 中只有 4 个指针和变址寄存器以及 BX 寄存器可以存放偏移地址,用于存储器寻址。在 80386 及其后继机型中,所有 32 位通用寄存器既可以存放数据,也可以存放地址。也就是说,这些寄存器都可以用于存储器寻址。在这 8 个通用寄存器中,每个寄存器的专用特性与 8086/8088 和 80286 的 AX,BX,CX,DX,SP,BP,DI,SI 是一一对应的。如 EAX 专用于乘、除法和 I/O 指令,ECX 的计数特性,EDI 和 ESI 作为串处理指令专用的地址寄存器等。

2. 专用寄存器

8086/8088 和 80286 的专用寄存器包括 IP、SP 和 FLAGS 3 个 16 位寄存器。

IP(instruction pointer)为指令指针寄存器,它用来存放代码段中的偏移地址。在程序运行的过程中,它始终指向下一条指令的首地址,它与段寄存器 CS 联用确定下一条指令的物理地址。当这一地址送到存储器后,控制器可以取得下一条要执行的指令,而控制器一旦取得这条指令就马上修改 IP 的内容,使它指向下一条指令的首地址。可见,计算机就是用 IP 寄存器来控制指令序列的执行流程的,因此 IP 寄存器是计算机中很重要的一个控制寄存器。

SP 为堆栈指针寄存器,它与堆栈段寄存器联用来确定堆栈段中栈顶的地址,也就是说 SP 用来存放栈顶的偏移地址。

FLAGS 为标志寄存器,又称程序状态寄存器(program status word,PSW)。这是一个存放条件码标志、控制标志和系统标志的寄存器。

80386 及其后继机型也有三个 32 位专用寄存器,它们是 EIP、ESP 和 EFLAGS。它们的作用和相应的 16 位寄存器相同。

下面介绍标志寄存器。图 2.4 说明了 80x86 中标志寄存器的内容,图中未标明的位暂不用。

(1) 条件码标志用来记录程序中运行结果的状态信息,它们是根据有关指令的运行结果由 CPU 自动设置的。由于这些状态信息往往作为后续条件转移指令的转移控制条件,所以称为条件码。它包括以下 6 位:

溢出标志(overflow flag,OF),在运算过程中,如操作数超出了机器能表示的范围称为溢出。此时 OF 位 1,否则置 0。

符号标志(sign flag,SF),记录运算结果的符号,结果为负时置 1,否则置 0。

零标志(zero flag,ZF),运算结果为 0 时 ZF 位 1,否则置 0。

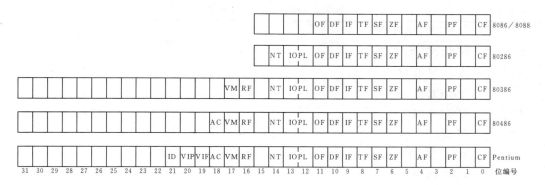

图 2.4 80x86 的标志寄存器

进位标志(carry flag,CF),记录运算时从最高有效位产生的进位值。例如,执行加法指令时,最高有效位有进位时置 1,否则置 0。

辅助进位标志(auxiliary carry flag,AF),记录运算时第 3 位(半个字节)产生的进位值。例如,执行加法指令第 3 位有进位时置 1,否则置 0。

奇偶标志(parity flag,PF),用来为机器中传送信息时可能产生的代码出错情况提供检验条件。当结果操作数中 1 的个数为偶数时置 1,否则置 0。

(2) 控制标志位为方向标志(direction flag,DF),在串处理指令中控制处理信息的方向用。当 DF 位为 1 时,每次操作后使变址寄存器 SI 和 DI 减小,这样就使串处理从高地址向低地址方向处理。当 DF 位为 0 时,则使 SI 和 DI 增大,使串处理从低地址向高地址方向处理。

(3) 系统标志位可以用于 I/O、可屏蔽中断、程序调试、任务切换和系统工作方式等的控制。一般应用程序不必关心或修改这些位的状态,只有系统程序员或需要编制低层 I/O 设备控制等程序时才需要访问其中的有关位。下面只简单介绍某些位的情况,其他各位由于在本书中不涉及,故不再加以说明。

陷阱标志(trap flag,TF),用于调试时的单步方式操作。当 TF 位为 1 时,每条指令执行完后产生陷阱,由系统控制计算机;当 TF 位为 0 时,CPU 正常工作,不产生陷阱。

中断标志(interrupt flag,IF),当 IF 位为 1 时,允许 CPU 响应可屏蔽中断请求,否则关闭中断。有关中断原理将于第 8 章说明。

I/O 特权级(I/O privilege level,IOPL),在保护模式下,用于控制对 I/O 地址空间的访问。有关内容将于第 8 章说明。

以上就是 EFLAGS 中主要标志位的含义。机器提供了设置某些状态信息的指令。必要时,程序员可使用这些指令来建立状态信息。

在调试程序 DEBUG 中提供了测试标志位的手段,它用符号表示某些标志位的值。表 2.2 说明这些标志位的符号表示。

表 2.2 标志位的符号表示

标 志 名		标志为 1	标志为 0
OF	溢出(是/否)	OV	NV
DF	方向(减量/增量)	DN	UP
IF	中断(允许/关闭)	EI	DI
SF	符号(负/正)	NG	PL
ZF	零(是/否)	ZR	NZ
AF	辅助进位(是/否)	AC	NA
PF	奇偶(偶/奇)	PE	PO
CF	进位(是/否)	CY	NC

3. 段寄存器

段寄存器也是一种专用寄存器,它们专用于存储器寻址,用来直接或间接地存放段地址。段寄存器的长度为 16 位,在 80286 以前的处理器中,只有代码段(code segment,CS)、数据段(data segment,DS)、堆栈段(stack segment,SS)和附加段(extra segment,ES)4 个寄存器。从 80386 起,增加了 FS 和 GS 两个段寄存器,它们也属于附加的数据段。有关段寄存器的使用将在 2.4 节中专门说明。

2.4 存 储 器

2.4.1 存储单元的地址和内容

计算机存储信息的基本单位是一个二进制位,一位可存储一个二进制数:0 或 1。每 8 位组成一个字节,位编号如图 2.5(1)所示。8086,80286 的字长为 16 位,由两个字节组成,位编号如图 2.5(2)所示。80386 到 Pentium II 机的字长为 32 位,由两个字即 4 个字节组成,在 80x86 系列中称其为双字,位编号如图 2.5(3)所示。此外,还有一种由 8 个字节即字长为 64 位组成的 4 字,位编号如图 2.5(4)所示。

在存储器里以字节为单位存储信息。为了正确地存放或取得信息,每一个字节单元给以一个惟一的存储器地址,称为物理地址。地址从 0 开始编号,顺序地每次加 1,因此存储器的物理地址空间是呈线性增长的。在机器里,地址也是用二进制数来表示的,当然它是无符号整数,书写格式为十六进制数。

既然每个字节单元有一个二进制数表示地址,那么 16 位二进制数可以表示多少个字节单元的地址呢?显然答案应该是有 2^{16} 个,所以它可以表示的地址范围应该是 0~65535,即 64K,其地址编号的范围用十六进制数表示为 0000~FFFFH。如下所示:

```
0000,0001,0002,0003,…,0009,000A,000B,000C,000D,000E,000F,
0010,0011,… …,     ,0019,001A,001B,001C,001D,001E,001F,
0020,0021,… …,                                    ,002F,
        ⋮
```

```
        ⋮
FFE0,FFE1,⋯ ⋯,                              ,FFEE,FFEF,
FFF0,FFF1,⋯ ⋯,                              ,FFFE,FFFF。
```

在 2.1 节中,曾提到 8086、8088 的地址总线为 20 位,那么它们可访问的字节单元地址范围为 00000H～FFFFFH;80286 的地址总线宽度为 24 位,它可访问的地址范围为 000000H～FFFFFFH;80386、80486 和 Pentium 的地址总线宽度为 32 位,相应的地址范围为 00000000H～FFFFFFFFH;而 Pentium Pro 和 Pentium Ⅱ 的地址总线宽度为 36 位,则相应的地址范围为 000000000H～FFFFFFFFFH。

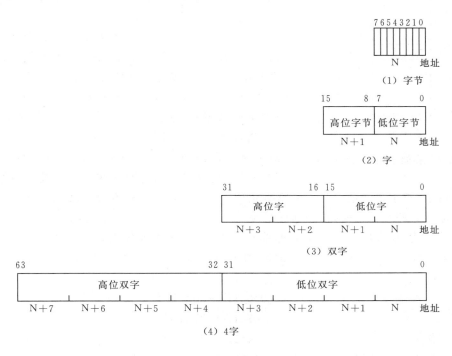

图 2.5　数据类型

一个存储单元中存放的信息称为该存储单元的内容,图 2.6 表示了存储器里存放信息的情况。可以看出,4 号字节单元中存放的信息为 78H。也就是说,4 单元中的内容为 78H,表示为:

$$(0004)=78H$$

但当机器字长是 16 位时,大部分数据都是以字为单位表示的。那么一个字怎样存入存储器呢?一个字存入存储器要占有相继的两个字节,存放时低位字节存入低地址,高位字节存入高地址(参见图 2.5)。也就是说,是以相反的次序存入的。这样二个字节单元就构成了一个字单元,字单元的地址采用它的低地址来表示。图 2.6 中 4 号字单元的内容为 5678H,表示为:

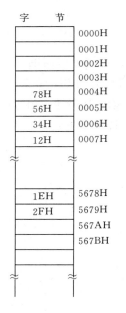

图 2.6　存储单元的地址和内容

$$(0004)=5678H$$

双字单元的存放方式与字单元类似，它被存放在相继的 4 个字节中，低位字存入低地址区，高位字存入高地址区。双字单元的地址由其最低字节的地址指定，因此 4 号双字单元的内容为：

$$(0004)=12345678H$$

80386 及其后继机型还可处理 4 字。4 字是一个 64 位的数，它由 4 个字即 8 个字节组成，它在存储器中被存放在相继的 8 个字节中，其表示方法和存放方式与以上类似，读者可自行推断。

可以看出，同一个地址既可看作字节单元的地址，又可看作字单元、双字单元或 4 字单元的地址，这要根据使用情况确定。字单元的地址可以是偶数，也可以是奇数。但是，在 8086 和 80286 中，访问存储器（要求取数或存数）都是以字为单位进行的，也就是说，机器是以偶地址访问存储器的。这样，对于奇地址的字单元，要取一个字需要访问二次存储器，当然这样做要花费较多的时间。在 80386 及其后继的 32 位处理机中，双字单元地址为 4 的整数倍时，访问存储器的速度可以较快。同样，4 字单元的地址为 8 的整数倍时，访问速度最快。

如上所述，如果用 X 表示某存储单元的地址，则 X 单元的内容可以表示为(X)；假如 X 单元中存放着 Y，而 Y 又是一个地址，则可用(Y)=((X))来表示 Y 单元的内容。如图 2.6 中

$$(0004H)=5678H$$

而

$$(5678H)=2F1EH$$

则也可记作

$$((0004H))=2F1EH$$

存储器有这样的特性：它的内容是取之不尽的。也就是说，从某个单元取出其内容后，该单元仍然保存着原来的内容不变，可以重复取出，只有存入新的信息后，原来保存的内容就自动丢失了。

2.4.2　实模式存储器寻址

80x86 中除 8086/8088 只能在实模式下工作外，其他微处理器均可在实模式或保护模式下工作。本节说明实模式存储器寻址，2.4.3 节将说明保护模式存储器寻址。

1. 存储器地址的分段

实模式下允许的最大寻址空间为 1MB。8086/8088 的地址总线宽度为 20 位，由于

$$2^{20}=1048576=1024K=1M$$

因而，其最大寻址空间正好是 1MB。而其他微处理器则在实模式下只能访问前 1MB 的

存储器地址。实际上,实模式就是为 8086/8088 而设计的工作方式,它要解决在 16 位字长的机器里怎么提供 20 位地址的问题,而解决的办法是采用存储器地址分段的方法。

程序员在编制程序时要把存储器划分成段,在每个段内地址空间是线性增长的。每个段的大小可达 64KB,这样段内地址可以用 16 位表示。实际上,可以根据需要来确定段的大小,它可以是 1B、100B、1000B 或在 64KB 范围内的任意个字节。段不能起始于任意地址,而必须从任一小段(paragraph)的首地址开始。机器规定:从 0 地址开始,每 16 个字节为一小段,下面列出了存储器最低地址区的三个小段的地址空间,每行为一小段。

$$00000,00001,00002,\cdots,0000E,0000F;$$
$$00010,00011,00012,\cdots,0001E,0001F;$$
$$00020,00021,00022,\cdots,0002E,0002F;$$
$$\vdots$$

其中,第一列就是每个小段的首地址。其特征是:在十六进制表示的地址中,最低位为 0(即 20 位地址的低 4 位为 0)。在 1MB 的地址空间里,共有 64K 个小段首地址,可表示如下:

$$00000H$$
$$00010H$$
$$\vdots$$
$$41230H$$
$$41240H$$
$$\vdots$$
$$FFFE0H$$
$$FFFF0H$$

在 1MB 的存储器里,每一个存储单元都有一个惟一的 20 位地址,称为该存储单元的物理地址。CPU 访问存储器时,必须先确定所要访问的存储单元的物理地址才能取得(或存入)该单元中的内容。20 位物理地址由 16 位段地址和 16 位偏移地址组成,段地址是指每一段的起始地址(又称段基地址),由于它必须是小段的首地址,所以其低 4 位一定是 0,这样,就可以规定段地址只取段起始地址的高 16 位值。偏移地址则是指在段内相对于段起始地址的偏移值。物理地址的计算方法,如图 2.7 所示:

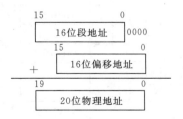

图 2.7 实模式存储器寻址时物理地址的计算方法

也就是说,把段地址左移 4 位再加上偏移地址值就形成物理地址。或写成:

16d×段地址＋偏移地址＝物理地址

图 2.8 是这种寻址方式的图示。显然,每个存储单元只有惟一的物理地址,但它却可由不同的段地址和不同的偏移地址组成。

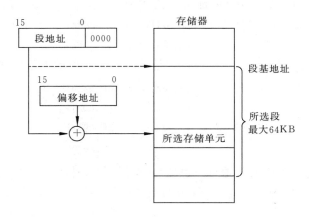

图 2.8 实模式存储器寻址

2. 段寄存器

在 8086~80286 中,有四个专门存放段地址的寄存器,称为段寄存器。它们是代码段 CS、数据段 DS、堆栈段 SS 和附加段 ES 寄存器。每个段寄存器可以确定一个段的起始地址,而这些段则各有各的用途。代码段存放当前正在运行的程序。数据段存放当前运行程序所用的数据,如果程序中使用了串处理指令,则其源操作数也存放在数据段中。堆栈段定义了堆栈的所在区域,堆栈是一种数据结构,它开辟了一个比较特殊的存储区,并以后进先出的方式来访问这一区域,在第 3 章里还会专门加以说明。附加段是附加的数据段,它是一个辅助的数据区,也是串处理指令的目的操作数存放区。程序员在编制程序时,应该按照上述规定把程序的各部分放在规定的段区之内。

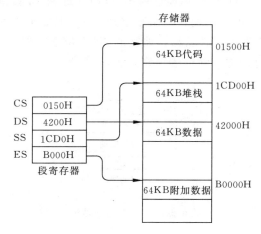

图 2.9 段分配方式之一

在80386及其后继的80x86中,除上述4个段寄存器外,又增加了2个段寄存器FS和GS,它们也是附加的数据段寄存器,所以8086~80286的程序允许4个存储段,而其他80x86程序可允许6个存储段。

除非专门指定,一般情况下,各段在存储器中的分配是由操作系统负责的。每个段可以独立地占用64KB存储区,如图2.9所示。各段也可以允许重叠,下面的例子就可以说明这种情况。

例2.1 如果代码段中的程序占有8KB(2000H)存储区,数据段占有2KB(800H)存储区,堆栈段只占有256个字节的存储区。此时段区的分配如图2.10所示。从图中可以看出,代码段的区域可以是02000~11FFFH,但由于程序区只需要8KB,所以程序区结束后的第一个小段的首地址就作为数据段的起始地址。也就是说,在这里,代码段和数据段可以重叠在一起。当然每个存储单元的内容是不允许发生冲突的,所谓的重叠只是指每个段区的大小允许根据实际需要来分配,而不一定要占有64KB的最大段空间。实际上,段区的分配工作是由操作系统完成的。但是,系统允许程序员在必要时可指定所需占用的内存区,指定方法将在第4章中说明。

如果程序中的四个段都在64KB的范围之内,而且程序运行时所需要的信息都在本程序所定义的区段之内,那么程序员只要在程序的首部设定各段寄存器的值就可以了。如果程序的某一段(例如数据段)在程序运行过程中会超过64KB空间,或者程序中可能访问除本身四个段以外的其他段区的信息,那么在程序中必须动态地修改段寄存器的内容,以保证所获得的信息的正确性,并不会因段区的划分而限制了程序空间。

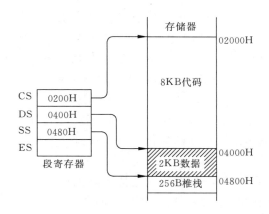

图2.10 段分配方式之二

这种存储器分段的方法虽然给程序设计带来一定的麻烦,但它可以扩大存储空间,而且对于程序的再定位也是很方便的。

在80x86中,段寄存器和与其对应存放偏移地址的寄存器之间有一种默认组合关系,如表2.3和表2.4所示:

表 2.3　8086/8088、80286 的段寄存器和相应存放偏移地址的寄存器之间的默认组合

段	偏移
CS	IP
SS	SP 或 BP
DS	BX、DI、SI 或一个 16 位数
ES	DI(用于串指令)

表 2.4　80386 及其后继机型的段寄存器和相应存放偏移地址的寄存器之间的默认组合

段	偏移
CS	EIP
SS	ESP 或 EBP
DS	EAX、EBX、ECX、EDX、EDI、ESI、一个 8 位数或一个 32 位数
ES	EDI(用于串指令)
FS	无默认
GS	无默认

在这种默认组合下,程序中不必专门指定其组合关系,但程序如用到非默认的组合关系,则必须用段跨越前缀加以说明。这一点将在第 3 章中说明。

2.4.3　保护模式存储器寻址

从 80286 起,就引出了保护模式的存储器寻址,其直接原因:首先是实模式的寻址空间为 1MB,8086/8088 本身只有 1MB 地址空间,这显然是不够用的;80286 提供了 16MB,80386 及其后继机型均提供 4GB 或更多的地址空间,那么系统要解决的首要问题就是如何寻址;其次,引出保护模式的更重要原因在于它使微机系统能支持多任务处理。

随着微机被广泛的使用,要求系统能提供多任务处理功能,即多个应用程序能同时在同一台计算机上运行,而且它们之间必须相互隔离,使一个应用程序中的缺陷或故障不会破坏系统,也不会影响其他应用程序的运行。为实现这样的要求,从 80286 起,系统就提供了保护模式存储器寻址。

在系统支持多任务功能的同时,系统也支持了虚拟存储器特性。虚拟存储器可支持程序员编写的程序具有比主存储器能提供的更大的空间。这样,即使主存储器提供的空间不够大,仍能在计算机上运行占有空间要大得多的程序。实际上,程序将存放在外存储器中,程序运行时,由操作系统进行管理,把正在执行的那部分程序调入主存储器。而保护模式寻址则对虚拟存储特性有很好地支持。由于有关计算机的多任务处理和虚拟存储器问题的讨论,均已超出本书所要讨论问题的范围,所以不在这里说明保护模式存储器寻址是如何支持上述两种功能的,只在本节简单介绍保护模式下的存储器寻址。

1. 逻辑地址

在实模式存储器寻址时,程序员只要在程序中给出存放在段寄存器中的段地址并在

指令中给出偏移地址,机器就会自动用段地址左移4位再加上偏移地址的方法,求得所选存储单元的物理地址,从而取得所要的存储单元的内容。因此,程序员在编程时并未直接指定所选存储单元的物理地址,而是给出了一个逻辑地址(即段地址:偏移地址),是机器自动用某种方法来取得所选物理地址的。

在保护模式存储器寻址中,仍然要求程序员在程序中指定逻辑地址,只是机器采用另一种比较复杂、或者说比较间接的方法来求得相应的物理地址。因此,对程序员编程来说,并未增加复杂性。在保护模式下,逻辑地址由选择器和偏移地址两部分组成,选择器存放在段寄存器中,但它不能直接表示段基地址,而由操作系统通过一定的方法取得段基地址,再和偏移地址相加,从而求得所选存储单元的物理地址。图2.11所表示的为保护模式存储器寻址的示意图。从图2.11中可以看出,它和实模式寻址的另一个区别是:偏移地址为32位长,最大段长可从64KB扩大到4GB。

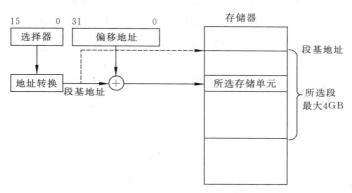

图2.11 保护模式存储器寻址示意图

2. 描述符

描述符用来描述段的大小、段在存储器中的位置及其控制和状态信息,它由基地址、界限、访问权和附加字段四部分组成。基地址(base)部分用来指定段的起始地址;界限(limit)部分存放着该段的段长度;访问权(access rights)部分用来说明该段在系统中的功能,并给出访问该段的一些控制信息;附加字段部分在386及其后继机型中存在,它用来表示该段的一些属性。描述符的内容是由系统设置的,而不是由用户建立的。

系统按选择器的内容,根据指定的途径可以找到所选段对应的描述符,从而可以根据其给出的基地址和界限值,确定所要找的存储单元所在的段,再加上逻辑地址中指定的偏移地址,就可以找到相应的存储单元。

2.5 外部设备

计算机运行时的程序和数据都要通过输入设备送入机器,程序运行的结果要通过输出设备送给用户,所以输入、输出设备是计算机必不可少的组成部分。大容量的外存储器(如磁盘、光盘)能存储大量信息,也是现代计算机不可缺少的一部分。对于外部设备的管理是汇编语言的重要使用场合之一。

从图2.1可见,外部设备与主机(CPU和存储器)的通信是通过外设接口进行的。每个接口包括一组寄存器。一般说来,这些寄存器有三种不同的用途:

(1) 数据寄存器,用来存放要在外设和主机间传送的数据,这种寄存器实际上起缓冲器的作用。

(2) 状态寄存器,用来保存外部设备或接口的状态信息,以便CPU在必要时测试外设状态,了解外设的工作情况。例如,每个设备都有忙闲位用来标志设备当前是否正在工作,是否有空接受CPU给予的新任务等。

(3) 命令寄存器,CPU给外设或接口的控制命令通过此寄存器送给外部设备。例如,CPU要启动磁盘工作,必须发出启动命令等。

各种外部设备都有以上三种类型的寄存器,只是每个接口所配备的寄存器数量是根据设备的需要确定的。例如,工作方式较简单、速度又慢的键盘只有一个8位的数据寄存器,并把状态和命令寄存器合二为一个控制寄存器。又如,工作速度快、工作方式又比较复杂的磁盘则需要多个数据、状态和命令寄存器。

为使主机访问外设方便起见,外设中的每个寄存器给予一个端口(port)地址(又称端口号),这样就组成了一个独立于内存储器的I/O地址空间。80x86机的I/O地址空间可达64KB,所以端口地址的范围是0000～FFFFH,用16位二进制代码来表示。端口可以是8位或16位的,386及其后继机型还可以有32位端口,但整个I/O空间不允许超过64KB。

主机与外设交换信息是通过输入、输出指令来完成的,在第3章中还要专门说明。主机与外设的信息传送方式可有直接、查询、中断、成组传送等,将在第8章说明。实际上,对外设的管理及信息传送是汇编语言最经常使用的场合,也是最复杂的一部分程序。

为了便于用户使用外设,80x86提供了两种类型的例行程序供用户调用。一种是基本输入输出系统(basic input/output system,BIOS),另一种是磁盘操作系统(disk operating system,DOS)功能调用。它们都是系统编制的子程序,通过中断方式转入所需要的子程序去执行,执行完后返回原来的程序继续执行。这些例行程序有的完成一次简单的外设信息传送,如从键盘输入一个字符,或送一个字符到显示器等;也有的完成相当复杂的一次外设操作,如磁盘读写一个文件等。总之,操作系统把一些复杂的外设操作编成例行程序,使用户用简单的中断指令(INT)就可以进入这些例行程序,完成所需要的外设操作。所以,用户应尽量利用这些系统所提供的工具来编写自己的程序,本书将在第9章中讨论它们的使用方法,第10章和第11章说明几种主要外设的典型应用程序。

BIOS和DOS功能调用虽然都是系统提供的例行程序,但它们之间又有差别。BIOS存放在机器的只读存储器ROM中,所以可把它看成是机器硬件的一个组成部分,它的层次比DOS更低,更接近硬件,因此它的语句要完成每一个对设备的直接命令或信息传送。DOS功能调用是操作系统DOS的一个组成部分,它在开机时由磁盘装入存储器,在它的例行程序中可以一次或多次调用BIOS,以完成比BIOS更高级的功能。用户需要使用外设时,应尽可能使用层次较高的DOS功能调用。但有时它不能满足要求,就需要直接调用BIOS,如果BIOS还不能解决问题,就只好自己编制中断处理程序了。

习 题

2.1 在 80x86 微机的输入输出指令中，I/O 端口号通常是由 DX 寄存器提供的，但有时也可以在指令中直接指定 00~FFH 的端口号。试问可直接由指令指定的 I/O 端口数。

2.2 有两个 16 位字 1EE5H 和 2A3CH 分别存放在 80x86 微机的存储器的 000B0H 和 000B3H 单元中，请用图表示出它们在存储器里的存放情况。

2.3 80x86 微机的存储器中存放信息如图 2.12 所示。试读出 30022H 和 30024H 字节单元的内容，以及 30021H 和 30022H 字单元的内容。

```
         存储器
          ⋮
30020  | 12H |
30021  | 34H |
30022  | ABH |
30023  | CDH |
30024  | EFH |
          ⋮
```

图 2.12 2.3 题的信息存放情况

2.4 在实模式下，段地址和偏移地址为 3017:000A 的存储单元的物理地址是什么？如果段地址和偏移地址是 3015:002A 和 3010:007A 呢？

2.5 如果在一个程序开始执行以前（CS）=0A7F0H（如十六进制数的最高位为字母，则应在其前加一个 0），(IP)=2B40H，试问该程序的第一个字的物理地址是多少？

2.6 在实模式下，存储器中每一段最多可有 10000H 个字节。如果用调试程序 DEBUG 的 r 命令在终端上显示出当前各寄存器的内容如下，请画出此时存储器分段的示意图，以及条件标志 OF、SF、ZF、CF 的值。

```
C>debug
-r
AX=0000   BX=0000   CX=0079   DX=0000   SP=FFEE   BP=0000   SI=0000
DI=0000
DS=10E4   ES=10F4   SS=21F0   CS=31FF   IP=0100   NV  UP  DI  PL  NZ
NA  PO  NC
```

2.7 下列操作可使用那些寄存器？

(1) 加法和减法

(2) 循环计数

(3) 乘法和除法

(4) 保存段地址

(5) 表示运算结果为 0

(6) 将要执行的指令地址
(7) 将要从堆栈取出数据的地址

2.8 那些寄存器可以用来指示存储器地址?

2.9 请将下列左边的项和右边的解释联系起来(把所选的字母写在括号内):

(1) CPU　　　　　（　）　A. 保存当前栈顶地址的寄存器。

(2) 存储器　　　　（　）　B. 指示下一条要执行的指令的地址。

(3) 堆栈　　　　　（　）　C. 存储程序、数据等信息的记忆装置,微机有 RAM 和 ROM 两种。

(4) IP　　　　　　（　）　D. 以后进先出方式工作的存储空间。

(5) SP　　　　　　（　）　E. 把汇编语言程序翻译成机器语言程序的系统程序。

(6) 状态标志　　　（　）　F. 惟一代表存储空间中每个字节单元的地址。

(7) 控制标志　　　（　）　G. 能被计算机直接识别的语言。

(8) 段寄存器　　　（　）　H. 用指令的助记符、符号地址、标号等符号书写程序的语言。

(9) 物理地址　　　（　）　I. 把若干个模块连接起来成为可执行文件的系统程序。

(10) 汇编语言　　　（　）　J. 保存各逻辑段的起始地址的寄存器,8086/8088 机有 4 个:CS、DS、SS、ES。

(11) 机器语言　　　（　）　K. 控制操作的标志,如 DF 位。

(12) 汇编程序　　　（　）　L. 记录指令操作结果的标志,共 6 位:OF、SF、ZF、AF、PF 和 CF。

(13) 连接程序　　　（　）　M. 分析、控制并执行指令的部件,由算术逻辑部件 ALU 和寄存器组等组成。

(14) 指令　　　　　（　）　N. 由汇编程序在汇编过程中执行的指令。

(15) 伪指令　　　　（　）　O. 告诉 CPU 要执行的操作(一般还要指出操作数地址),在程序运行时执行。

第 3 章　80x86 的指令系统和寻址方式

　　计算机是通过执行指令序列来解决问题的,因而每种计算机都有一组指令集供给用户使用,这组指令集就称为计算机的指令系统。目前,一般小型或微型计算机的指令系统可以包括几十种或百余种指令。本章介绍 80x86 的指令系统以及在指令中为取得操作数地址所使用的寻址方式(addressing mode)。

　　计算机中的指令由操作码字段和操作数字段两部分组成。操作码字段指示计算机所要执行的操作,而操作数字段则指出在指令执行操作的过程中所需要的操作数。例如,加法指令除需要指定做加法操作外,还需提供加数和被加数。操作数字段可以是操作数本身,也可以是操作数地址或是地址的一部分,还可以是指向操作数地址的指针或其他有关操作数的信息。

　　指令的格式一般是:

| 操作码 | 操作数 | …… | 操作数 |

　　操作数字段可以有一个、二个或三个,通常称为一地址、二地址或三地址指令。例如,单操作数指令就是一地址指令,它只需要指定一个操作数,如加 1 指令只需要指出需要加 1 的操作数。大多数运算型指令可使用三地址指令:除给出参加运算的两个操作数外,还指出运算结果的存放地址。也可使用二地址指令,此时分别称两个操作数为源操作数和目的操作数。尽管在指令执行前这两个操作数都是输入操作数,但指令执行后将把运算结果存放到目的操作数的地址之中。也就是说,经过运算后,参加运算的一个操作数将会丢失,但一般不关心这个问题。如果在以后的运算中还会用到这个操作数的话,则应在运算之前先为它准备一个副本(即提前存储起来)。80x86 的大多数运算型指令就采用这种二地址指令,少数采用三地址指令。

　　指令的操作码字段在机器里的表示比较简单,只需对每一种操作指定确定的二进制代码就可以了。指令的操作数字段的情况就比较复杂了,如果操作数存放在寄存器中,则由于寄存器的数量较少,因而需要指定的操作数地址的位数就较少;但如果操作数存放在存储器里,那么一个存储单元的地址对 8086 就需要 20 位,对 80386 及其后继机型则需要 32 位。怎样设法使它在指令的操作数字段的表示中减少位数呢?另外,从程序运行时的数据结构来看,操作数常常不是单个的数,往往是成组的以表格或数组的形式存放在存储器的某一区中,在这种情况下,指令用什么方式来指定操作数地址更好呢?这在计算机的设计中是一个很重要的问题,它会影响机器运行的速度和效率。3.1 节将专门讨论这个问题。

　　计算机只能识别二进制代码,所以机器指令是由二进制代码组成的。为便于人们使用而采用汇编语言来编写程序。汇编语言是一种符号语言,它用助记符来表示操作码,用符号或符号地址来表示操作数或操作数地址,它与机器指令是一一对应的。在本书中,将使用汇编语言格式来表示指令。

3.1 80x86 的寻址方式

3.1.1 与数据有关的寻址方式

这种寻址方式用来确定操作数地址从而找到操作数。在 80x86 系列中,8086 和 80286 的字长是 16 位,一般情况下只处理 8 位和 16 位数,只是在乘、除指令中才会有 32 位数;80386 及其后继机型其字长为 32 位,因此它除可处理 8 位和 16 位操作数外,还可处理 32 位操作数,在乘除法情况下可产生 64 位数。本节下面所述例子,如处理的是 32 位数,则适用于 386 及其后继机型。数据寻址方式的讨论中均以 MOV d,s 为例,这是传送指令,第 1 操作数为目的操作数 d,第 2 操作数为源操作数 s,指令执行的结果应把 s 送到 d。

1. 立即寻址方式(immediate addressing)

操作数直接存放在指令中,紧跟在操作码之后,它作为指令的一部分存放在代码段里,这种操作数称为立即数。立即数可以是 8 位的或 16 位的。对于 386 及其后继机型则可以是 8 位或 32 位的。如果是 16 位数,则高位字节存放在高地址中,低位字节存放在低地址中;如果是 32 位数,则高位字在高地址中,低位字在低地址中。这种寻址方式如图 3.1(1)所示。

立即寻址方式用来表示常数,它经常用于给寄存器赋初值,并且只能用于源操作数字段,不能用于目的操作数字段,且源操作数长度应与目的操作数长度一致。

例 3.1　　MOV　　AL,5

则指令执行后,(AL)=05H

例 3.2　　MOV　　AX,3064H

则指令执行后,(AX)=3064H,可用图 3.2 表示。图中指令存放在代码段中,OP 表示该指令的操作码部分,3064 则为立即数,它是指令的一个组成部分。

例 3.3　　MOV　　EAX,12345678H

则指令执行后,(EAX)=12345678H。

2. 寄存器寻址方式(register addressing)

操作数在寄存器中,指令指定寄存器号。对于 16 位操作数,寄存器可以是 AX,BX,CX,DX,SI,DI,SP 和 BP;对于 8 位操作数,寄存器可以是 AL,AH,BL,BH,CL,CH,DL,DH;对于 386 及其后继机型还可以有 32 位操作数,寄存器可以是 EAX,EBX,ECX,EDX,ESI,EDI,ESP 和 EBP。这种寻址方式由于操作数就在寄存器中,不需要访问存储器来取得操作数,因而可以得到较高的运算速度。这种方式如图 3.1(2) 所示。

例 3.4(a)　　MOV　　AX,BX

如指令执行前(AX)=3064H,(BX)=1234H;则指令执行后,(AX)=1234H,(BX)

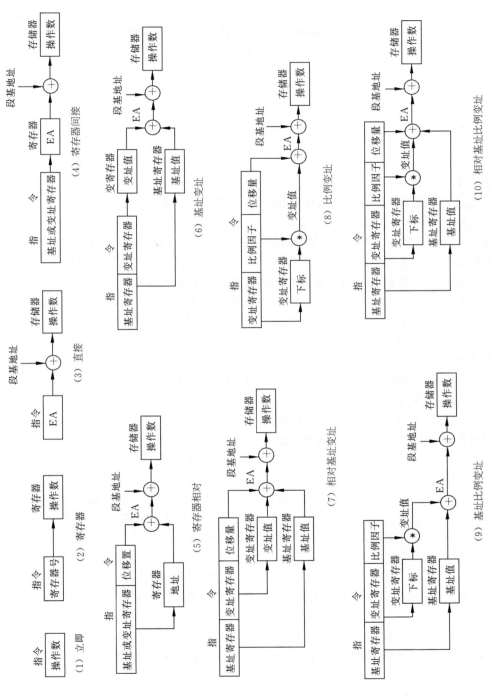

图3.1 与数据有关的寻址方式

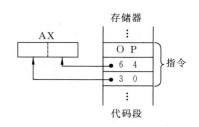

图 3.2 例 3.2 的执行情况

保持不变。

例 3.4(b) MOV ECX,EDX

如执行前(ECX)＝01237541H,(EDX)＝12345678H；则指令执行后,(ECX)＝12345678H,(EDX)保持不变。

除上述两种寻址方式外,以下各种寻址方式的操作数都在除代码段以外的存储区中。通过不同寻址方式求得操作数地址,从而取得操作数。从第 2 章里已经知道,操作数地址是由段基地址和偏移地址相加而取得的。段基地址在实模式和保护模式下可从不同途径取得。在这一节里要解决的问题是如何取得操作数的偏移地址。在 80x86 里,把操作数的偏移地址称为有效地址(effective address,EA),所以下述各种寻址方式即为求得有效地址(EA)的不同途径。

有效地址可以由以下四种成分组成：

(1) 位移量(displacement)是存放在指令中的一个 8 位、16 位或 32 位的数,但它不是立即数,而是一个地址。

(2) 基址(base)是存放在基址寄存器中的内容。它是有效地址中的基址部分,通常用来指向数据段中数组或字符串的首地址。

(3) 变址(index)是存放在变址寄存器中的内容。它通常用来访问数组中的某个元素或字符串中的某个字符。

(4) 比例因子(scale factor)是 386 及其后继机型新增加的寻址方式中的一个术语,其值可为 1,2,4 或 8。在寻址中,可用变址寄存器的内容乘以比例因子来取得变址值。这类寻址方式对访问元素长度为 2,4,8 字节的数组特别有用。

有效地址的计算可以下式表示：

$$EA = 基址 + (变址 \times 比例因子) + 位移量 \qquad (3.1)$$

这四个成分中,除比例因子是固定值外,其他三个成分都可正可负,以保证指针移动的灵活性。

8086/80286 只能使用 16 位寻址,而 80386 及其后继机型则既可用 32 位寻址,也可用 16 位寻址。在这两种情况下,对以上四种成分的组成有不同的规定,表 3.1 说明了这一规定。

注意：选择寻址方式所对应的寄存器时,必须符合这一规定。

表 3.1 16/32 位寻址时有效地址四种成分的组成

四种成分	16 位寻址	32 位寻址
位移量	0,8,16 位	0,8,32 位
基址寄存器	BX,BP	任何 32 位通用寄存器(包括 ESP)
变址寄存器	SI,DI	除 ESP 以外的 32 位通用寄存器
比例因子	无	1,2,4,8

表 3.2 则说明了各种访存类型下所对应的段的默认选择。实际上，在某些情况下，80x86 允许程序员用段跨越前缀来改变系统所指定的默认段，如允许数据存放在除 DS 段以外的其他段中，此时程序中应使用段跨越前缀。有关段跨越前缀的使用方法，本节的很多例子中将会加以说明。但在以下三种情况下，不允许使用段跨越前缀，它们是：

（1）串处理指令的目的串必须用 ES 段；

（2）PUSH 指令的目的和 POP 指令的源必须用 SS 段；

（3）指令必须存放在 CS 段中。

现在，可以来看访问存储器的数据寻址方式了。式 3.1 中的四种成分可以任意组合使用，在各种不同组合下其中每一种成分均可空缺，但比例因子只能与变址寄存器同时使用，这样可以得到 8 种不同组合的寻址方式。其中有关比例因子的三种组合只能用于 80386 及其后继机型。

表 3.2 默认段选择规则

访存类型	所用段及段寄存器	缺省选择规则
指　　令	代码段　　CS 寄存器	用于取指
堆　　栈	堆栈段　　SS 寄存器	所有的堆栈的进栈和出栈， 任何用 ESP 或 EBP 作为基址寄存器的访存
局部数据	数据段　　DS 寄存器	除相对于堆栈以及串处理指令的目的串以外的所有数据访问
目的串	附加数据段　ES 寄存器	串处理指令的目的串

3. 直接寻址方式(direct addressing)

操作数的有效地址只包含位移量一种成分，其值就存放在代码段中指令的操作码之后。位移量的值即操作数的有效地址，如图 3.1(3)所示。

例 3.5　MOV　　AX,[2000H]

如(DS)=3000H,则执行情况如图 3.3 所示。(这里用实模式来计算物理地址,本章中其他例子也如此处理。)

执行结果为:(AX)=3050H

在汇编语言指令中，可以用符号地址代替数值地址，如：

MOV　　AX,VALUE

此时 VALUE 为存放操作数单元的符号地址。如写成：

MOV　　AX,[VALUE]

也是可以的,两者是等效的。如 VALUE 在附加段中,则应指定段跨越前缀如下:
  MOV   AX,ES:VALUE
或 MOV   AX,ES:[VALUE]

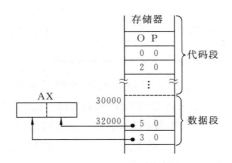

图 3.3 例 3.5 的执行情况

例 3.6 MOV   EAX,DATA

指令中的 DATA 为符号地址,其中存放着 32 位操作数,故目的操作数也应使用 32 位寄存器。

直接寻址方式适用于处理单个变量。例如,要处理某个存放在存储器里的变量,可以用直接寻址方式把该变量先取到一个寄存器中,然后再作进一步处理。

80x86 中为了使指令字不要过长,规定双操作数指令的两个操作数中,只能有一个使用存储器寻址方式,这就是一个变量常常先要送到寄存器的原因。

4. 寄存器间接寻址方式(register indirect addressing)

操作数的有效地址只包含基址寄存器内容或变址寄存器内容一种成分。因此,有效地址就在某个寄存器中,而操作数则在存储器中,如图 3.1(4)所示。根据表 3.1 中的规定,在 16 位寻址时可用的寄存器是 BX,BP,SI 和 DI;在 32 位寻址时可用 EAX,EBX,ECX,EDX,ESP,EBP,ESI 和 EDI 等 8 个通用寄存器。又根据表 3.2 中的规定,凡使用 BP,ESP 和 EBP 时,其默认段为 SS 段。其他寄存器的默认段为 DS 寄存器。

例 3.7 MOV   AX,[BX]

如果   (DS)=2000H,(BX)=1000H
则   物理地址 = 20000+1000=21000H
执行情况如图 3.4 所示。执行结果为:(AX)=50A0H

指令中也可指定段跨越前缀来取得其他段中的数据。如:
  MOV   AX,ES:[BX]

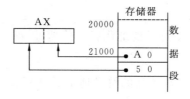

图 3.4 例 3.7 的执行情况

例 3.8 MOV   ECX,[EDX]

指令把数据段中有效地址存放在 EDX 寄存器中的 32 位操作数传送到 ECX 寄存器中。

这种寻址方式可以用于表格处理,执行完一条指令后,只需修改寄存器内容就可以取出表格的下一项。

5. 寄存器相对寻址方式(register relative addressing)(或称直接变址寻址方式)

操作数的有效地址为基址寄存器或变址寄存器的内容和指令中指定的位移量之和,所以有效地址由两种成分组成。这种寻址方式如图 3.1(5)所示。它所允许使用的寄存器及其对应的默认段情况与"4. 寄存器间接寻址方式"中所说明的相同,这里不再赘述。

例 3.9 MOV   AX,COUNT[SI]

(也可表示为 MOV   AX,[COUNT+SI])

其中 COUNT 为 16 位位移量的符号地址。

如果 (DS)=3000H, (SI)=2000H, COUNT=3000H

则物理地址=30000+2000+3000=35000H

指令执行情况如图 3.5 所示。执行结果是: (AX)=1234H

类似地,可有 MOV   EAX,TABLE[ESI]

TABLE 为 32 位位移量的符号地址,ESI 的内容指向此表格中的一项。

这种寻址方式同样可用于表格处理,表格的首地址可设置为位移量,利用修改基址或变址寄存器的内容来取得表格中的值。

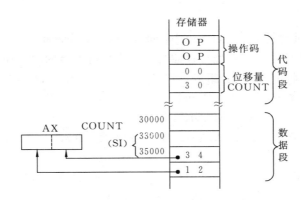

图 3.5 例 3.9 的执行情况

直接变址寻址方式也可以使用段跨越前缀。例如:

MOV   DL,ES:STRING[SI]

6. 基址变址寻址方式(based indexed addressing)

操作数的有效地址是一个基址寄存器和一个变址寄存器的内容之和,所以有效地址由两种成分组成。这种寻址方式图示如图 3.1(6)。它所允许使用的寄存器及其对应的默认段见表 3.1 和表 3.2。

例 3.10 MOV   AX,[BX][DI]

(或写为:MOV   AX,[BX+DI])

如(DS)=2100H,(BX)=0158H,(DI)=10A5H

则 EA=0158+10A5=11FDH,物理地址=21000+11FD=221FDH

指令执行情况如图3.6所示。执行结果(AX)=1234H。

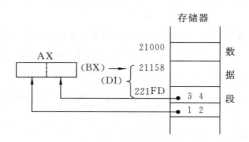

图3.6 例3.10的执行情况

类似地,对于32位寻址方式可有:

　　MOV　　EDX,[EBX][ESI]

这种寻址方式同样适用于数组或表格处理,首地址可存放在基址寄存器中,而用变址寄存器来访问数组中的各个元素。由于两个寄存器都可以修改,所以它比直接变址方式更加灵活。

此种寻址方式使用段跨越前缀时的格式为:

　　MOV　　AX,ES：[BX][SI]

7. 相对基址变址寻址方式(relative based indexed addressing)

操作数的有效地址是一个基址寄存器与一个变址寄存器的内容和指令中指定的位移量之和,所以有效地址由三种成分组成。这种寻址方式如图3.1(7)所示。它所允许使用的寄存器及其对应的默认段见表3.1和表3.2。

例3.11　　MOV　　AX,MASK[BX][SI]

(也可写成 MOV　　AX,MASK[BX+SI] 或 MOV　　AX,[MASK+BX+SI])。

如(DS)=3000H, (BX)=2000H, (SI)=1000H,MASK=0250H,

则 物理地址 = 16d×(DS)+(BX)+(SI)+MASK
　　　　　　　= 30000+2000+1000+0250 = 33250H

指令执行情况如图3.7所示。执行结果(AX)=1234H。

类似地,对于32位寻址方式可有:

　　MOV　　EAX,ARRAY[EBX][ECX]

这种寻址方式通常用于对二维数组的寻址。例如,存储器中存放着由多个记录组成的文件,则位移量可指向文件之首,基址寄存器指向某个记录,变址寄存器则指向该记录中的一个元素。这种寻址方式也为堆栈处理提供了方便,一般(BP)可指向栈顶,从栈顶到数组的首址可用位移量表示,变址寄存器可用来访问数组中的某个元素。

这种寻址方式及以下几种寻址方式使用段跨越前缀的方式与以前所述类似,不再

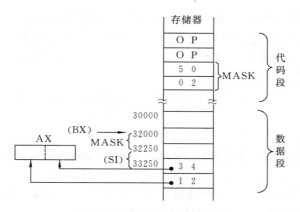

图 3.7 例 3.11 的执行情况

赘述。

以下三种寻址方式均与比例因子有关,这些寻址方式只能用在 80386 及其后继机型中,8086/80286 不支持这几种寻址方式。

8. 比例变址寻址方式(scaled indexed addressing)

操作数的有效地址是变址寄存器的内容乘以指令中指定的比例因子再加上位移量之和,所以有效地址由三种成分组成。这种寻址方式如图 3.1(8)所示。它所允许使用的寄存器及相应的默认段见表 3.1 和表 3.2。

这种寻址方式与相对寄存器寻址相比,增加了比例因子,其优点在于:对于元素大小为 2,4,8 字节的数组,可以在变址寄存器中给出数组元素下标,而由寻址方式控制直接用比例因子把下标转换为变址值。

例 3.12 MOV   EAX, COUNT[ESI * 4]

如要求把双字数组 COUNT 中的元素 3 送到 EAX 中,用这种寻址方式可直接在 ESI 中放入 3,选择比例因子 4(数组元素为 4 字节长)就可以方便地达到目的(见图 3.8),而不必像在相对寄存器寻址方式中要把变址值直接装入寄存器中。

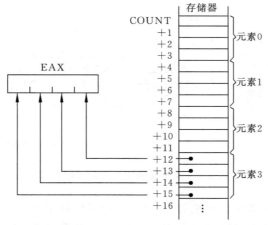

图 3.8 例 3.12 的执行情况

9. 基址比例变址寻址方式(based scaled indexed addressing)

操作数的有效地址是变址寄存器的内容乘以比例因子再加上基址寄存器的内容之和,所以有效地址由三种成分组成。这种寻址方式如图 3.1(9)所示。它所允许使用的寄存器及相应的默认段见表 3.1 和表 3.2。

这种寻址方式与基址变址寻址方式相比,增加了比例因子,其优点是很明显的,读者可自行推断。

例 3.13   MOV   ECX,[EAX][EDX*8]

10. 相对基址比例变址寻址方式(relative based scaled indexed addressing)

操作数的有效地址是变址寄存器的内容乘以比例因子,加上基址寄存器的内容,再加上位移量之和,所以有效地址由四种成分组成。这种寻址方式如图 3.1(10)所示。它所允许使用的寄存器及相应的默认段见表 3.1 和表 3.2。

这种寻址方式比相对基址变址方式增加了比例因子,便于对元素为 2,4,8 字节的二维数组的处理。

例 3.14   MOV   EAX,TABLE[EBP][EDI*4]

3.1.2 与转移地址有关的寻址方式

这种寻址方式用来确定转移指令及 CALL 指令的转向地址。

1. 段内直接寻址(intrasegment direct addressing)

转向的有效地址是当前 IP 寄存器的内容和指令中指定的 8 位或 16 位位移量之和,如图 3.9(1)所示。

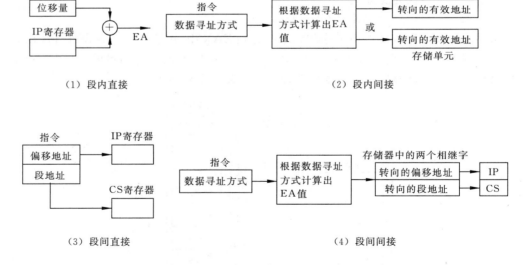

图 3.9 与转移地址有关的寻址方式

这种方式的转向有效地址用相对于当前 IP 值的位移量来表示,所以它是一种相对寻址方式。指令中的位移量是转向的有效地址与当前 IP 值之差,所以当这一程序段在内存中的不同区域运行时,这种寻址方式的转移指令本身不会发生变化,这是符合程序的再定位要求的。这种寻址方式适用于条件转移及无条件转移指令,但是当它用于条件转移指令时,位移量只允许8位(386及其后继机型条件转移指令的位移量可为8位或32位)。无条件转移指令在位移量为8位时称为短跳转,位移量为16位时则称为近跳转。

指令的汇编语言格式表示为:

JMP　　NEAR PTR PROGIA
JMP　　SHORT QUEST

其中,PROGIA 和 QUEST 均为转向的符号地址,在机器指令中,用位移量来表示。在汇编指令中,如果位移量为16位,则在符号地址前加操作符 NEAR PTR;如果位移量为8位,则在符号地址前加操作符 SHORT。

对于386及其后继机型,代码段的偏移地址存放在 EIP 中,同样用相对寻址的段内直接方式,只是其位移量为8位或32位。8位对应于短跳转;32位对应于近跳转。由于位移量本身是个带符号数,所以8位位移量的跳转范围在$-128\sim +127$的范围内;16位位移量的跳转范围为$\pm 32K$,32位位移量的跳转范围在$\pm 2G$。所有机型的汇编格式均相同。

2. 段内间接寻址(intrasegment indirect addressing)

转向有效地址是一个寄存器或是一个存储单元的内容。这个寄存器或存储单元的内容可以用数据寻址方式中除立即数以外的任何一种寻址方式取得,所得到的转向的有效地址用来取代 IP 寄存器的内容。这种寻址方式如图3.9(2)所示。

这种寻址方式以及以下的两种段间寻址方式都不能用于条件转移指令。也就是说,条件转移指令只能使用段内直接寻址的8位位移量(386及其后继机型允许8位或32位位移量),而 JMP 和 CALL 指令则可用四种寻址方式中的任何一种。

段内间接寻址转移指令的汇编格式可以表示为:

JMP　　BX
JMP　　WORD PTR[BP+TABLE]

等。其中 WORD PTR 为操作符,用以指出其后的寻址方式所取得的转向地址是一个字的有效地址,也就是说它是一种段内转移。

以上两种寻址方式均为段内转移,所以直接把求得的转向的有效地址送到 IP 寄存器就可以了。

下面举例说明在段内间接寻址方式的转移指令中,转向的有效地址的计算方法。

假设:　　　(DS)= 2000H,(BX)=1256H,(SI)= 528FH,
　　　　　位移量= 20A1H,(232F7H)= 3280H,(264E5H)= 2450H。

例 3.15　JMP　　BX
　　　　则执行该指令后(IP)= 1256H

例 3.16　JMP　　TABLE[BX]
　　　　则执行该指令后(IP)=(16d×(DS)+(BX)+位移量)
　　　　　　　　　　　　=(20000+1256+20A1)
　　　　　　　　　　　　=(232F7)

$$= 3280H$$

例 3.17　JMP　　[BX][SI]

$$则指令执行后(IP) = (16d \times (DS) + (BX) + (SI))$$
$$= (20000 + 1256 + 528F)$$
$$= (264E5)$$
$$= 2450H$$

以上说明及举例均针对 8086 的 16 位寻址来分析的,对于 386 及其后继机型除 16 位寻址方式外,还可使用 32 位寻址方式,就方法而言是与 16 位寻址完全相同的。

例 3.18　JMP　　ECX

例 3.19　JMP　　WORD PTR TABLE[ESI]

3. 段间直接寻址(intersegment direct addressing)

在指令中直接提供了转向段地址和偏移地址,所以只要用指令中指定的偏移地址取代 IP 寄存器的内容,用指令中指定的段地址取代 CS 寄存器的内容就完成了从一个段到另一个段的转移操作,如图 3.9(3)所示。

指令的汇编语言格式可表示为:

JMP　　FAR PTR NEXTROUTINT

其中,NEXTROUTINT 为转向的符号地址,FAR PTR 则是表示段间转移的操作符。

对于 386 及其后继机型,段间转移应修改 CS 和 EIP 的内容,方法仍然和 16 位寻址时一致。

4. 段间间接寻址(intersegment indirect addressing)

用存储器中的两个相继字的内容来取代 IP 和 CS 寄存器中的原始内容,以达到段间转移的目的。这里,存储单元的地址是由指令指定除立即数方式和寄存器方式以外的任何一种数据寻址方式取得,如图 3.9(4)所示。

这种指令的汇编语言格式可表示为:

JMP　　DWORD PTR[INTERS+BX]

其中,[INTERS+BX]说明数据寻址方式为直接变址寻址方式,DWORD PTR 为双字操作符,说明转向地址需取双字为段间转移指令。

对于 386 及其后继机型,除 16 位寻址方式外,还可使用 32 位寻址方式,方法上也与 16 位寻址相同。如:

JMP　　DWORD PTR [EDI]

3.2　程序占有的空间和执行时间

当我们用汇编语言编写的汇编语言程序输入计算机后,由机器提供的"汇编程序"将它翻译成由机器指令组成的机器语言程序,才能由计算机识别并执行。

一方面,80x86 的机器指令是可变字节指令,即不同指令或不同寻址方式的机器指令长度不同。一条 16 位格式指令的长度可为 1～7 个字节,32 位指令则可达 14 个字节,如计入前缀字节(如段跨越前缀等)长度还会增加。这样,一个程序一旦装入计算机,它就会占有一定的存储空间。程序量越大,占有的存储空间也越大。

另一方面,当程序在计算机上运行时,访问存储器取得操作数或存放结果需要时间,运算器执行指令也需要时间。尽管从8086到Pentium的15年左右时间内,计算机的生产厂商在提高计算机的运行速度上下了很大功夫,除时钟频率已从原来的5MHz提高到300MHz外,又在体系结构方面采用了如数据预取、高速缓冲、流水线等多项重叠或并发技术,使指令的执行速度有了很大的提高,但程序运行仍是需要时间的。

完成同样功能的不同程序,可能在占有存储空间和执行时间上有很大差别。因此,在编制程序时,如果对程序所占有的空间或程序的执行时间要求不高,那就只要根据题意编制出合乎要求的程序就可以了,当然也应该尽量在空间和时间上提高程序运行的效率。有时,对于程序所占有的存储空间或者对于程序执行的时间要求很高,在这种情况下,应仔细斟酌程序的算法、数据结构以及指令与寻址方式的选用,以便编制出符合要求的程序。

3.3 80x86的指令系统

80x86的指令系统可以分为以下6组:

数据传送指令　　　　　串处理指令
算术指令　　　　　　　控制转移指令
逻辑指令　　　　　　　处理机控制指令

下面分别加以说明。

3.3.1 数据传送指令

数据传送指令负责把数据、地址或立即数传送到寄存器或存储单元中。它又可分为五种,分别说明如下:

1. 通用数据传送指令

MOV(move)　　　　　　　　　　　　　传送
MOVSX(move with sign-extend)　　　　带符号扩展传送
MOVZX(move with zero-extend)　　　　带零扩展传送
PUSH(push onto the stack)　　　　　　进栈
POP(pop from the stack)　　　　　　　出栈
PUSHA/PUSHAD(push all registers)　　 所有寄存器进栈
POPA/POPAD(pop all registers)　　　　所有寄存器出栈
XCHG(exchange)　　　　　　　　　　　交换

(1) MOV 传送指令

格式为:MOV　　DST,SRC

执行操作:(DST)←(SRC)

其中 DST 表示目的操作数,SRC 表示源操作数。

MOV 指令的机器语言可以有 7 种格式：

① MOV   mem/reg1，mem/reg2

当然，双操作数指令不允许两个操作数都使用存储器，因而两个操作数中必须有一个是寄存器。这种方式不允许指定段寄存器。

② MOV   reg，data

其中 reg 指定寄存器，data 为立即数。当然，这种方式也不允许指定段寄存器。

③ MOV   ac，mem

其中 ac 为累加器。

④ MOV   mem，ac

⑤ MOV   segreg，mem/reg

其中 segreg 指定段寄存器，但不允许使用 CS 寄存器。此外，这条指令执行完后不响应中断，要等下一条指令执行完后才可能响应中断(有关中断处理问题，将在第 8 章说明。)。

⑥ MOV   mem/reg，segreg

⑦ MOV   mem/reg，data

这种方式的目的操作数只用存储器寻址方式而不用寄存器方式。

以上 7 种方式说明 MOV 指令可以在 CPU 内或 CPU 和存储器之间传送字或字节，386 及其后继机型还可以传送双字(80x86 系统中凡 32 位指令均为 386 及其后继机型可用，这一点以后不另加说明。)。它传送的信息可以从寄存器到寄存器，立即数到寄存器，立即数到存储单元，从存储单元到寄存器，从寄存器到存储单元，从寄存器或存储单元到除 CS 外的段寄存器(注意，立即数不能直接送段寄存器。)，从段寄存器到寄存器或存储单元。但是 MOV 指令的目的操作数不允许用立即数方式，也不允许用 CS 寄存器，而且除源操作数为立即数的情况外，两个操作数中必须有一个是寄存器。也就是说，不允许用 MOV 指令在两个存储单元之间直接传送数据。此外，也不允许在两个段寄存器间直接传送信息。还应该注意的是 MOV 指令不影响标志位。

例 3.20   MOV   AX，DATA_SEG
             MOV   DS，AX

段地址必须通过寄存器(如 AX 寄存器)送到 DS 寄存器。

例 3.21   MOV   AL，′E′

把立即数(字符 E 的 ASCII 码)送到 AL 寄存器。

例 3.22   MOV   BX，OFFSET TABLE

把 TABLE 的偏移地址(不是内容)送到 BX 寄存器。其中 OFFSET 为属性操作符，表示应把其后跟着的符号地址的值(不是内容)作为操作数。

例 3.23   MOV   AX，Y[BP][SI]

把堆栈段中有效地址为(BP)+(SI)+位移量 Y 的存储单元的内容送给 AX 寄存器。

例 3.24   MOV   EAX，[EBX+ECX*4]

该指令为 386 及其后继机型可用的 32 位指令，它可把 DS 段中有效地址为(EBX)+(ECX)*4 的存储单元的 32 位内容送给 EAX 寄存器。

(2) MOVSX 带符号扩展传送指令(386 及其后继机型可用)

格式为：MOVSX   DST，SRC

执行操作：(DST)←符号扩展(SRC)

本指令可以有两种格式：

  MOVSX reg1, reg2
  MOVSX reg, mem

该指令的源操作数可以是 8 位或 16 位的寄存器或存储单元的内容，而目的操作数则必须是 16 位或 32 位寄存器，传送时把源操作数符号扩展送入目的寄存器。可以是 8 位符号扩展到 16 位或 32 位，也可以是 16 位符号扩展到 32 位。MOVSX 不影响标志位。

  例 3.25 MOVSX EAX, CL

把 CL 寄存器中的 8 位数，符号扩展为 32 位数，送到 EAX 寄存器中。

  例 3.26 MOVSX EDX, [EDI]

把 DS 段中由 EDI 内容指定地址的 16 位数符号扩展为 32 位数，送到 EDX 寄存器中。

(3) MOVZX 带零扩展传送指令(386 及其后继机型可用)

  格式为：MOVZX DST, SRC

  执行操作：(DST)←零扩展(SRC)

  本指令可以有两种格式：

  MOVZX reg1, reg2
  MOVZX reg, mem

有关源操作数和目的操作数以及对标志位的影响均和 MOVSX 相同。它们的差别只是 MOVSX 的源操作数是带符号数，所以作符号扩展；而 MOVZX 的源操作数应是无符号整数，所以作零扩展(不管源操作数的符号位是否为 1，高位均扩展为零)。MOVSX 和 MOVZX 指令与一般双操作数指令的差别是：一般双操作数指令的源操作数和目的操作数的长度是一致的，但 MOVSX 和 MOVZX 的源操作数长度一定要小于目的操作数长度。

  例 3.27 MOVZX DX, AL

把 AL 寄存器中的 8 位数，零扩展成 16 位数，送到 DX 寄存器中。

  例 3.28 MOVZX EAX, DATA

把 DATA 单元中的 16 位数，零扩展为 32 位数，送到 EAX 寄存器中。

(4) PUSH 进栈指令

  格式为：PUSH SRC

  执行操作：

    16 位指令：(SP)←(SP)－2
         ((SP)+1, (SP))←(SRC)

    32 位指令：(ESP)←(ESP)－4
         ((ESP)+3, (ESP)+2, (ESP)+1, (ESP))←(SRC)

(5) POP 出栈指令

  格式为：POP DST

  执行操作：

16 位指令：$(DST) \leftarrow ((SP)+1, (SP))$
$(SP) \leftarrow (SP)+2$

32 位指令：$(DST) \leftarrow ((ESP)+3, (ESP)+2, (ESP)+1, (ESP))$
$(ESP) \leftarrow (ESP)+4$

这是两条堆栈的进栈和出栈指令。堆栈是以"后进先出"方式工作的一个存储区，它必须存在于堆栈段中，因而其段地址存放于 SS 寄存器中。它只有一个出入口，所以只有一个堆栈指针寄存器。当堆栈地址长度为 16 位时用 SP，堆栈地址长度为 32 位时用 ESP。SP 或 ESP 的内容在任何时候都指向当前的栈顶，所以 PUSH 和 POP 指令都必须根据当前 SP 或 ESP 的内容来确定进栈或出栈的存储单元，而且必须及时修改指针，以保证 SP 或 ESP 指向当前的栈顶。

堆栈的存取在 16 位指令中必须以字为单位（不允许字节堆栈），在 32 位指令中必须以双字为单位，所以 PUSH 和 POP 指令只能作字或双字操作。

PUSH 指令可以有四种格式：

PUSH　　reg
PUSH　　mem
PUSH　　data
PUSH　　segreg

也就是说，它可使用所有的寻址方式，但 8086 不允许 PUSH 指令使用立即数寻址方式。

POP 指令允许的格式有：

POP　　reg
POP　　mem
POP　　segreg

即 POP 指令不允许使用立即数寻址方式。还应该说明的是，POP 指令的目的为段寄存器时，不允许使用 CS 寄存器。

PUSH 和 POP 指令在操作数长度为 16 位时，SP 或 ESP 均为∓2，进栈或出栈的是字；而操作数长度为 32 位时，SP 或 ESP 均为∓4，进栈或出栈的是双字。也就是说，地址长度为 16 位时，使用 SP 作为堆栈指针，进出栈的可以是字，也可以是双字；同样，地址长度为 32 位时，使用 ESP 作为堆栈指针，进出栈的可以是双字，也可以是字。PUSH 和 POP 指令执行的操作，如表 3.3 中的说明。

表 3.3　PUSH/POP 指令执行的操作

操 作 数 长 度（位）	地 址 长 度（位）	执行的操作	
16	16	SP←SP±2	字出栈或进栈
16	32	ESP←ESP±2	字出栈或进栈
32	16	SP←SP±4	双字出栈或进栈
32	32	ESP←ESP±4	双字出栈或进栈

需要说明的特殊情况是：8086 中的 PUSH　　SP 指令入栈的是该指令已修改了的 SP 新值，而 PUSH　　ESP 指令入栈的却是 ESP 在执行该指令之前的旧值。此外，

PUSH 和 POP 指令在使用与存储器有关的寻址方式且用 ESP 作为基址寄存器时,PUSH 指令使用该指令执行前的 ESP 内容,POP 指令则使用该指令执行后的 ESP 内容来计算基地址。

PUSH 和 POP 指令均不影响标志位。

例 3.29　PUSH　　AX

指令执行情况如图 3.10 所示。

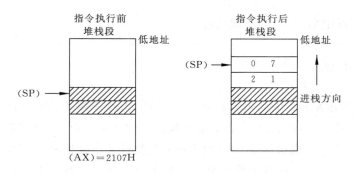

图 3.10　PUSH AX 指令的执行情况

例 3.30　POP　　AX

指令执行情况如图 3.11 所示。

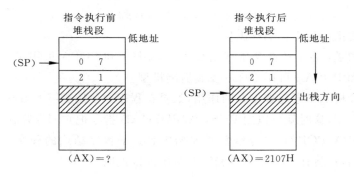

图 3.11　POP AX 指令的执行情况

例 3.31　PUSH　　[EAX]

该指令为 386 及其后继机型可用的 32 位指令,它可把 DS 段中以 EAX 内容为指针所指向的存储单元内容进入堆栈,指令执行后 ESP 内容应减 4,仍指向当前栈顶。

例 3.32　PUSH　　12H

286 及其后继机型允许立即数进栈。

(6) PUSHA/PUSHAD　所有寄存器进栈指令

格式为:PUSHA
　　　　PUSHAD

执行操作：

  PUSHA：16 位通用寄存器依次进栈，进栈次序为：AX，CX，DX，BX，指令执行前的 SP，BP，SI，DI。指令执行后(SP)←(SP)－16 仍指向栈顶。

  PUSHAD：32 位通用寄存器依次进栈，进栈次序为：EAX，ECX，EDX，EBX，指令执行前的 ESP，EBP，ESI 和 EDI。指令执行后(SP)←(SP)－32。

(7) POPA/POPAD 所有寄存器出栈指令

格式为：POPA

    POPAD

执行操作：

  POPA：16 位通用寄存器依次出栈，出栈次序为：DI，SI，BP，SP，BX，DX，CX，AX，指令执行后(SP)←(SP)＋16 仍指向栈顶。应该说明的是：SP 的出栈只是修改了指针使其后的 BX 能顺利出栈，而堆栈中原先由 PUSHA 指令存入的 SP 的原始内容被丢弃，并未真正送到 SP 寄存器中去。

  POPAD：32 位通用寄存器依次出栈，出栈次序为：EDI，ESI，EBP，ESP，EBX，EDX，ECX，EAX。指令执行后(ESP)←(ESP)＋32 仍指向栈顶。与 POPA 相同，堆栈中存放的原 ESP 的内容被丢弃而不装入 ESP 寄存器。

这两条堆栈指令均不影响标志位。

应该说明的是：在上述两条堆栈指令中，PUSHA 和 POPA 可用于 286 及其后继机型；PUSHAD 和 POPAD 可用于 386 及其后继机型。

堆栈在计算机工作中起着重要的作用，如果在程序中要用到某些寄存器，但它们的内容却在将来还有用，这时就可以用 PUSHA/PUSHAD 指令把它们保存在堆栈中，然后在需要时再用 POPA/POPAD 指令恢复其原始内容。子程序结构的程序和中断程序中就经常会用到它们，这将在以后第 6 章和第 8 章中加以说明。

例 3.33 PUSHAD

指令执行情况如图 3.12 所示。

(8) XCHG 交换指令

格式为：XCHG OPR1，OPR2

执行操作：(OPR1)↔(OPR2)

其中 OPR 表示操作数。该指令的两个操作数中必须有一个在寄存器中，因此它可以在寄存器之间或者在寄存器与存储器之间交换信息，但不允许使用段寄存器。指令允许字或字节操作，386 及其后继机型还允许双字操作。该指令可用除立即数外的任何寻址方式，且不影响标志位。

例 3.34 XCHG BX，[BP＋SI]

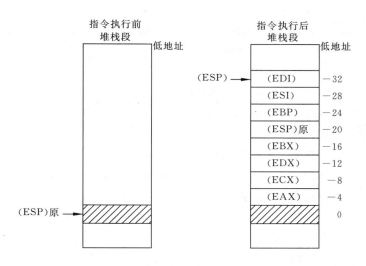

图 3.12 PUSHAD 指令的执行情况

如指令执行前：

(BX)=6F30H，(BP)=0200H，(SI)=0046H，(SS)=2F00H，(2F246H)=4154H

OPR2 的物理地址=2F000+0200+0046=2F246

则指令执行后：

(BX)=4154H　　(2F246H)=6F30H

例 3.35　XCHG　　EAX，EBX

EAX 和 EBX 寄存器的内容互换。

2. 累加器专用传送指令

IN(input)　　　　　输入
OUT(output)　　　 输出
XLAT(translate)　　换码

这组指令只限于使用累加器 EAX，AX 或 AL 传送信息。

(1) IN 输入指令

长格式为：　　　IN　　AL，PORT(字节)
　　　　　　　　IN　　AX，PORT(字)
　　　　　　　　IN　　EAX，PORT(双字)

执行的操作：　　(AL)←(PORT)(字节)
　　　　　　　　(AX)←(PORT+1，PORT)(字)
　　　　　　　　(EAX)←(PORT+3，PORT+2，PORT+1，PORT)(双字)

短格式为：　　　IN　　AL，DX(字节)
　　　　　　　　IN　　AX，DX(字)
　　　　　　　　IN　　EAX，DX(双字)

执行的操作：　　(AL)←((DX))(字节)

$(AX) \leftarrow ((DX)+1,(DX))$(字)

$(EAX) \leftarrow ((DX+3,(DX)+2,(DX)+1,(DX))$(双字)

(2) OUT 输出指令

长格式为：　　　OUT　PORT, AL(字节)

　　　　　　　OUT　PORT, AX(字)

　　　　　　　OUT　PORT, EAX(双字)

执行的操作：　$(PORT) \leftarrow (AL)$(字节)

　　　　　　$(PORT+1, PORT) \leftarrow (AX)$(字)

　　　　　　$(PORT+3, PORT+2, PORT+1, PORT) \leftarrow (EAX)$(双字)

短格式为：　　　OUT　DX, AL(字节)

　　　　　　　OUT　DX, AX(字)

　　　　　　　OUT　DX, EAX(双字)

执行的操作：　$((DX)) \leftarrow (AL)$(字节)

　　　　　　$((DX+1,(DX)) \leftarrow (AX)$(字)

　　　　　　$((DX)+3,(DX)+2,(DX)+1,(DX)) \leftarrow (EAX)$(双字)

在 80x86 里，所有 I/O 端口与 CPU 之间的通信都由 IN 和 OUT 指令来完成。其中 IN 完成从 I/O 到 CPU 的信息传送，而 OUT 则完成从 CPU 到 I/O 的信息传送。CPU 只能用累加器(AL 或 AX 或 EAX)接收或发送信息。外部设备最多可有 65536 个 I/O 端口，端口号(即外部设备的端口地址)为 0000～FFFFH。其中前 256 个端口(0～FFH)可以直接在指令中指定，这就是长格式中的 PORT，此时机器指令用两个字节表示，第 2 个字节就是端口号。所以用长格式时可以在指令中直接指定端口号，但只限于外设的前 256 个端口。当端口号≥256 时，只能使用短格式，此时必须先把端口号放到 DX 寄存器中(端口号可以从 0000～0FFFFH)，然后再用 IN 或 OUT 指令来传送信息。

必须注意：这里的端口号或 DX 的内容均为地址，而传送的是端口中的信息，而且在用短格式时 DX 内容就是端口号本身，不需要由任何段寄存器来修改它的值。

IN 和 OUT 指令提供了双字、字和字节三种使用方式，选用哪一种，则取决于外设端口宽度。如端口宽度只有 8 位，则只能用字节指令传送信息。

输入、输出指令不影响标志位。

例 3.36　　IN　　　AX, 28H

　　　　　MOV　　DATA_WORD, AX

这两条指令把端口 28 的内容经过 AX 传送到存储单元 DATA_WORD 中。

例 3.37　　MOV　　DX, 3FCH

　　　　　IN　　　EAX, DX

从端口 03FCH 送一个双字到 EAX 寄存器。

例 3.38　　OUT　　5, AL

从 AL 寄存器输出一个字节到端口 5。

（3）XLAT   换码指令

格式为：XLAT   OPR

或 XLAT

执行的操作：

16 位指令：(AL)←((BX)+(AL))

32 位指令：(AL)←((EBX)+(AL))

经常需要把一种代码转换为另一种代码。例如，把字符的扫描码转换成 ASCII 码，或者把数字 0～9 转换成 7 段数码管所需要的相应代码等，XLAT 就是为这种用途所设置的指令。在使用这条指令之前，应先建立一个字节表格，表格的首地址提前存入 BX 或 EBX 寄存器，需要转换的代码应该是相对于表格首地址的位移量也提前存放在 AL 寄存器中，表格的内容则是所要换取的代码，该指令执行后就可在 AL 中得到转换后的代码。该指令可用 XLAT 或 XLAT OPR 两种格式中的任一种，使用 XLAT OPR 时，OPR 为表格的首地址（一般为符号地址），但在这里的 OPR 只是为提高程序的可读性而设置的，指令执行时只使用预先已存入 BX 或 EBX 中的表格首地址，而并不用汇编格式中指定的值。该指令不影响标志位。

例 3.39   如(BX)=0040H,(AL)=0FH,(DS)=F000H

所建立的表格如图 3.13 所示。

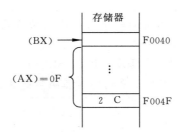

图 3.13 例 3.39 所用的表格

指令   XLAT

把 F0000+0040+0F=F004F 的内容送 AL（请注意，相加时 AL 的内容应零扩展到 16 位或 32 位），所以指令执行后：(AL)=2CH，即指令把 AL 中的代码 0FH 转换为 2CH。

必须注意，由于 AL 寄存器只有 8 位，所以表格的长度不能超过 256。

3．地址传送指令

LEA（load effective address）   有效地址送寄存器

LDS（load DS with pointer）   指针送寄存器和 DS

LES（load ES with pointer）   指针送寄存器和 ES

LFS（load FS with pointer）   指针送寄存器和 FS

LGS（load GS with pointer）   指针送寄存器和 GS

LSS（Load SS with Pointer）   指针送寄存器和 SS

这一组指令完成把地址送到指定寄存器的功能。

(1) LEA 有效地址送寄存器指令

格式为：LEA   REG,SRC

执行的操作：(REG)←SRC

指令把源操作数的有效地址送到指定的寄存器中。

该指令的目的操作数可使用 16 位或 32 位寄存器,但不能使用段寄存器。源操作数可使用除立即数和寄存器外的任一种存储器寻址方式。由于存在操作数长度和地址长度的不同,该指令执行的操作如表 3.4 所示。该指令不影响标志位。

表 3.4 LEA 指令执行的操作

操作数长度	地址长度	执行的操作
16	16	计算得的 16 位有效地址存入 16 位目的寄存器
16	32	计算得的 32 位有效地址,截取低 16 位存入 16 位目的寄存器
32	16	计算得的 16 位有效地址,零扩展后存入 32 位目的寄存器
32	32	计算得的 32 位有效地址存入 32 位目的寄存器

例 3.40 LEA   BX,[BX+SI+0F62H]

如指令执行前(BX)=0400H,(SI)=003CH,

则指令执行后(BX)=0400+003C+0F62=139EH。

必须注意：在这里 BX 寄存器得到的是有效地址而不是该存储单元的内容。如果指令为：

MOV   BX,[BX+SI+0F62H]

则 BX 中得到的是偏移地址为 139EH 单元的内容而不是其偏移地址。

例 3.41 LEA   BX,LIST

MOV   BX,OFFSET   LIST

可以看出,这两条指令在功能上是相同的,BX 寄存器中都可得到符号地址 LIST 的值,而且此时 MOV 指令的执行速度会比 LEA 指令更快。但是,OFFSET 只能与简单的符号地址相连,而不能和诸如 LIST[SI]或[SI]等复杂操作数相连(其原因将在第 4 章中说明)。因此,LEA 指令在取得访问变量的工具方面是很有用的。本组的其他指令也是为同一目的服务的。

(2) LDS、LES、LFS、LGS 和 LSS 指针送寄存器和段寄存器指令

格式以 LDS 为例为：LDS   REG,SRC

其他指令格式与 LDS 指令格式相同,仅指定的段寄存器不同。

执行的操作：(REG)←(SRC)

(SREG)←(SRC+2)

或   (SREG)←(SRC+4)

该组指令的源操作数只能用存储器寻址方式,根据任一种存储器寻址方式找到一个存储

单元。当指令指定的是 16 位寄存器时,把该存储单元中存放的 16 位偏移地址(即(SRC))装入该寄存器中,然后把(SRC+2)中的 16 位数装入指令指定的段寄存器中;当指令指定的是 32 位寄存器时,把该存储单元中存放的 32 位偏移地址装入该寄存器中,然后把(SRC+4)中的 16 位数装入指令指定的段寄存器中。

本组指令的目的寄存器不允许使用段寄存器,LFS、LGS 和 LSS 只能用于 386 及其后继机型中。本组指令不影响标志位。

例 3.42　　LES　　DI,[BX]

如指令执行前(DS)=B000H,(BX)=080AH,(0B080AH)=05AEH,(0B080CH)=4000H,

则指令执行后(DI)=05AEH,(ES)=4000H。

例 3.43　　LSS　　ESP,MEM

把 MEM 单元中存放的 48 位地址分别装入 ESP 和 SS 寄存器中。

4. 标志寄存器传送指令

```
LAHF(load AH with flags)              标志送 AH
SAHF(store AH into flags)             AH 送标志寄存器
PUSHF/PUSHFD(push the flags or eflags)  标志进栈
POPF/POPFD(pop the flags or eflags)     标志出栈
```

(1) LAHF 标志送 AH 指令

格式为:LAHF

执行的操作:(AH)←(FLAGS 的低字节)

(2) SAHF　AH 送标志寄存器指令

格式为:SAHF

执行的操作:(FLAGS 的低字节)←(AH)

(3) PUSHF/PUSHFD　标志进栈指令

格式为:PUSHF
　　　　PUSHFD

执行的操作:

PUSHF: (SP)←(SP)−2
　　　　((SP)+1,(SP))←(FLAGS)

PUSHFD: (ESP)←(ESP)−4
　　　　((ESP)+3,(ESP)+2,(ESP)+1,(ESP))←(EFLAGS AND 0FCFFFFH)(清除 VM 和 RF 位)

(4) POPF/POPFD　标志出栈指令

格式为:POPF
　　　　POPFD

执行的操作:

POPF: (FLAGS)←((SP)+1,(SP))

(SP)←(SP)+2
POPFD：(EFLAGS)←((ESP)+3,(ESP)+2,(ESP)+1,(ESP))
(ESP)←(ESP)+4

这组指令中的 LAHF 和 PUSHF/PUSHFD 不影响标志位。SAHF 和 POPF/POPFD 则由装入的值来确定标志位的值,但 POPFD 指令不影响 VM,RF,IOPL,VIF 和 VIP 的值。

5. 类型转换指令

CBW(convert byte to word)　　　　　　字节转换为字
CWD/CWDE(convert word to double word)　字转换为双字
CDQ(convert double to quad)　　　　　双字转换为 4 字
BSWAP(byte swap)　　　　　　　　　字节交换

(1) CBW　字节转换为字指令

格式：CBW

执行的操作：AL 的内容符号扩展到 AH,形成 AX 中的字。即如果(AL)的最高有效位为 0,则(AH)=0；如(AL)的最高有效位为 1,则(AH)=0FFH。

(2) CWD/CWDE　字转换为双字指令

格式：CWD

执行的操作：AX 的内容符号扩展到 DX,形成 DX：AX 中的双字。即如果(AX)的最高有效位为 0,则(DX)=0；如(AX)的最高有效位为 1,则(DX)=0FFFFH。

格式：CWDE

执行的操作：AX 的内容符号扩展到 EAX,形成 EAX 中的双字。

(3) CDQ　双字转换为 4 字指令

格式：CDQ

执行的操作：EAX 的内容符号扩展到 EDX,形成 EDX：EAX 中的 4 字。

(4) BSWAP　字节交换指令

格式：BSWAP　　r32

该指令只能用于 486 及其后继机型。r32 指 32 位寄存器。

执行的操作：使指令指定的 32 位寄存器的字节次序变反。具体操作为：1、4 字节互换,2、3 字节互换。

例 3.44　　BSWAP　　EAX

如指令执行前(EAX)=11223344H,则指令执行后(EAX)=44332211,字节次序变反。

本组指令均不影响标志位。

3.3.2　算术指令

80x86 的算术运算指令包括二进制运算和十进制运算指令。算术指令用来执行算术运算,它们中有双操作数指令,也有单操作数指令。如前所述,双操作数指令的两个操作数中除源操作数为立即数的情况外,必须有一个操作数在寄存器中。单操作数指令不允许使用立即数方式。算术指令的寻址方式均遵循这一规则。

1. 加法指令

ADD(add)　　　　　　　　加法
ADC(add with carry)　　　带进位加法
INC(increment)　　　　　　加1
XADD(exchange and add)　交换并相加

(1) ADD 加法指令

格式：ADD　　DST，SRC

执行的操作：(DST)←(SRC)+(DST)

(2) ADC 带进位加法指令

格式：ADC　　DST，SRC

执行的操作：(DST)←(SRC)+(DST)+CF

其中 CF 为进位位的值。

(3) INC 加1指令

格式：INC　　OPR

执行的操作：(OPR)←(OPR)+1

以上三条指令都可作字或字节运算,386 及其后继机型还可作双字操作,而且除 INC 指令不影响 CF 标志外,它们都影响条件标志位。

(4) XADD 交换并相加指令

格式：XADD　　DST，SRC

执行的操作：TEMP←(SRC)+(DST)
　　　　　　(SRC)←(DST)
　　　　　　(DST)←TEMP

该指令只能用于 486 及其后继机型,它把目的操作数装入源,并把源和目的操作数之和送目的地址。该指令的源操作数只能用寄存器寻址方式,目的操作数则可用寄存器或任一种存储器寻址方式。指令可作双字、字或字节运算。它对标志位的影响和 ADD 指令相同。

条件标志(或称条件码)位中最主要的是 CF,ZF,SF,OF 四位,分别表示进位、结果为零、符号和溢出的情况。其中 ZF 和 SF 位的设置比较简单,在第 2 章中已经说明,这里不再赘述。这里将进一步分析 CF 和 OF 位的设置情况。

执行加法指令时,CF 位是根据最高有效位是否有向高位的进位设置的。有进位时 CF=1,无进位时 CF=0。OF 位则根据操作数的符号及其变化情况来设置:若两个操作数的符号相同,而结果的符号与之相反时 OF=1,否则 OF=0。溢出位 OF 既然是根据数的符号及其变化来设置的,当然它是用来表示带符号数的溢出的,从其设置条件来看结论也是明显的。那么,进位位 CF 的意义是什么呢？在第 1 章说明补码运算规则的过程中,曾经提到在补码加、减法中,从最高有效位向高位的进位说明模运算中的进位自动丢失现象,对结果并没有影响。那么,为什么还要保存 CF 位呢？

CF 位可以用来表示无符号数的溢出。一方面,由于无符号数的最高有效位只有数值意义而无符号意义,所以从该位产生的进位应该是结果的实际进位值,但是在有限数位的范围内就说明了结果的溢出情况;另一方面,它所保存的进位值有时是有用的。例如,双字长数运算时,可以利用进位值把低位字的进位计入高位字中等。这可以根据不同情况在程序中加以处理。

8 位二进制数可以表示十进制数的范围是:无符号数为 0～255,带符号数为 －128～＋127。16 位二进制数可以表示十进制数的范围是:无符号数为 0～65535,带符号数为 －32768～＋32767。

下面以 8 位数为例分析一下数的溢出情况。

(1) 带符号数和无符号数都不溢出

```
   二进制加法           看作无符号数         看作带符号数
    0000  0100              4                 ＋ 4
  ＋0000  1011            ＋ 11              ＋(＋11)
    0000  1111              15                ＋15
                          CF＝0               OF＝0
```

(2) 无符号数溢出

```
   二进制加法           看作无符号数         看作带符号数
    0000  0111              7                 ＋7
  ＋1111  1011            ＋ 251             ＋(－5)
    0000  0010             258                ＋2
  ↙
  1                       CF＝1               OF＝0
                        现为 2,结果错
```

(3) 带符号数溢出

```
   二进制加法           看作无符号数         看作带符号数
    0000  1001              9                 ＋9
  ＋0111  1100            ＋ 124             ＋(＋124)
    1000  0101             133                ＋133
                          CF＝0               OF＝1
                                            现为－123,结果错。
```

(4) 带符号数和无符号数都溢出

```
   二进制加法           看作无符号数         看作带符号数
    1000  0111             135                (－121)
  ＋1111  0101            ＋ 245             ＋(－11)
    0111  1100             380                －132
  ↙
  1                       CF＝1               OF＝1
                        现为 124,结果错。    现为 124,结果错。
```

上面的 4 个例子清楚地说明了 OF 位可以用来表示带符号数的溢出,CF 位则可用来表示无符号数的溢出。

注意:如果(2)和(4)中的进位值以 $2^8＝256$ 为其权值考虑在内时,得到的运算结果应该是正确的。

ADC 及 INC 对条件码的设置方法与 ADD 指令相同,但 INC 指令不影响 CF 位标志。

例 3.45　ADD　DX,0F0F0H

如指令执行前　(DX)=4652H 则

```
   4652           0100 0110 0101 0010
 + F0F0  ⇒      + 1111 0000 1111 0000
                  0011 0111 0100 0010
                       ↙
                       1
```

指令执行后(DX)=3742H,ZF=0,SF=0,CF=1,OF=0,结果正确。

例 3.46　下列指令序列可在 8086 和 80286 中实现两个双精度数的加法。设目的操作数存放在 DX 和 AX 寄存器中,其中 DX 存放高位字。源操作数存放在 BX、CX 中,其中 BX 存放高位字。如指令执行前:(DX)=0002H,(AX)=0F365H,(BX)=0005H,(CX)=0E024H

指令序列为　ADD　AX,CX
　　　　　　ADC　DX,BX

则第一条指令执行后　(AX)=0D389H,SF=1,ZF=0,CF=1,OF=0
第二条指令执行后　(DX)=0008H,SF=0,ZF=0,CF=0,OF=0
因此该指令序列执行完后　(DX)=0008H,(AX)=0D389H　结果正确。

可以看出,为实现双精度加法,必须用两条指令分别完成低位字和高位字的加法,而且在高位字相加时,应该使用 ADC 指令,以便把前一条 ADD 指令作低位字加法所产生的进位值加入高位字之内。另外,带符号的双精度数的溢出,应该根据 ADC 指令的 OF 位来判断,而作低位加法用的 ADD 指令的溢出是无意义的。

在 80386 及其后继机型中,因机器字长为 32 位,故可直接用双字指令实现 32 位字的相加。如 ADD EAX,ECX 不必再用上述两条指令的指令序列来实现,但它可以用同样的方法来实现 64 位字的相加。

例 3.47　XADD　BL,DL

如指令执行前　(BL)=12H,(DL)=02H,
则指令执行后　(BL)=14H,(DL)=12H。

2. 减法指令

SUB(subtract)　　　　　　　　　　　　　　　减法
SBB(subtract with borrow)　　　　　　　　　带借位减法
DEC(decrement)　　　　　　　　　　　　　　减 1
NEG(negate)　　　　　　　　　　　　　　　　求补
CMP(compare)　　　　　　　　　　　　　　　比较
CMPXCHG(compare and exchange)　　　　　比较并交换
CMPXCHG8B(compare and exchange 8 byte)　比较并交换 8 字节

(1) SUB 减法指令
格式:SUB　DST,SRC
执行的操作:(DST)←(DST)-(SRC)

(2) SBB 带借位减法指令

格式：SBB DST，SRC

执行的操作：(DST)←(DST)－(SRC)－CF

其中 CF 为进位位的值。

(3) DEC 减 1 指令

格式：DEC OPR

执行的操作：(OPR)←(OPR)－1

(4) NEG 求补指令

格式：NEG OPR

执行的操作：(OPR)←－(OPR)

亦即把操作数按位求反后末位加 1，因而执行的操作也可表示为：

    (OPR)←0FFFFH－(OPR)＋1

(5) CMP 比较指令

格式：CMP OPR1，OPR2

执行的操作：(OPR1)－(OPR2)

该指令与 SUB 指令一样执行减法操作，但它并不保存结果，只是根据结果设置条件标志位。CMP 指令后往往跟着一条条件转移指令，根据比较结果产生不同的程序分支。

(6) CMPXCHG 比较并交换指令

格式：CMPXCHG DST，SRC

该指令只能用于 486 及其后继机型。SRC 只能用 8 位、16 位或 32 位寄存器。DST 则可用寄存器或任一种存储器寻址方式。

执行的操作：累加器 AC 与 DST 相比较，

  如   (AC)=(DST)

  则   ZF←1,(DST)←(SRC)

  否则， ZF←0,(AC)←(DST)

累加器可为 AL、AX 或 EAX 寄存器。该指令对其他标志位的影响与 CMP 指令相同。

(7) CMPXCHG8B 比较并交换 8 字节指令

格式：CMPXCHG8B DST

该指令只能用于 Pentium 及其后继机型。源操作数为存放于 EDX，EAX 中的 64 位字，目的操作数可用存储器寻址方式确定一个 64 位字。

执行的操作：EDX，EAX 与 DST 相比较，

  如   (EDX, EAX)=(DST)

  则   ZF←1,(DST)←(ECX, EBX)

  否则， ZF←0,(EDX, EAX)←(DST)

该指令影响 ZF 位，但不影响其他标志位。

以上前六种指令均可作字或字节运算，386 及其后继机型还可作双字运算，而且除 DEC 不影响 CF 标志外，它们都影响条件标志位。

减法运算的条件码情况与加法类似。CF 位说明无符号数相减的溢出，同时它又确实

是被减数的最高有效位向高位的借位值。OF 位则说明带符号数的溢出,这里不再详细讨论,只说明其设置方法。减法的 CF 值反映无符号数运算中的借位情况,因此当作为无符号数运算时,若减数>被减数,此时有借位,则 CF=1;否则 CF=0。或者,也可以简单地用二进制减法运算中最高有效位向高位的进位的情况来判别:有进位时 CF=0,无进位时 CF=1。减法的 OF 位的设置方法为:若两个数的符号相反,而结果的符号与减数相同,则 OF=1;除上述情况外 OF=0。OF=1 说明带符号数的减法溢出,结果是错误的。

这里再简单说明一下 NEG 指令的条件码设置情况:NEG 指令的条件码按求补后的结果设置,只有当操作数为 0 时,求补运算的结果使 CF=0,其他情况均为 1。所以,只有当字节运算时对 -128 求补,以及字运算时对 -32768 求补和双字运算时对 -2^{31} 求补的情况下 OF=1,其他则均为 0。

例 3.48　SUB　[SI+14H],0136H

如指令执行前　(DS)=3000H,(SI)=0040H,(30054H)=4336H,

则指令执行后

```
    4336           0100  0011  0011  0110
   -0136    ⇒    -0000  0001  0011  0110
                         ⇓
                  0100  0011  0011  0110
                 +1111  1110  1100  1010
                  0100  0010  0000  0000
                 ↙
                1
```

所以,(30054H)=4200H,SF=0,ZF=0,CF=0,OF=0。

例 3.49　SUB　DH,[BP+4]

如指令执行前(DH)=41H,(SS)=0000H,(BP)=00E4H,(000E8)=5AH

则指令执行后

```
    41           0100  0001           0100  0001
   -5A    ⇒    -0101  1010    ⇒    +1010  0110
                                     1110  0111
```

所以,(DH)=0E7H,SF=1,ZF=0,CF=1,OF=0。

例 3.50　设 X、Y、Z 均为双精度数,它们分别存放在地址为 X,X+2;Y,Y+2;Z,Z+2 的存储单元中,存放时高位字在高地址中,低位字在低地址中。在 8086 和 80286 中可用下列指令序列实现　　　　　　w←x+y+24-z

并用 w 和 w+2 单元存放运算结果。

```
    MOV    AX,X        ;  ⎫
    MOV    DX,X+2      ;  ⎬ X+Y
    ADD    AX,Y        ;  ⎪
    ADC    DX,Y+2      ;  ⎭
    ADD    AX,24       ;  ⎱ +24
    ADC    DX,0        ;  ⎰
```

```
SUB    AX,Z       ;  ⎤
SBB    DX,Z+2     ;  ⎦ −Z
MOV    W,AX       ;  ⎤
MOV    W+2,DX     ;  ⎦ 结果存入 W,W+2
```

在 386 及其后继机型中就更加简单了,下列指令序列就可实现上述运算。

```
MOV    EAX,X
ADD    EAX,Y
ADD    EAX,24
SUB    EAX,Z
MOV    W,EAX
```

但上述 8086 中处理双字的方法可用于 386 及其后继机型中对 4 字的处理。

例 3.51　CMPXCHG　　CX,DX

如指令执行前:(AX)=2300H,(CX)=2300H,(DX)=2400H

则指令执行后:因(CX)=(AX),故(CX)=2400H,ZF=1。

如指令执行前:(AX)=2500H,(CX)=2300H,(DX)=2400H

则指令执行后:因(CX)≠(AX),故(AX)=2300H,ZF=0。

3. 乘法指令

MUL　(unsigned multiple)无符号数乘法

IMUL　(signed multiple)带符号数乘法

(1) MUL　无符号数乘法指令

格式:MUL　　SRC

执行的操作:

　　　　字节操作数:(AX)←(AL)∗(SRC)

　　　　字操作数:(DX,AX)←(AX)∗(SRC)

　　　　双字操作数:(EDX,EAX)←(EAX)∗(SRC)

(2) IMUL 带符号数乘法指令

格式:IMUL　　SRC

执行的操作:与 MUL 相同,但必须是带符号数,而 MUL 是无符号数。

在乘法指令里,目的操作数必须是累加器,字运算为 AX,字节运算为 AL。两个 8 位数相乘得到的是 16 位乘积存放在 AX 中,两个 16 位数相乘得到的是 32 位乘积,存放在 DX,AX 中。其中 DX 存放高位字,AX 存放低位字。386 及其后继机型可作双字运算。累加器为 EAX,两个 32 位数相乘得到 64 位乘积存放于 EDX,EAX 中。EDX 存放高位双字,EAX 存放低位双字。指令中的源操作数可以使用除立即数方式以外的任何一种寻址方式。

MUL 指令和 IMUL 指令的使用条件是由数的格式决定的。很明显(11111111b)∗(11111111b)当把它看作无符号数时应为 255d×255d=65025d;而把它看作带符号数时则为(−1)×(−1)=1。因此,必须根据所要相乘数的格式来决定选用哪一种指令。

乘法指令对除 CF 位和 OF 位以外的条件码位无定义。

注意:"无定义"的意义和"不影响"不同,"无定义"是指指令执行后这些条件码位的状态不定;而"不影响"则是指该指令的结果并不影响条件码,因而条件码应保持原状态不变。

对于 MUL 指令,如果乘积的高一半为 0,即字节操作的(AH)或字操作的(DX)或双字操作的(EDX)为 0,则 CF 位和 OF 位均为 0;否则(即字节操作时的(AH)或字操作的(DX)或双字操作的(EDX) 0),则 CF 位和 OF 位均为 1。这样的条件码设置可以用来检查字节相乘的结果是字节还是字,或者可以检查字相乘的结果是字还是双字,双字相乘的结果是双字还是 4 字。对于 IMUL 指令,如果乘积的高一半是低一半的符号扩展,则 CF 位和 OF 位均为 0,否则就均为 1。

例 3.52 如(AL)=0B4H,(BL)=11H,求执行指令 IMUL   BL 和 MUL   BL 后的乘积值。

(AL)=0B4H 为无符号数的 180D,带符号数的-76D,

(BL)=11H 为无符号数的 17D,带符号数的 17D,

则执行 IMUL   BL 的结果为

(AX)=0FAF4H=-1292D,CF=OF=1

执行 MUL   BL 的结果为

(AX)=0BF4H=3060D,CF=OF=1

对于 80286 及其后继机型,IMUL 除上述的单操作数指令(累加器是隐含的)外,还增加了双操作数和三操作数指令格式:

格式:IMUL   REG,SRC

执行的操作:

  字操作数: (REG16)←(REG16) * (SRC)

  双字操作数: (REG32)←(REG32) * (SRC)

其中,目的操作数必须是 16 位或 32 位寄存器,而源操作数则可用任一种寻址方式取得和目的操作数长度相同的数;但如源操作数为立即数时,除相应地用 16 位或 32 位立即数外,指令中也可指定 8 位立即数,在运算时机器会自动把该数符号扩展成与目的操作数长度相同的数。

格式:IMUL   REG,SRC,IMM

执行的操作:

  字操作数:(REG16)←(SRC) * IMM

  双字操作数:(REG32)←(SRC) * IMM

其中,目的操作数必须是 16 位或 32 位寄存器;源操作数可用除立即数外的任一种寻址方式取得和目的操作数长度相同的数;IMM 表示立即数,它可以是 8、16 或 32 位数,但其长度必须和目的操作数一致,如长度为 8 位时,运算时将符号扩展成与目的操作数长度一致的数。

读者可能已经注意到,IMUL 的双操作数和三操作数指令格式与单操作数指令的一个很大的区别是:后者的乘积字长是源和目的操作数字长的二倍,因而即使 OF 位为 1,乘积结果也是正确的,不存在溢出问题。但前两种格式乘积的字长和源与目的操作数的字长是一致的,即字操作数相乘得到的乘积也是字,双字操作数相乘得到的乘积也是双字,这样就可能产生溢出。机器规定:16 位操作数相乘得到的乘积在 16 位之内或 32 位操作数相乘得到的乘积在 32 位之内时,OF 位和 CF 位置 0,否则置 1,这时的 OF 位为 1 是说明溢出的。其他标志位也是无定义的。

例 3.53　IMUL　　DX,TWORD

DX 中的内容乘以 TWORD 单元中的 16 位字,结果送 DX 寄存器中。

例 3.54　IMUL　　EBX,ARRAY[ESI * 4],7

根据存储器寻址方式从 ARRAY 中取出相应单元的 32 位字乘以常数 7,结果送 EBX 寄存器。

4. 除法指令

DIV (unsigned divide)　无符号数除法

IDIV(signed divide)　带符号数除法

(1) DIV　无符号数除法指令

格式:DIV　　SRC

执行的操作:

　　字节操作:16 位被除数在 AX 中,8 位除数为源操作数,结果的 8 位商在 AL 中,8 位余数在 AH 中。表示为:

　　　　　　(AL)←(AX)/(SRC)的商

　　　　　　(AH)←(AX)/(SRC)的余数

　　字操作:32 位被除数在 DX,AX 中。其中 DX 为高位字;16 位除数为源操作数,结果的 16 位商在 AX 中,16 位余数在 DX 中。表示为:

　　　　　　(AX)←(DX,AX)/(SRC)的商

　　　　　　(DX)←(DX,AX)/(SRC)的余数

　　双字操作:64 位被除数在 EDX,EAX 中。其中 EDX 为高位双字;32 位除数为源操作数,结果的 32 位商在 EAX 中,32 位余数在 EDX 中。表示为:

　　　　　　(EAX)←(EDX,EAX)/(SRC)的商

　　　　　　(EDX)←(EDX,EAX)/(SRC)的余数

　　商和余数均为无符号数。

(2) IDIV　带符号数除法指令

格式:IDIV　　SRC

执行的操作:与 DIV 相同,但操作数必须是带符号数,商和余数也都是带符号数,且余数的符号和被除数的符号相同。

除法指令的寻址方式和乘法指令相同。其目的操作数必须存放在 AX 或 DX,AX 或 EDX,EAX 中;而其源操作数可以用除立即数以外的任一种寻址方式。

除法指令对所有条件码位均无定义。

由于除法指令的字节操作要求被除数为 16 位,字操作要求被除数为 32 位,双字操作要求被除数为 64 位,因此往往需要用符号扩展的方法取得除法指令所需要的被除数格式。此类指令的使用方法请参阅 3.3.1 节中"5. 类型转换指令"。

在使用除法指令时,还需要注意一个问题,除法指令要求字节操作时商为 8 位,字操作时商为 16 位,双字操作时商为 32 位。如果字节操作时,被除数的高 8 位绝对值≥除数的绝对值;或者字操作时,被除数的高 16 位绝对值≥除数的绝对值;或者双字操作时,被除数高 32 位绝对值≥除数的绝对值,则商就会产生溢出。在 80x86 中这种溢出是由系统直接转入 0 型中断处理的(有关中断处理问题将在第 8 章专门说明)。为避免出现这种情况,必要时程序应进行溢出判断及处理。

除法运算举例如下:

例 3.55 设(AX)=0400H,(BL)=0B4H
即(AX)为无符号数的 1024D,带符号数的+1024D
(BL)为无符号数的 180D,带符号数的−76D
执行 DIV　　BL 的结果是:
　　(AH)=7CH=124D　　　余数
　　(AL)=05H=5D　　　　商
执行 IDIV　　BL 的结果是:
　　(AH)=24H=36D　　　余数
　　(AL)=0F3H=−13D　　商

例 3.56 算术运算综合举例,计算:(V−(X*Y+Z−540))/X
其中 X,Y,Z,V 均为 16 位带符号数,已分别装入 X,Y,Z,V 单元中,要求上式计算结果的商存入 AX,余数存入 DX 寄存器。编制程序如下:

```
mov      ax,x           ;multiply x
imul     y              ;   by y and
mov      cx,ax          ;     store product
mov      bx,dx          ;       in BX,CX
mov      ax,z           ;add sign_extended z
cwd                     ;    to the
add      cx,ax          ;      product
adc      bx,dx          ;        in BX,CX
sub      cx,540         ;substract 540
sbb      bx,0           ;   from BX,CX
mov      ax,v           ;substract (BX,CX)
cwd                     ;    from sign_extended
sub      ax,cx          ;      and
sbb      dx,bx          ; divide by x leave
idiv     x              ;    quotient in AX
                        ;      and remainder in DX
```

5. 十进制调整指令

前面提到的所有算术运算指令都是二进制数的算术运算指令,但是人们最常用的是十进制数。这样,当用计算机进行计算时,必须先把十进制数转换成二进制数,然后再进行二进制数的计算,计算结果又转换成十进制数输出。为了便于十进制数的计算,计算机还提供了一组十进制数调整指令,这组指令在二进制计算的基础上,给予十进制调整,可以直接得到十进制的结果。

二-十进制(binary code decimal,BCD)是一种用二进制编码的十进制数。它是用4位二进制数表示一个十进制数码的,由于这4位二进制数的权为8421,所以BCD码又称8421码。可以用压缩的BCD码和非压缩的BCD码两种格式来表示一个十进制数。压缩的BCD码(packed BCD format)用4位二进制数表示一个十进制数位,整个十进制数形成为一个顺序的以4位为一组的数串。非压缩的BCD码(unpacked BCD format)则以8位为一组表示一个十进制数位,8位中的低4位是以8421码表示的十进制数位,而高4位则没有意义。可以看出,数字的ASCII码是一种非压缩的BCD码。因为数字的ASCII码的高4位值为0011,而低4位是以8421码表示的十进制数位,这符合非压缩BCD码高4位无意义的规定。

80x86的十进制调整指令分为两组,下面分别加以说明:

(1) 压缩的BCD码调整指令

  DAA(decimal adjust for addition) 加法的十进制调整指令
  DAS(decimal adjust for subtraction) 减法的十进制调整指令

(2) 非压缩的BCD码调整指令

  AAA(ASCII adjust for addition) 加法的ASCII调整指令
  AAS(ASCII adjust for subtraction) 减法的ASCII调整指令
  AAM(ASCII adjust for multiplication) 乘法的ASCII调整指令
  AAD(ASCII adjust for division) 除法的ASCII调整指令

由于篇幅有限,本书对本组指令不加详细说明。读者如有需要,请查阅手册或《IBM-PC汇编语言程序设计》(清华大学出版社,1991年版)或《80x86汇编语言程序设计》(清华大学出版社,2001年版)。

3.3.3 逻辑指令

1. 逻辑运算指令

  AND(and) 逻辑与
  OR (or) 逻辑或
  NOT (not) 逻辑非
  XOR (exclusive or) 异或
  TEST (test) 测试

逻辑运算指令可以对字或字节执行逻辑运算,386及其后继机型还可执行双字操作。由于逻辑运算是按位操作的,因此一般说来,其操作数应该是位串而不是数。

(1) AND 逻辑与指令

格式：AND　　DST，SRC

执行的操作：(DST)←(DST)∧(SRC)

(2) OR　逻辑或指令

格式：OR　　DST，SRC

执行的操作：(DST)←(DST)∨(SRC)

(3) NOT　逻辑非指令

格式：NOT　　OPR

执行的操作：(OPR)←($\overline{\text{OPR}}$)

(4) XOR　异或指令

格式：XOR　　DST，SRC

执行的操作：(DST)←(DST)∀(SRC)

(5) TEST　测试指令

格式：TEST　　OPR1，OPR2

执行的操作：(OPR1)∧(OPR2)

两个操作数相与的结果不保存,只根据其特征置条件码。

在以上五种指令中,NOT 不允许使用立即数,其他 4 条指令除非源操作数是立即数,至少有一个操作数必须存放在寄存器中,另一个操作数则可以使用任意寻址方式。它们对标志位的影响是：NOT 指令不影响标志位,其他 4 种指令将使 CF 位和 OF 位为 0,AF 位无定义,而 SF 位、ZF 位和 PF 位则根据运算结果设置。

这些指令对处理操作数的某些位很有用,例如可屏蔽某些位(将这些位置 0),或使某些位置 1 或测试某些位等,下面举例说明。

例 3.57　要求屏蔽 0、1 两位,可用 AND 指令并设置常数 0FCH。

MOV　　AL,0BFH
AND　　AL,0FCH

这两条指令执行的结果使(AL)=0BCH

```
          1011  1111
    AND   1111  1100
          1011  1100
```

所以用 AND 指令可以使操作数的某些位被屏蔽。只需要把 AND 指令的源操作数设置成一个立即数,并把需要屏蔽的位设为 0,这样指令执行的结果就可把操作数的相应位置 0,其他各位保持不变。

例 3.58　要求第 5 位置 1,可用 OR 指令

MOV　　AL,43H
OR　　AL,20H

这两条指令执行后,(AL)=63H

```
          0100  0011
    OR    0010  0000
          0110  0011
```

所以用 OR 指令可以使操作数的某些位置 1,其他位则保持不变。只需把 OR 指令的源操作数设置为一个立即数,并把需要置 1 的位设为 1,就可达到目的。

例 3.59 要测试操作数的某些位是否为 0,可用 TEST 指令,同样把 TEST 指令的源操作数设置成一个立即数,其中需要测试的位应设为 1。

  MOV AL,40H
  TEST AL,0AFH

这两条指令执行后

         0100 0000
     AND 1010 1111
         0000 0000

这里要求测试第 0,1,2,3,5,7 位是否为 0,根据测试的结果设置条件码为 CF=OF=0,SF=0,ZF=1,说明所需测试的位均为 0。如果在这两条指令之后跟一条条件转移指令 JNZ,结果如不是 0 则转移,结果如是 0 则顺序往下执行,这样就可根据测试的情况产生不同的程序分支,转向不同的处理。

例 3.60 要测试操作数的某位是否为 1,可先把该操作数求反,然后用 TEST 指令测试。如要测试 AL 寄存器中第 2 位是否为 1,若为 1 则转移到 EXIT 去执行,可用下列指令序列:

  MOV DL,AL
  NOT DL
  TEST DL,0000 0100B
  JE   EXIT

如 AL 寄存器的内容为 0FH,为避免破坏操作数的原始内容,把它复制到 DL 去测试,执行完 TEST 指令后,因结果为全 0 而有 ZF=1,说明操作数的第 2 位为 1 而引起转移到 EXIT 去执行。

例 3.61 要使操作数的某些位变反,可以使用 XOR 指令,只要把源操作数的立即数字段的相应位置成 1 就可以达到目的。如果求第 0,1 位变反,可使用如下指令:

  MOV AL,11H
  XOR AL,3

则指令执行后,(AL)=12H,达到第 0,1 位变反而其他位不变的目的。

例 3.62 XOR 指令还可以用来测试某一操作数是否与另一确定的操作数相等。这种操作在检查地址是否匹配时是常用的。

  XOR AX,042EH
  JZ   MATCH

这两条指令用来检查 AX 的内容是否等于 042EH,若相等则转到 MATCH 去执行匹配时要做的工作,否则执行 JZ 指令下面的程序。

2. 位测试并修改指令

386 及其后继机型增加了本组指令。

BT(bit test)　　　　　　　　　　　位测试
BTS(bit test and set)　　　　　　　位测试并置 1
BTR(bit test and reset)　　　　　　位测试并置 0
BTC(bit test and complement)　　　位测试并变反

(1) BT　位测试指令

格式：BT　　　DST，SRC

执行的操作：把目的操作数中由源操作数所指定位的值送往标志位 CF。

(2) BTS　位测试并置 1 指令

格式：BTS　　　DST，SRC

执行的操作：把目的操作数中由源操作数所指定位的值送往标志位 CF，并将目的操作数中的该位置 1。

(3) BTR　位测试并置 0 指令

格式：BTR　　　DST，SRC

执行的操作：把目的操作数中由源操作数所指定位的值送往标志位 CF，并将目的操作数中的该位置 0。

(4) BTC　位测试并变反指令

格式：BTC　　　DST，SRC

执行的操作：把目的操作数中由源操作数所指定位的值送往标志位 CF，并将目的操作数中的该位变反。

本组指令中的 SRC 可以使用寄存器方式或立即数方式，即可以在指令中用 8 位立即数直接指出目的操作数所要测试位的位位置，(注意，在指定位位置的方式上本组指令与 3.3.3 小节中的"逻辑运算指令"是不同的。)也可用任一字寄存器或双字寄存器的内容给出同一个值。目的操作数则可用除立即数外的任一种寻址方式指定一个字或双字。由于目的操作数的字长最大为 32 位，所以位位置的范围应是 0～31。

本组指令影响 CF 位的值，其他标志位则无定义。

例 3.63　　BT　　AX，4

指令测试 AX 寄存器的位 4。如指令执行前(AX)=1234H，则指令执行后(CF)=1；如指令执行前(AX)=1224H，则指令执行后(CF)=0。

3. 位扫描指令

386 及其后继机型增加了本组指令。

BSF(bit scan forward)　　　正向位扫描
BSR(bit scan reverse)　　　反向位扫描

(1) BSF　正向位扫描指令

格式：BSF　　　REG，SRC

执行的操作：指令从位 0 开始自右向左扫描源操作数，目的是检索第一个为 1 的位。如遇到第一个为 1 的位则将 ZF 位置 0，并把该位的位位置装入目的寄存器中；如源操作数为 0，则将 ZF 位置 1，目的寄存器无定义。

该指令的源操作数可以用除立即数以外的任一种寻址方式指定字或双字，目的操作数则必须用字或双字寄存器。

该指令影响 ZF 位,其他标志位无定义。

(2) BSR 反向位扫描指令

格式:BSR　　REG,SRC

执行的操作:指令从最高有效位开始自左向右扫描源操作数,目的是检索第一个为 1 的位。该指令除方向与 BSF 相反外,其他规定均与 BSF 相同。因此它们之间的差别是 BSF 指令检索从低位开始第一个出现的 1,而 BSR 则检索从高位开始第一个出现的 1。

例 3.64　BSF　　ECX,EAX
　　　　　　BSR　　EDX,EAX

如指令执行前(EAX)=60000000H,可见该数中有两个 1 位并出现于位位置为 30 和 29 之处。BSF 执行后,(ECX)=29D;BSR 执行后,(EDX)=30D,ZF 位应为 0。

4. 移位指令

SHL(shift logical left)　　　　　　逻辑左移
SAL(shift arithmetic left)　　　　　算术左移
SHR(shift logical right)　　　　　　逻辑右移
SAR(shift arithmetic right)　　　　 算术右移
ROL(rotat left)　　　　　　　　　　循环左移
ROR(rotate right)　　　　　　　　　循环右移
RCL(rotate left through carry)　　 带进位循环左移
RCR(rotate right through carry)　　带进位循环右移
SHLD(shift left double)　　　　　　双精度左移
SHRD(shift right double)　　　　　 双精度右移

这十条指令可分为三组加以说明。

(1) 移位指令

① SHL　逻辑左移指令

格式:SHL　　OPR,CNT

执行的操作如图 3.14(1) 所示。

其中 OPR 用除立即数外的任何寻址方式。移位次数由 CNT 决定,在 8086 中它可以是 1 或 CL。CNT 为 1 时只移一位;如需要移位的次数大于 1,则可以在该移位指令前把移位次数置于 CL 寄存器中,而移位指令中的 CNT 写为 CL 即可。在其他机型中可使用 CL 和 CNT,且 CNT 的值除可用 1 外,还可以用 8 位立即数指定范围从 1 到 31 的移位次数。有关 OPR 及 CNT 的规定适用于以下的所有移位指令。

② SAL　算术左移指令

格式:SAL　　OPR,CNT

执行的操作:与 SHL 相同。

③ SHR　逻辑右移指令

格式:SHR　　OPR,CNT

执行的操作:如图 3.14(2)所示。

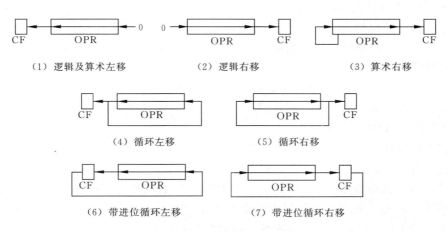

图 3.14 移位指令的操作

④ SAR 算术右移指令

格式：SAR　　OPR,CNT

执行的操作：如图 3.14(3)所示。

这里最高有效位右移,同时再用它自身的值填入,即如原来是 0 则仍为 0,原来是 1 则仍为 1。

(2) 循环移位指令

① ROL 循环左移指令

格式：ROL　　OPR,CNT

执行的操作：如图 3.14(4)所示。

② ROR 循环右移指令

格式：ROR　　OPR,CNT

执行的操作：如图 3.14(5)所示。

③ RCL 带进位循环左移指令

格式：RCL　　OPR,CNT

执行的操作：如图 3.14(6)所示。

④ RCR 带进位循环右移指令

格式：RCR　　OPR,CNT

执行的操作：如图 3.14(7)所示。

以上(1)和(2)两组指令都可以作字或字节操作,386 及其后继机型还可作双字操作。它们对条件码的影响是：CF 位根据各条指令的规定设置。OF 位只有当 CNT=1 时才是有效的,否则该位无定义。当 CNT=1 时,在移位后最高有效位的值发生变化时(原来为 0,移位后为 1;或原来为 1,移位后为 0。)OF 位置 1,否则置 0。循环移位指令不影响除 CF 和 OF 以外的其他条件标志。而移位指令则根据移位后的结果设置 SF、ZF 和 PF 位,AF 位则无定义。可以看出,循环移位指令可以改变操作数中所有位的位置,在程序中还是很有用的。移位指令则常常用来作乘以 2 或除以 2 的操作。其中算术移位指令适用于带符号数运算,SAL 用来乘以 2,SAR 用来除以 2;而逻辑移位指令则用于无符号数运算,

SHL 用来乘以 2,SHR 用来除以 2。

(3) 双精度移位指令

386 及其后继机型可以使用本组指令。

① SHLD　双精度左移指令

格式：SHLD　　DST, REG, CNT

执行的操作：如图 3.15(1)所示。

② SHRD　双精度右移指令

格式：SHRD　　DST, REG, CNT

执行的操作：如图 3.15(2)所示。

这是一组三操作数指令,其中 DST 可以用除立即数以外的任一种寻址方式指定字或双字操作数。源操作数则只能使用寄存器方式指定与目的操作数相同长度的字或双字。第三个操作数 CNT 用来指定移位次数,它可以是一个 8 位的立即数,也可以是 CL,用其内容存放移位计数值。移位计数值的范围应为 1 到 31,对于大于 31 的数,机器则自动取模 32 的值来取代。

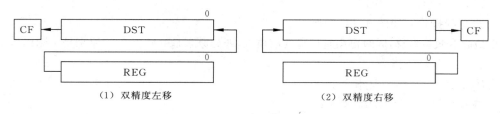

(1) 双精度左移　　　　　　　　(2) 双精度右移

图 3.15　双精度移位指令的操作

注意：① 这组指令可以取两个字作移位操作而得到是一个字的结果；也可以取两个双字作移位操作而得到的是一个双字的结果。在移位中,作为源操作数的寄存器提供移位值,以补目的操作数因移位引起的空缺,而指令执行完后,只取目的操作数作为移位的结果,源操作数寄存器则保持指令执行前的值不变。

② 这组指令在边界不对准的位串传送中很有用。如所要传送的位串的第一个位的位位置不在字节地址的边界上时,就可用这组指令来处理。

③ 这组指令当移位次数为 0 时,不影响标志位；否则,根据移位后的结果设置 SF、ZF、PF 和 CF 值。OF 的设置情况是：当移位次数为 1 时,如移位后引起符号位改变则 OF 位为 1,否则为 0；当移位次数大于 1 时 OF 位无定义。AF 位除移位次数为 0 外均无定义。

例 3.65　MOV　　CL, 5

　　　　　　SAR　　[DI], CL

如指令执行前：(DS)=0F800H,(DI)=180AH,(0F980A)=0064H,

则指令执行后：(0F980A)=0003H,CF=0,相当于 100d/32d=3d。

例3.66　MOV　　CL, 2
　　　　　　SHL　　SI, CL

如指令执行前：(SI)=1450H,

则指令执行后：(SI)=5140H,CF=0,相当于5200d×4d=20800d。

例3.67　如(AX)=0012H,(BX)=0034H,要求把它们装配在一起形成(AX)=1234H,可编制程序如下：

MOV　　CL, 8
ROL　　AX, CL
ADD　　AX, BX

例3.68　SHR　　EDX, 10

如指令执行前：(EDX)=12345678H,

则指令执行后：(EDX)=00048D15H,CF=1。

例3.69　SHLD　　EBX, ECX, 16

如指令执行前：(EBX)=12345678H,(ECX)=13572468H,

则指令执行后：(EBX)=56781357H,(ECX)=13572468H,CF=0。

3.3.4　串处理指令

MOVS(move string)　　　　　　　　　　串传送
CMPS(compare string)　　　　　　　　　串比较
SCAS(scan string)　　　　　　　　　　　串扫描
LODS(load from string)　　　　　　　　从串取
STOS(store in to string)　　　　　　　　存入串
INS(input from port to string)　　　　　串输入
OUTS(output string to port)　　　　　　串输出

与上述基本指令配合使用的前缀有：

REP(repeat)　　　　　　　　　　　　　　重复
REPE/REPZ(repeat while equal/zero)　　　　相等/为零则重复
REPNE/REPNZ(repeat while not equal/not zero)　　不相等/不为零则重复

串处理指令处理存放在存储器里的数据串,所有串指令都可以处理字节或字,386及其后继机型还可处理双字。下面,我们分两组来说明：

1. 与REP相配合工作的MOVS,STOS,LODS,INS和OUTS指令

(1) REP重复串操作直到计数寄存器Count Reg的内容为0为止。

　　格式：REP　　string primitive

其中string primitive可为MOVS,LODS,STOS,INS和OUTS指令。

　　执行的操作：

①　如(Count Reg)=0,则退出REP,否则往下执行。

② (Count Reg)←(Count Reg)－1
③ 执行其后的串指令
④ 重复①～③

其中,地址长度为 16 位时,用 CX 作为 Count Reg;地址长度为 32 位时,用 ECX 作为 Count Reg。

(2) MOVS 串传送指令

格式:可有四种

 MOVS DST,SRC
 MOVSB (字节)
 MOVSW (字)
 MOVSD (双字)(386 及其后继机型可用)

其中后三种格式明确地注明是传送字节、字或双字,第一种格式则应在操作数中表明是双字、字还是字节操作,例如:

   MOVS ES:BYTE PTR[DI], DS:[SI]

实际上 MOVS 的寻址方式是隐含的(这在下面所执行的操作中可以看到),所以这种格式中的 DST 及 SRC 只提供给汇编程序作类型检查用,并且不允许用其他寻址方式来确定操作数。

执行的操作:

① ((Destination-index))←((Source-index))
② 字节操作:
 (Source-index)←(Source-index)±1
 (Destination-index)←(Destination-index)±1
③ 字操作:
 (Source-index)←(Source-inedx)±2
 (Destination-index)←(Destination-index)±2
④ 双字操作:
 (Source-index)←(Source-index)±4
 (Destination-index)←(Destination-index)±4

在上述操作中,当方向标志 DF＝0 时用＋,DF＝1 时用－。

其中,Source-index 为源变址寄存器,当其地址长度为 16 位时用 SI 寄存器,当地址长度为 32 位时用 ESI 寄存器;Destination-index 为目的变址寄存器,当地址长度为 16 位时用 DI 寄存器,当地址长度为 32 位时用 EDI 寄存器。这一有关源和目的变址寄存器的说明适用于以下所有的串处理指令。

该指令不影响条件码。

MOVS 指令可以把由源变址寄存器指向的数据段中的一个字(或双字,或字节)传送到由目的变址寄存器指向的附加段中的一个字(或双字,或字节)中去,同时根据方向标志及数据格式(字、双字或字节)对源变址寄存器和目的变址寄存器进行修改。当该指令与前缀 REP 联用时,则可将数据段中的整串数据传送到附加段中去。这里,源串必须在数据段中,目的串必须在附加段中,但源串允许使用段跨越前缀来修改。在与 REP 联用时还必须先把数据串的长度值送到计数寄存器中,以便控制指令结束。因此在执行该指令前,应该先做好以下准备工作:

① 把存放在数据段中的源串首地址(如反向传送则应是末地址)放入源变址寄存器中;

② 把将要存放数据串的附加段中的目的串首地址(或反向传送时的末地址)放入目的变址寄存器中;

③ 把数据串长度放入计数寄存器;

④ 建立方向标志。

在完成这些准备工作后,就可以使用串指令传送信息了。

为了建立方向标志,这里介绍两条指令:

① CLD(clear direction flag)该指令使 DF=0,在执行串处理指令时可使地址自动增量;

② STD(set direction flag)该指令使 DF=1,在执行串处理指令时可使地址自动减量。

下面举例说明这一组串指令的使用方法:

例 3.70 在数据段中有一个字符串,其长度为 17,要求把它们转送到附加段中的一个缓冲区中。编制程序如下所示。

```
;****************************************************************
datarea    segment                              ;define data segment
    mess1          db         'personal computer $'
datarea    ends
;****************************************************************
extra      segment                              ;define extra segment
    mess2          db         17 dup(?)
extra      ends
;****************************************************************
code       segment                              ;define code segment
    assume cs: code, ds: datarea, es: extra
                         ⋮
;set DS register to current data segment
        mov    ax, datarea              ;datarea segment addr
        mov    ds, ax                   ;    into DS register
;set ES register to current extra segment
        mov    ax, extra                ;extra segment addr
        mov    es, ax                   ;    into ES register
                         ⋮
        lea    si, mess1
        lea    di, mess2
```

```
            mov    cx,17
            cld
            rep    movsb
                   ⋮
     code   ends                              ;end of code segment
```

程序中 SEGMENT，ENDS 为定义段的伪操作，DB 为定义字节数据的伪操作（这在第 4 章中将专门说明）。在这里，读者只要知道 MESS1 为源串，存放在数据段中从符号地址 MESS1 开始的存储区内，每个字符占有一个字节；MESS2 为目的串，存放在附加段中从符号地址 MESS2 开始空出 17 个字节的存储区内。后面的程序则存放在代码段中。程序运行过程可以用图 3.16 来说明。通过 MOVS 指令可总结一下串处理指令的特性，如表 3.5 所示。这些特性适用于所有的串处理指令。

表 3.5 串处理指令的特性

数据类型	字节	字	双字
方向(DF)	向前(DF=0)		向后(DF=1)
串长度	1		重复 REP (count reg) 次
源串(source-index)	在数据段中(可用段跨越前缀修改)		
目的串(destination-index)	在附加段中		

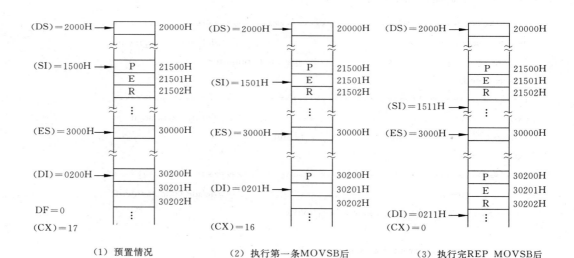

(1) 预置情况　　　(2) 执行第一条 MOVSB 后　　　(3) 执行完 REP MOVSB 后

图 3.16　REP　MOVSB 指令的执行情况

(3) STOS　存入串指令

格式：STOS　　DST
　　　STOSB　（字节）
　　　STOSW　（字）
　　　STOSD　（双字）(386 及其后继机型可用)

执行的操作：

  字节操作：((Destination-index))←(AL)，
      (Destnation-index)←(Destnation-index)±1
  字操作：((Destination-index))←(AX)，
      (Destnation-index)←(Destnation-index)±2
  双字操作：((Destination-index))←(EAX)，
      (Destnation-index)←(Destnation-index)±4

该指令把 AL、AX 或 EAX 的内容存入由目的变址寄存器指向的附加段的某单元中，并根据 DF 的值及数据类型修改目的变址寄存器的内容。当它与 REP 联用时，可把 AL、AX 或 EAX 的内容存入一个长度为(Count Reg)的缓冲区中。上述有关串处理指令的特性也适用于 STOS 指令，该指令也不影响标志位。

  STOS 指令在初始化某一缓冲区时很有用。

例 3.71 如果要把附加段中的五个字节缓冲区置为 20H 值，则可根据串处理指令的要求预置指针及方向标志，并用 REP STOSB 指令执行结果如图 3.17 所示。

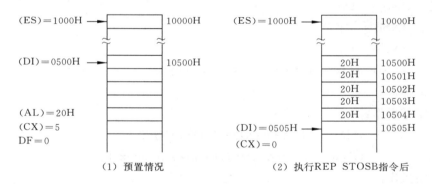

图 3.17 REP STOSB 指令的执行情况

(4) LODS 从串取指令

格式： LODS SRC
    LODSB (字节)
    LODSW (字)
    LODSD (双字)(386 及其后继机型可用)

执行的操作：

  字节操作：(AL)←((Source-index))，(Source-index)←(Source-index)±1
  字操作：(AX)←((Source-index))，(Source-index)←(Source-index)±2
  双字操作：(EAX)←((Source-index))，(Source-index)←(Source-index)±4

该指令把由源变址寄存器指向的数据段中某单元的内容送到 AL、AX 或 EAX 中，并根据方向标志和数据类型修改源变址寄存器的内容。指令允许使用段跨越前缀来指定非数据段的存储区。该指令也不影响条件码。

一般说来,该指令不和 REP 联用。有时缓冲区中的一串字符需要逐次取出来测试时,可使用本指令。

使用 LODSW 指令的预置情况及执行结果如图 3.18 所示。

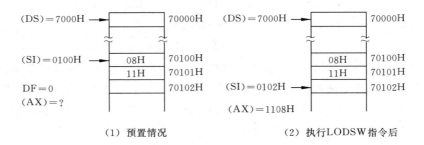

(1) 预置情况　　　　　　(2) 执行LODSW指令后

图 3.18　LODSW 指令的执行情况

(5) INS　串输入指令

格式：　INS　　DST，DX

　　　　INSB　（字节）

　　　　INSW　（字）

　　　　INSD　（双字）(386 及其后继机型可用)

执行的操作：

　字节操作：

　　　　　　((Destination-index))←((DX))（字节）

　　　　　　(Destination-index)←(Destination-index)±1

　字操作：

　　　　　　((Destination-index))←((DX))（字）

　　　　　　(Destination-index)←(Destination-index)±2

　双字操作：

　　　　　　((Destination-index))←((DX))（双字）

　　　　　　(Destination-index)←(Destination-index)±4

该指令把端口号在 DX 寄存器中的 I/O 空间的字节、字或双字传送到附加段中的由目的变址寄存器所指向的存储单元中,并根据 DF 的值和数据类型修改目的变址寄存器的内容。当它与 REP 联用时,可以把成组的字节、字或双字输入到长度为(Count Reg)的缓冲区中。

必须注意：I/O 端口的处理速度必须与 REP INS 指令的执行速度相匹配。该指令不影响标志位。

(6) OUTS　串输出指令

格式：　OUTS　　DX，SRC
　　　　OUTSB　（字节）
　　　　OUTSW　（字）
　　　　OUTSD　（双字）(386 及其后继机型可用)

执行的操作：

字节操作：

$$((DX))\leftarrow((\text{Source-index}))（字节）$$
$$(\text{Source-index})\leftarrow(\text{Source-index})\pm 1$$

字操作：

$$((DX))\leftarrow((\text{Source-index}))（字）$$
$$(\text{Source-index})\leftarrow(\text{Source-index})\pm 2$$

双字操作：

$$((DX))\leftarrow((\text{Source-index}))（双字）$$
$$(\text{Source-index})\leftarrow(\text{Source-index})\pm 4$$

该指令把由源变址寄存器所指向的存储器中的字节、字或双字传送到端口号在 DX 寄存器中的 I/O 端口中去，并根据 DF 的值及数据类型修改源变址寄存器的内容。当它与 REP 联用时，可以把存储器中长度为(Count Reg)的字节、字或双字成组地传送到 I/O 空间。

必须注意：与 INS 指令一样，I/O 端口处理速度必须与 REP OUTS 指令的执行速度相匹配。该指令不影响标志位。

---

2. 与 REPE/REPZ 和 REPNE/REPNZ 联合工作的 CMPS 和 SCAS 指令

(1) REPE/REPZ　当相等/为零时重复串操作

　　格式：　REPE（或 REPZ）　　String Primitive

其中 String Primitive 可为 CMPS 或 SCAS 指令。

执行的操作：

① 如(Count Reg)＝0 或 ZF＝0（即某次比较的结果两个操作数不等）时退出，否则往下执行。

② (Count Reg)←(Count Reg)－1

③ 执行其后的串指令

④ 重复①～③

有关 Count Reg 的规定和 REP 相同。实际上 REPE 和 REPZ 是完全相同的，只是表达的方式不同而已。与 REP 相比，除满足(Count Reg)＝0 的条件可结束操作外，还增加了 ZF＝0 的条件。也就是说，只要两数相等就可继续比较，如果遇到两数不相等时可提前结束操作。

(2) REPNE/REPNZ 当不相等/不为零时重复串操作

格式：REPNE(或 REPNZ)   String Primitive

其中 String Primitive 可为 CMPS 和 SCAS 指令。

执行的操作：

除退出条件为(Count Reg)＝0 或 ZF＝1 外，其他操作与 REPE 完全相同。也就是说，只要两数比较不相等，就可继续执行串处理指令，如某次两数比较相等或(Count Reg)＝0 时，就可结束操作。

(3) CMPS 串比较指令

格式：　CMPS　　SRC，DST
　　　　CMPSB　（字节）
　　　　CMPSW　（字）
　　　　CMPSD　（双字）(386 及其后继机型可用)

执行的操作：

① ((Source-index))－((Destination-index))

② 字节操作：

(Source-index)←(Source-index)±1，

(Destination-index)←(Destination-index)±1

字操作：

(Source-index)←(Source-index)±2，

(Destination-index)←(Destination-index)±2

双字操作：

(Source-index)←(Source-index)±4，

(Destination-index)←(Destination-index)±4

指令把由源变址寄存器指向的数据段中的一个字节、字或双字与由目的变址寄存器所指向的附加段中的一个字节、字或双字相减，但不保存结果，只根据结果置条件码。指令的其他特性和 MOVS 指令的规定相同。

(4) SCAS 串扫描指令

格式：　SCAS　　DST
　　　　SCASB　（字节）
　　　　SCASW　（字）
　　　　SCASD　（双字）(386 及其后继机型可用)

执行的操作：

字节操作：

(AL)－((Destination-index))，(Destination-index)←(Destination-index)±1

字操作：

(AX)−((Destination-index)), (Destination-index)←(Destination-index)±2

双字操作：

(EAX)−((Destination-index)), (Destination-index)←(Destination-index)±4

指令把 AL、AX 或 EAX 的内容与由目的变址寄存器指向的在附加段中的一个字节、字或双字进行比较,并不保存结果,只根据结果设置条件码。指令的其他特性和 MOVS 的规定相同。

以上(3)和(4)两条串处理指令和 REPE/REPZ 或 REPNE/REPNZ 相结合,可以用来比较两个数据串,或从一个字符串中查找一个指定的字符。

例 3.72 要求从一个字符串中查找一个指定的字符,可用指令 REPNZ SCASB。下面用图 3.19 来表示预置及找到后的情况。由图 3.19 可见,(AL)中指定的字符为 space(空格),其 ASCII 码为 20H。开始比较时,因(DI)指定的字符与(AL)不符合而不断往下比较,当(DI)=1508H 时,比较结果相符,因此 ZF=1,在修改(DI)值后指令停止比较而提前结束,此时

(DI)是相匹配字符的下一个地址；

(CX)是剩下还未比较的字符个数。

所以根据(DI)和(CX)的值可以很方便地找到所需查找字符的位置。

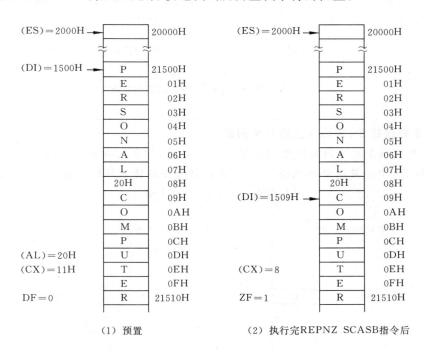

图 3.19 REPNZ SCASB 指令的执行情况

例 3.73 要求比较两个字符串,找出它们不相匹配的位置。这里可以使用指令 REPE CMPSB。下面用图 3.20 表示其预置及找到后的情况。

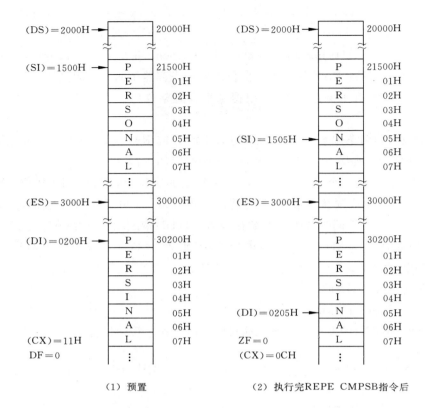

（1）预置　　　　　　（2）执行完REPE CMPSB指令后

图 3.20　REPE CMPSB 指令的执行情况

对于串处理指令,需要注意的几个问题:

① 串处理指令在不同的段之间传送或比较数据,如果需要在同一段内处理数据,可以在 DS 和 ES 中设置同样的地址,或者在源操作数字段使用段跨越前缀来实现,例如:
　　　　　　MOVS　　[DI],ES:[SI]。

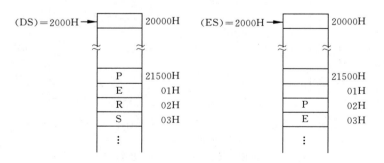

（1）REP MOVSB 指令执行前　　（2）REP MOVSB 指令执行后

图 3.21　反向传送举例

② 当使用重复前缀时,(Count Reg)是每次减1的,因此对于字或双字指令来说,预置时设置的值应该是字或双字的个数而不是字节数。

③ 上面所举的例子都把 DF 设置为0,作正向传送或比较。实际上反向传送也是很有用的,有些情况下必须反向处理,例如需要把数据缓冲区向前(地址增加的方向)错一个字,此时为避免信息的丢失,不可能用正向传送而必须使用反向传送(DF=1)的方法(见图 3.21)。这里的(DS)和(ES)应设置为同一个值。

3.3.5 控制转移指令

一般情况下指令是顺序地逐条执行的,但实际上程序不可能全部顺序执行而经常需要改变程序的执行流程,这里要介绍的控制转移指令就是用来控制程序的执行流程的。

1. 无条件转移指令

JMP （jmp） 跳转指令

无条件地转移到指令指定的地址去执行从该地址开始的指令。可以看出 JMP 指令必须指定转移的目标地址(或称转向地址)。

总的说来,转移可以分成两类:段内转移和段间转移。段内转移是指在同一段的范围内进行转移,此时只需改变 IP 或 EIP 寄存器的内容,即用新的转移目标地址代替原有的 IP 或 EIP 的值就可达到转移的目的。段间转移则是要转到另一个段去执行程序,此时不仅要修改 IP 或 EIP 寄存器内容,还需要修改 CS 寄存器的内容才能达到目的。因此,此时的转移目标地址应由新的段地址和偏移地址两部分组成。有关转移的寻址方式,已在3.1.2 小节中作了介绍,这里只简单给出 JMP 指令的各种格式及其执行的操作。

(1) 段内直接短转移

格式:JMP　　SHORT OPR

执行的操作:(IP)←(IP)+8 位位移量

　　　　386 及其后继机型则为:(EIP)←(EIP)+8 位位移量

　　　　如操作数长度为16位,则还需(EIP)←(EIP) AND 0000FFFFH。

其中8位位移量是由目标地址 OPR 确定的。转移的目标地址在汇编格式中可直接使用符号地址,而在机器执行时则是当前的 IP 或 EIP 的值(即 JMP 指令的下一条指令的地址)与指令中指定的8位位移量之和。位移量需要满足向前或向后转移的需要,因此它是一个带符号数,也就是说这种转移格式只允许在-128～+127 字节的范围内转移。

(2) 段内直接近转移

格式:JMP　　NEAR PTR OPR

执行的操作:(IP)←(IP)+16 位位移量

　　　　386 及其后继机型为:(EIP)←(EIP)+32 位位移量

　　　　如操作数长度为16位,则(EIP)←(EIP) AND 0000FFFFH。

可以看出除位移量为16位或32位外,它和段内短转移一样,也采用相对寻址方式,在汇编格式中 OPR 也只需要使用符号地址。在8086及其他机型的实模式下段长为

64KB,所以16位位移量可以转移到段内的任一个位置。在386及其后继机型的保护模式下,段的大小可达4GB,32位位移量可转移到段内的任何位置。

(3) 段内间接近转移

格式:JMP　　WORD PTR OPR

执行的操作:(IP)←(EA)

386及其后继机型则为:(EIP)←(EA)

如操作数长度为16位,则(EIP)←(EIP) AND 0000FFFFH。

其中有效地址EA值由OPR的寻址方式确定。它可以使用除立即数方式以外的任一种寻址方式。如果指定的是寄存器,则把寄存器的内容送到IP或EIP寄存器中;如果指定的是存储器中的一个字或双字,则把该存储单元的内容送到IP或EIP寄存器中去。在3.1.2小节中已经说明,这里不再赘述。

(4) 段间直接远转移

格式:JMP　　FAR PTR OPR

执行的操作:(IP)←OPR的段内偏移地址

(CS)←OPR所在段的段地址

386及其后继机型则为:

(EIP)←OPR的段内偏移地址

(CS)←OPR所在段的段地址

如操作数长度为16位,则(EIP)←(EIP) AND 0000FFFFH。

在这里使用的是直接寻址方式。在汇编格式中OPR可使用符号地址,而机器语言中则要指定转向的偏移地址和段地址。

(5) 段间间接远转移

格式:JMP　　DWORD PTR OPR

执行的操作:(IP)←(EA)

(CS)←(EA+2)

386及其后继机型则为:

(EIP)←(EA)

(CS)←(EA+4)

如操作数长度为16位,则(EIP)←(EIP) AND 0000FFFFH。

其中EA由OPR的寻址方式确定,它可以使用除立即数及寄存器方式以外的任何存储器寻址方式,根据寻址方式求出EA后,把指定存储单元的字内容送到IP或EIP寄存器,并把下一个字的内容送到CS寄存器,这样就实现了段间跳转。例如:

JMP　　DWORD PTR ALPHA[SP][DI]

等。

最后需要说明的是,JMP指令不影响条件码。

2. 条件转移指令

条件转移指令根据上一条指令所设置的条件码来判别测试条件,每一种条件转移指令有它的测试条件,满足测试条件则转移到由指令指定的转向地址去执行那里的程序;如不满足条件则顺序执行下一条指令。因此,当满足条件时:IP 或 EIP 与 8 位、16 位或 32 位位移量相加得到转向地址;如不满足测试条件,则 IP 或 EIP 值不变。可见条件转移指令使用了相对寻址方式,它只可用 JMP 中的短转移和近转移两种格式。在汇编格式中,OPR 应指定一个目标地址,在 8086 和 80286 中只提供短转移格式,目标地址应在本条转移指令下一条指令地址的 $-128 \sim +127$ 个字节的范围之内。在 386 及其后继机型中,除短转移格式外,还提供了近转移格式,这样它就可以转移到段内的任何位置。但条件转移不提供段间远转移格式,如有需要可采用转换为 JMP 指令的办法来解决。另外,所有的条件转移指令都不影响条件码。下面我们把条件转移指令分为四组来介绍。

(1) 根据单个条件标志的设置情况转移。这组包括 10 种指令。它们一般适用于测试某一次运算的结果并根据其不同特征产生程序分支作不同处理的情况。

① JZ (或 JE) (jump if zero,or equal)结果为零(或相等)则转移。

格式:JZ(或 JE)　　OPR

测试条件:ZF=1

② JNZ(或 JNE)(jump if not zero,or not equal)结果不为零(或不相等)则转移。

格式:JNZ(或 JNE)　　OPR

测试条件:ZF=0

③ JS(jump if sign)结果为负则转移。

格式:JS　　OPR

测试条件:SF=1

④ JNS(jump if not sign)结果为正则转移。

格式:JNS　　OPR

测试条件:SF=0

⑤ JO(jump if overflow)溢出则转移。

格式:JO　　OPR

测试条件:OF=1

⑥ JNO(jump if not overflow)不溢出则转移。

格式:JNO　　OPR

测试条件:OF=0

⑦ JP(或 JPE)(jump if parity,or parity even) 奇偶位为 1 则转移。

格式:JP(或 JPE)　　OPR

测试条件:PF=1

⑧ JNP(或 JPO)(jump if not parity,or parity odd)奇偶位为 0 则转移。

格式:JNP(或 JPO)　　OPR

测试条件:PF=0

⑨ JB(或 JNAE,或 JC)(jump if below, or not above or equal,or carry)低于,或者不高于或等于,或进位为 1 则转移。

格式:JB(或 JNAE,或 JC)　　OPR

测试条件:CF=1

⑩ JNB(或 JAE,或 JNC)(jump if not below, or above or equal,or not carry) 不低于,或者高于或等于,或进位为零则转移。

格式：JNB(或 JAE,或 JNC)   OPR

测试条件：CF=0

最后两种指令在这一组指令中可以只看作 JC 和 JNC,它们只用 CF 的值来判别是否转移。

例 3.74 如果需要根据一次加法运算的结果实行不同的处理,如结果为 0 做动作 2,否则做动作 1,程序框图如图 3.22 所示。编制程序如下：

```
          ⋮
        ADD     AX,TEMP
        JZ      ACTION_2
          ⋮                    ⎫
                               ⎬ ACTION_1
ACTION_2:                      ⎫
          ⋮                    ⎬ ACTION_2
          ⋮                    ⎭
```

或者：

```
          ⋮
        ADD     AX,TEMP
        JNZ     ACTION_1
          ⋮                    ⎫
                               ⎬ ACTION_2
ACTION_1:                      ⎫
          ⋮                    ⎬ ACTION_1
          ⋮                    ⎭
```

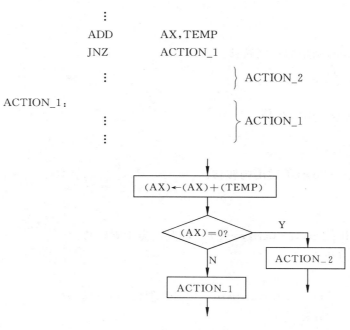

图 3.22 例 3.74 的程序框图

例 3.75 比较两个数是否相等,如相等做动作 1,否则做动作 2。可编写程序为：

```
          ⋮
        CMP     AX,BX
```

```
                    JE       ACTION_1
                     ⋮
ACTION_1:
                     ⋮
                     ⋮
```

或者:
```
                     ⋮
                    CMP      AX,BX
                    JNE      ACTION_2
                     ⋮
ACTION_2:
                     ⋮
                     ⋮
```

(2) 比较两个无符号数,并根据比较的结果转移。

① JB(或 JNAE,或 JC)低于,或者不高于或等于,或进位位为 1 则转移。

② JNB(或 JAE,或 JNC)不低于,或者高于或等于,或进位位为 0 则转移。

以上两种指令与(1)组指令中的⑨和⑩两种完全相同。

③ JBE(或 JNA)(jump if below or equal, or not above)低于或等于,或不高于则转移。

格式:JBE(或 JNA)   OPR

测试条件:CF∨ZF=1

④ JNBE(或 JA)(jump if not below or equal, or above)不低于或等于,或高于则转移。

格式:JNBE(或 JA)   OPR

测试条件:CF∨ZF=0

(3) 比较两个带符号数,并根据比较结果转移。

① JL(或 JNGE)(jump if less, or not greater or equal)小于,或者不大于或等于则转移。

格式:JL(或 JNGE)   OPR

测试条件:SF∀OF=1

② JNL(或 JGE)(jump if not less, or greater or equal)不小于,或者大于或等于则转移。

格式:JNL(或 JGE)   OPR

测试条件:SF∀OF=0

③ JLE(或 JNG)(jump if less or equal, or not greater)小于或等于,或者不大于则转移。

格式:JLE(或 JNG)   OPR

测试条件：(SF∀OF)∨ZF=1

④ JNLE(或 JG)(jump if not less or equal, or greater)不小于或等于,或者大于则转移。

格式：JNLE(或 JG)　　OPR

测试条件：(SF∀OF)∨ZF=0

(2)和(3)两组条件转移指令用于对两个数进行比较,并根据比较结果的<、≥、≤、>几种不同情况来判断是否转移。其中(2)组 JB、JAE、JBE、JA 四种指令适用于判断无符号数的比较情况,如用于地址比较或双精度数的低位字的比较等；而(3)组 JL、JGE、JLE、JG 四种指令则适用于判断带符号数的比较情况。在使用时必须严格地加以区别,否则会引起错误的结果。例如,11111111 和 00000000 两个数相比较,如果把它们看作无符号数则前者大于后者(255>0),但是如果把它们看作带符号数则前者小于后者(-1<0)。因而要根据数的不同类型来选择条件转移指令的道理是很明显的。

(2)和(3)两组指令的测试条件是完全不同的,为什么要有这种差别呢？下面分析一下它们的测试条件。

无符号数比较的转移指令的测试条件是易于理解的。前面已经分析过,当两个无符号数相减时,CF 位的情况说明了是否有借位的问题。有借位时,CF=1,当然它对应于<的条件；CF=0 时没有借位,就必然对应于≥的条件了,这就是 JB 和 JAE 指令的测试条件的建立原因。既然如此,要求≤的 JBE 指令的测试条件为 CF∨ZF=1 以及要求>的 JA 指令的测试条件为 CF∨ZF=0 也是显而易见的了。

带符号数比较的转移指令的测试条件比较复杂,这里先分析要求<的 JL 指令的情况,它的测试条件为 SF ∀OF=1。

16 位带符号数可以表示数的范围为-32768～+32767,相应的十六进制表示为 8000～7FFF,可以用图 3.23 表示。由图可见：如正数溢出则进入负数区,而负数溢出则进入正数区。在溢出的情况下,运算结果都是错误的。

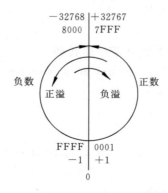

图 3.23　16 位二进制补码表示数的范围

异或运算的真值表如表 3.6 所示。下面我们根据真值表表示的四种情况来分析该测试条件的含义。

表 3.6 SF ∨ OF 的真值表

序 号	SF	OF	SF ∨ OF
a	0	0	0
b	0	1	1
c	1	0	1
d	1	1	0

例如： MOV　　AX，A
　　　　CMP　　AX，B
　　　　JL　　　X

如果(A)=α,(B)=β,如 α<β 则转移到 X 去执行。

① α-β 的结果使 SF=0,OF=0。这说明差值为正,且未溢出,结果在右半区,可以判断 α≥β 不满足转移条件 SF ∨ OF=0。

② α-β 的结果使 SF=0,OF=1。说明差值为正,但有溢出,这种情况必为负溢。现以 4 位二进制数为例,说明如下：

　　如　　α=1100,β=0101
　　　　α-β=1100-0101=0111(溢出,结果错。)
　　　　SF=0,OF=1。

可见此时 α<β,应满足测试条件：SF ∨ OF=1。

③ α-β 的结果使 SF=1,OF=0。差值为负且未溢出,说明 α<β 应满足测试条件 SF ∨ OF=1。

④ α-β 的结果使 SF=1,OF=1。差值为负但又溢出,说明是正溢。仍以 4 位二进制数为例说明如下：

　　如　　α=0101,β=1100
　　　　α-β=0101-1100=1001(溢出,结果错。)
　　　　SF=1,OF=1。

此时 α>β 不满足测试条件：SF ∨ OF=0。

从上面的分析中可以看出,这里不是简单地用差值的符号来判断两个数的大小,而是更全面地考虑到了可能存在的差值溢出的情况。虽然差值溢出了,但我们只关心数的比较而并不关心差值的具体情况,所以使用了 SF ∨ OF=1 的测试条件就更全面地反映了两数的比较情况。

既然 JL(<)的测试条件为 SF ∨ OF=1,则 JGE(≥)的测试条件为 SF ∨ OF=0 就是显而易见的了。而且 JLE(≤)的测试条件为(SF ∨ OF)∨ ZF=1 以及 JG(>)的测试条件为(SF ∨ OF)∨ ZF=0 也都是显而易见的了。

(4) 测试 CX 或 ECX 的值为 0 则转移指令。

① JCXZ(jump if CX register is zero)　　CX 寄存器的内容为零则转移。
　　格式：JCXZ　　OPR
　　测试条件：(CX)=0

② JECXZ(jump if ECX register is zero)　　ECX 寄存器的内容为零则转移。

格式：JECXZ　　　OPR　　（386 及其后继机型可用）

测试条件：(ECX)=0

(4)组两条指令与前面所有条件转移指令不同，它不是根据条件码的测试而是根据对 CX 或 ECX 寄存器的内容是否为零的测试来确定是否转移，CX 或 ECX 寄存器经常用来设置计数值，因此(4)组指令是根据计数值的变化情况来产生两个不同的程序分支的。还应该说明的是，这组指令只能提供 8 位位移量，也就是说它们只能作短转移，而且它们也不影响条件码。

下面举例说明转移指令的使用方法。

例 3.76　设 X、Y 均为存放在 X 和 Y 单元中的 16 位操作数，先判 X>50 否，如满足条件则转移到 TOO_HIGH 去执行，然后做 X－Y，如溢出则转移到 OVERFLOW 去执行，否则计算 |X－Y|，并把结果存入 RESULT 中。程序如下：

```
            MOV     AX, X
            CMP     AX, 50
            JG      TOO_HIGH
            SUB     AX, Y
            JO      OVERFLOW
            JNS     NONNEG
            NEG     AX
NONNEG:     MOV     RESULT, AX
                    ⋮
TOO_HIGH:
                    ⋮
OVERFLOW:
                    ⋮
```

例 3.77　α、β 为两个双精度数，分别存储于 DX、AX 及 BX、CX 中。要求编制一程序使 α>β 时转向 X 执行，否则转向 Y 执行。程序如下：

```
            CMP     DX, BX
            JG      X
            JL      Y
            CMP     AX, CX
            JA      X
Y:                  ⋮
X:                  ⋮
```

例 3.78　在存储器中有一个首地址为 ARRAY 的 N 字数组，要求测试其中正数、0 及负数的个数。正数的个数放在 DI 中，0 的个数放在 SI 中，并根据 N－(DI)－(SI)求得负数的个数放在 AX 中，如果有负数则转移到 NEG_VAL 中去执行。

实现例 3.78 的程序段

```
                mov     cx, N               ; load array size to CX
                mov     bx, 0               ; clear BX
                mov     di, bx              ; use DI to count elements＞0
                mov     si, bx              ; use SI to count elements＝0
again:          cmp     array[bx], 0        ; branch to less_or_eq
                jle     less_or_eq          ; if element less or
                                            ;     equal 0
                inc     di                  ; otherwise, increment DI
                jmp     short next          ; and branch to next
less_or_eq:
                jl      next                ; if element＜0, go to next
                inc     si                  ; otherwise, increment SI
next:           add     bx, 2               ; increment offset
                dec     cx                  ; decrement loop counter
                jnz     again               ; and repeat if nonzero
                mov     ax, N               ; subtract DI and SI
                sub     ax, di              ;   from N to get no. of
                sub     ax, si              ;      element＜0
                jz      skip                ; are there negative value?
                jmp     near ptr neg_val    ; if so, branch
                                            ;    to neg_val
skip:
                    ⋮
neg_val:
                    ⋮
```

3. 条件设置指令

这里所讨论的是 386 及其后继机型才提供的一组指令。这组指令本来不应该属于 3.3.5 小节所讨论的控制转移的范围之内，但是它的测试条件与条件转移指令中的测试条件是完全一致的，为方便起见，就把它放在这里介绍。

条件转移指令是根据上一条刚执行完的指令所设置的条件码情况，来判断是否产生程序分支的。有时并不希望在这一点就产生程序分支，而是希望在其后运行的程序的另一个位置，根据这一点条件码设置来产生程序分支，这样就要求把这一点的条件码情况保存下来，以便在其后使用，条件设置指令就是根据这样的要求建立的。

指令格式：SETcc　　DST

执行的操作：DST 可使用寄存器或任一种存储器寻址方式，但只能指定一个字节单元。指令根据所指定的条件码情况，如满足条件则把目的字节置为 1；如不满足条件则把目的字节置为 0。指令本身不影响标志位。

条件设置指令可分为三组：

(1) 根据单个条件标志的值把目的字节置1。

① SETZ(或 SETE)(set byte if zero, or equal)结果为零(或相等)则目的字节置1。

格式：SETZ(或 SETE)　　DST

测试条件：ZF＝1

② SETNZ(或 SETNE)(set byte if not zero, or not equal)结果不为零(或不相等)则目的字节置1。

格式：SETNZ(或 SETNE)　　DST

测试条件：ZF＝0

③ SETS(set byte if sign)结果为负则目的字节置1。

格式：SETS　　DST

测试条件：SF＝1

④ SETNS(set byte if not sign)结果为正则目的字节置1。

格式：SETNS　　DST

测试条件：SF＝0

⑤ SETO(set byte if overflow)溢出则目的字节置1。

格式：SETO　　DST

测试条件：OF＝1

⑥ SETNO(set byte if not overflow)不溢出则目的字节置1。

格式：SETNO　　DST

测试条件：OF＝0

⑦ SETP(或 SETPE)(set byte if parity, or parity even)奇偶位为1则目的字节置1。

格式：SETP(或 SETPE)　　DST

测试条件：PF＝1

⑧ SETNP(或 SETPO)(set byte if not parity, or parity odd)奇偶位为0则目的字节置1。

格式：SETNP(或 SETPO)　　DST

测试条件：PF＝0

⑨ SETC(或 SETB, 或 SETNAE)(set byte if carry, or below, or not above or equal)进位位为1,或低于,或不高于或等于则目的字节置1。

格式：SETC(或 SETB, 或 SETNAE)　　DST

测试条件：CF＝1

⑩ SETNC(或 SETNB, 或 SETAE)(set byte if not carry, or not below, or above or equal)进位位为0,或不低于,或高于或等于则目的字节置1。

格式：SETNC(或 SETNB, 或 SETAE)　　DST

测试条件：CF＝0

(2) 比较两个无符号数,并根据比较的结果把目的字节置1。

① SETB(或 SETNAE,或 SETC)低于,或不高于或等于,或进位位为1则目的字节置1。

② SETNB(或 SETAE,或 SETNC)不低于,或高于或等于,或进位位为 0 则目的字节置 1。

①和②两种指令与(1)组指令中的⑨和⑩两种完全相同。

③ SETBE(或 SETNA)(set byte if below or equal, or not above)低于或等于,或不高于则目的字节置 1。

格式：SETBE(或 SETNA)　　　DST
测试条件：CF ∨ ZF＝1

④ SETNBE(或 SETA)(set byte if not below or equal, or above)不低于或等于,或高于则目的字节置 1。

格式：SETNBE(或 SETA)　　　DST
测试条件：CF ∨ ZF＝0

(3) 比较两个带符号数,并根据比较结果把目的字节置 1。

① SETL(或 SETNGE)(set byte if less, or not greater or equal)小于,或不大于或等于则目的字节置 1。

格式：SETL(或 SETNGE)　　　DST
测试条件：SF ∀ OF＝1

② SETNL(或 SETGE)(set byte if not less, or greater or equal)不小于,或大于或等于则目的字节置 1。

格式：SETNL(或 SETGE)　　　DST
测试条件：SF ∀ OF＝0

③ SETLE(或 SETNG)(set byte if less or equal, or not greater)小于或等于,或不大于则目的字节置 1。

格式：SETLE(或 SETNG)　　　DST
测试条件：(SF ∀ OF) ∨ ZF＝1

④ SETNLE(或 SETG)(set byte if not less or equal, or greater)不小于或等于,或大于则目的字节置 1。

格式：SETNLE(或 SETG)　　　DST
测试条件：(SF ∀ OF) ∨ ZF＝0

4. 循环指令

例 3.78 的测试数组中正数、0 及负数个数的程序实际上已经是一个循环程序了。它每次测试数组中的一个数,测第二个数时仍然使用这组指令,只是对它的地址作了修改,每循环一次测试一个数。为了控制循环的结束,在进入循环前先建立循环次数,以后每循环一次就修改计数值,当计数值为 0 时就结束循环,这一过程用如图 3.24 的框图表示。

例 3.79　循环结构的实现方法是：

```
        MOV    CX,N
              ⋮
AGAIN:
              ⋮       循环体
              ⋮
```

```
        DEC    CX
        JNZ    AGAIN
```

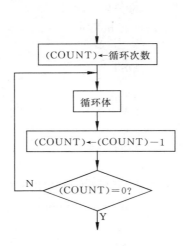

图 3.24　循环程序框图

80x86 为了简化循环程序的设计(第 5 章还会专门说明),设计了如下一组循环指令:
LOOP(loop)循环
LOOPZ/LOOPE(loop while zero, or equal)当为零或相等时循环
LOOPNZ/LOOPNE(loop while nonzero, or not equal)当不为零或不相等时循环

(1) LOOP 循环指令。

格式:LOOP　　OPR

测试条件:(Count Reg)≠0

(2) LOOPZ/LOOPE　当为零或相等时循环指令。

格式:LOOPZ(或 LOOPE)　　OPR

测试条件:ZF=1 且(Count Reg)≠0

(3) LOOPNZ/LOOPNE　当不为零或不相等时循环指令。

格式:LOOPNZ(或 LOOPNE)　　OPR

测试条件:ZF=0 且(Count Reg)≠0

这三条指令的执行步骤是:

① (Count Reg)←(Count Reg)−1

② 检查是否满足测试条件,如满足且操作数长度为 16 位,则(IP)←(IP)+D8 的符号扩展;如满足且操作数长度为 32 位,则(EIP)←(EIP)+D8 的符号扩展。

可见这里使用的是相对方式,在汇编格式中 OPR 必须指定一个表示转向地址的标号(符号地址),而在机器指令里则用 8 位位移量 D_8 来表示转向地址与当前 IP(或 EIP)值的差。由于位移量只有 8 位,所以转向地址必须在该循环指令的下一条指令地址的−128~+127 字节的范围之内。当满足测试条件时就转向由 OPR 指定的转向地址去执行,即实行循环;如不满足测试条件,则 IP(或 EIP)值不变,即退出循环,程序继续顺序执行。

上述说明中的 Count Reg 为计数寄存器,如地址长度为 16 位,则用 CX 寄存器,否则用 ECX 寄存器。

循环指令不影响条件码。

有了循环指令后,例 3.79 的程序结构可以这样实现:

```
            MOV     CX, N
              ⋮
AGAIN:
              ⋮
            LOOP    AGAIN
```

即用一条循环指令代替了原有的修改循环计数及判断转移条件两条指令。下面举几个例子说明循环指令的用法。

例 3.80 有一个首地址为 ARRAY 的 M 字数组,试编写一个程序:求出该数组的内容之和(不考虑溢出),并把结果存入 TOTAL 中。

```
              MOV     CX, M           ; put count in CX
              MOV     AX, 0           ; Zero AX
              MOV     SI, AX          ; and SI
START_LOOP:
              ADD     AX, ARRAY[SI]   ; Add next element to AX
              ADD     SI, 2           ; Increment index by two
              LOOP    START_LOOP      ; Repeat loop if CX nonzero
              MOV     TOTAL, AX       ; Store result in TOTAL
```

除 LOOP 指令外,另外两条 LOOPZ 和 LOOPNZ 指令提供了提前结束循环的可能性。有时需要在字符串中查找一个字符,在找到后就可提前结束循环而不需要一直查找到底了。此时,就可以用 LOOPZ 或 LOOPNZ 来处理。

例 3.81 有一串 L 个字符的字符串存储于首地址为 ASCII_STR 的存储区中。如要求在字符串中查找"空格"(ASCII 码为 20H)字符,找到则继续执行;如未找到则转到 NOT_FOUND 去执行,编制实现这一要求的程序如下:

```
              MOV     CX, L             ; Put array size in CX
              MOV     SI, -1            ; Initialize index, and
              MOV     AL, 20H           ; put code for space in AL
NEXT:         INC     SI                ; Increment index
              CMP     AL, ASCII_STR[SI] ; Test for space
              LOOPNE  NEXT              ; Loop if not space and count is nonzero
              JNZ     NOT_FOUND         ; If space not found
                                        ;   branch to NOT_FOUND
                ⋮
NOT_FOUND:
                ⋮
```

在程序执行过程中,有两种可能性:

① 在查找中找到了"空格",此时 ZF=1,因此提前结束循环。在执行 JNZ 指令时,因

不满足测试条件而顺序地继续执行。

② 如一直查找到字符串结束还未找到"空格"字符,此时因(CX)=0 而结束循环,但在执行 JNZ 指令时因 ZF=0 而转移到 NOT_FOUND 去执行。

在程序继续执行的部分还可以打印出找到"空格",并打印出(SI)以便确定它在字符串中的位置。在 NOT_FOUND 后可以打印出未找到"空格"等信息。

这个例子和例 3.72 的要求是完全相同的,只是所选择的指令和程序结构不同,同样都可完成所要求的任务。读者可比较选用。

5. 子程序(subroutine)

子程序结构相当于高级语言中的过程(procedure)。为便于模块化程序设计,往往把程序中某些具有独立功能的部分编写成独立的程序模块,称为子程序。程序中可由调用程序(或称主程序)调用这些子程序,而在子程序执行完后又返回调用程序继续执行。为实现这一功能,80x86 提供了以下指令:

CALL(call)　　　调用
RET(return)　　　返回

由于子程序与调用程序可以在一个段中,也可以不在同一段中,因此这两条指令的格式如下:

(1) CALL 调用指令

① 段内直接近调用

格式:CALL　　DST

执行的操作:当操作数长度为 16 位时,Push(IP)

$(IP) \leftarrow (IP) + D16$

或$(EIP) \leftarrow ((EIP) + D16)$ AND 0000FFFFH

当操作数长度为 32 位时,Push(EIP)

$(EIP) \leftarrow (EIP) + D32$

可以看出,这条指令的第一步操作是把子程序的返回地址(即调用程序中 CALL 指令的下一条指令的地址)存入堆栈中,以便子程序返回主程序时使用。第二步操作则是转移到子程序的入口地址去继续执行。指令中 DST 给出转向地址(即子程序的入口地址,亦即子程序的第一条指令的地址),D16 即为机器指令中的位移量,它是转向地址和返回地址之间的差值。

② 段内间接近调用

格式:CALL　　DST

执行的操作:当操作数长度为 16 位时,Push(IP)

$(IP) \leftarrow (EA)$

或$(EIP) \leftarrow (EA)$ AND 0000FFFFH

当操作数长度为 32 位时,Push(EIP)

$(EIP) \leftarrow (EA)$

指令中的 DST 可使用寄存器寻址方式或任一种存储器寻址方式,由指定的寄存器或存储单元的内容给出转向地址。如操作数长度为 16 位时,则有效地址 EA 应为 16 位;操作数长度为 32 位时,EA 应为 32 位。

①和②两种方式均为近调用,转向地址中只包含其偏移地址部分,段地址是保持不变的。

③ 段间直接远调用

格式：CALL　　DST

执行的操作：当操作数长度为 16 位时,Push（CS）
　　　　　　　　　　　　　　　　　　　Push（IP）
　　　　　　　　　　　　　　　　　　　(IP)←DST 指定的偏移地址
　　　　　　　　　　　　　　　　　　　(CS)←DST 指定的段地址
　　　　　　　　当操作数长度为 32 位时,Push（CS）
　　　　　　　　　　　　　　　　　　　Push（EIP）
　　　　　　　　　　　　　　　　　　　(EIP)←DST 指定的偏移地址
　　　　　　　　　　　　　　　　　　　(CS)←DST 指定的段地址

它同样是先保留返回地址,然后转移到由 DST 指定的转向地址去执行。由于调用程序和子程序不在同一段内,因此返回地址的保存以及转向地址的设置都必须把段地址考虑在内。

④ 段间间接远调用

格式：CALL　　DST

执行的操作：当操作数长度为 16 位时,Push（CS）
　　　　　　　　　　　　　　　　　　　Push（IP）
　　　　　　　　　　　　　　　　　　　(IP)←(EA)
　　　　　　　　　　　　　　　　　　　(CS)←(EA+2)
　　　　　　　　当操作数长度为 32 位时,Push（CS）
　　　　　　　　　　　　　　　　　　　Push（EIP）
　　　　　　　　　　　　　　　　　　　(EIP)←(EA)
　　　　　　　　　　　　　　　　　　　(CS)←(EA+4)

其中 EA 是由 DST 的寻址方式确定的有效地址,这里可使用任一种存储器寻址方式来取得 EA 值。

在上述 CALL 指令的格式中,并未加上如 NEAR PTR 或 FAR PTR 格式的属性操作符,在实际使用时,读者可根据具体情况加上它。使用方法可参考 JMP 指令的说明。

在保护方式下操作方式比较复杂,但由于是程序员不可见的,这里就不进一步说明了。

(2) RET　　返回指令

RET 指令放在子程序的末尾,它使子程序在功能完成后返回调用程序继续执行,而返回地址是调用程序调用子程序(或称转子)时存放在堆栈中的,因此 RET 指令的操作是返回地址出栈送 IP 或 EIP 寄存器(段内或段间)和 CS 寄存器(段间)。

① 段内近返回

格式：RET

执行的操作：当操作数长度为 16 位时,(IP)←Pop（ ）
　　　　　　　　386 及其后继机型：(EIP)←(EIP) AND 0000FFFFH

当操作数长度为 32 位时，(EIP)←Pop( )

② 段内带立即数近返回

格式：RET　　EXP

执行的操作：在完成与①的 RET 完全相同的操作后，还需要修改堆栈指针：

(SP 或 ESP)←(SP 或 ESP)+D16

其中 EXP 是一个表达式，根据它的值计算出来的常数成为机器指令中的位移量 D16。这种指令允许返回地址出栈后修改堆栈的指针，这就便于调用程序在用 CALL 指令调用子程序以前把子程序所需要的参数入栈，以便子程序运行时使用这些参数。当子程序返回后，这些参数已不再有用，就可以修改指针使其指向参数入栈以前的值。这种指令的使用方法将在第 6 章里说明。

③ 段间远返回

格式：RET

执行的操作：当操作数长度为 16 位时，(IP)←Pop( )

386 及其后继机型：(EIP)←(EIP) AND 0000FFFFH

(CS)←Pop( )

当操作数长度为 32 位时，(EIP)←Pop( )

(CS)←Pop( )（32 位数出栈，高 16 位废除。）

④ 段间带立即数远返回

格式：RET　　EXP

```
        MAIN                              PRO_A
        ────                              ────
        ────                              ────
        ────               (IP)=2500   CALL NEAR PTR PRO_B
     CALL FAR PTR PRO_A                   ────
(IP)=1000  ────                           ────
        ────                              ────
        ────               (IP)=3700   CALL NEAR PTR PRO_C
        ────                              ────
(CS)=0500                                 RET


                                          PRO_B
                                          ────
                                          ────
                           (IP)=4000   CALL NEAR PTR PRO_C
                                          ────
                                          ────
                                          RET

                                          PRO_C
                                          ────
                                          ────
                                          ────
                                          ────
                                          RET
```

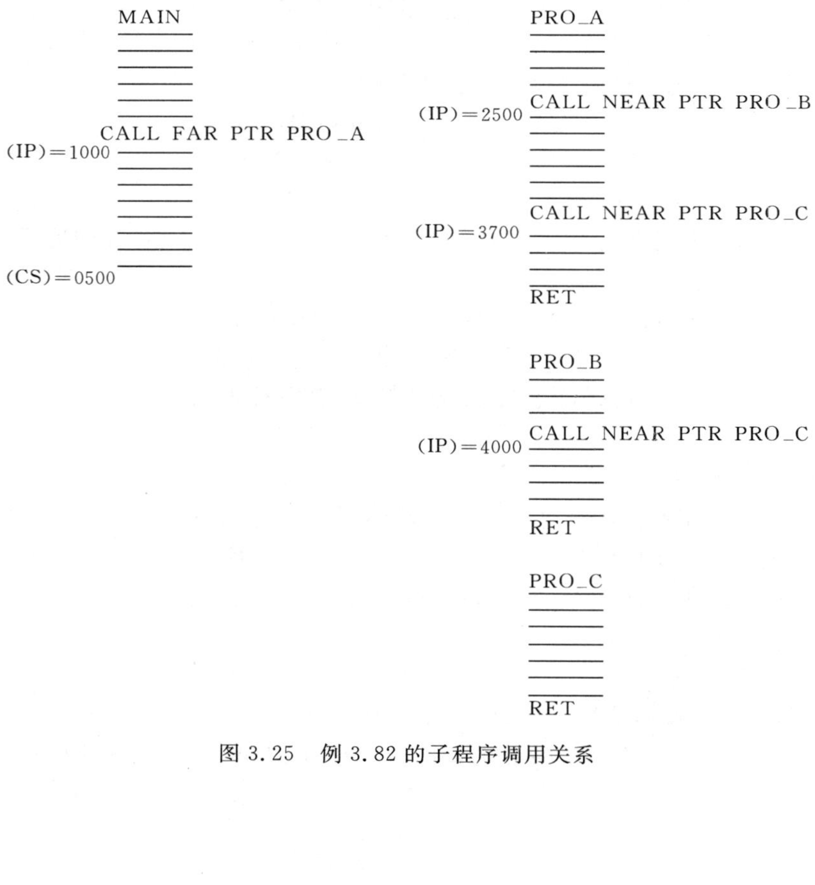

图 3.25　例 3.82 的子程序调用关系

执行的操作：在完成与③的 RET 完全相同的操作后，还要修改堆栈指针：

$$(SP 或 ESP) \leftarrow (SP 或 ESP) + D16$$

这里 EXP 的含义及使用情况与带立即数近返回指令相同。

CALL 和 RET 指令都不影响条件码。

为了说明指令的使用及堆栈情况，举例如下。（但这里并不讨论子程序的设计，有关这方面的问题将在第 6 章说明。）

例 3.82 主程序 MAIN 在一个代码段中，子程序 PRO_A、PRO_B、PRO_C 在另一个代码段中。如果这些程序之间的调用关系如图 3.25 所示，则在程序运行时，堆栈情况如图 3.26 所示。读者可以看到，在出栈后，堆栈中的内容并未破坏，但如果有新的内容进栈时，原有的内容便自动丢失了。

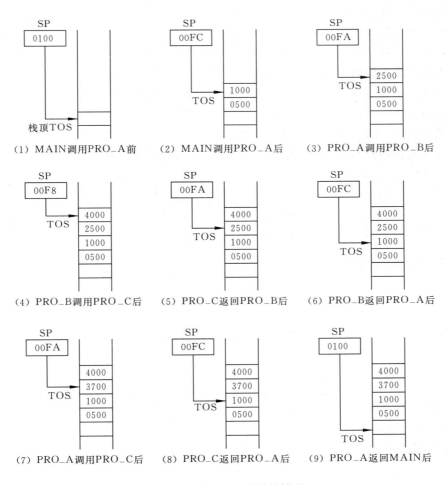

图 3.26 例 3.82 的堆栈情况

6. 中断

有时当系统运行或者程序运行期间在遇到某些特殊情况时，需要计算机自动执行一组专门的例行程序来进行处理。这种情况称为中断(interrupt),所执行的这组程序称为中断例行程序(interrupt routine)或中断子程序。中断分为内部中断和外部中断两类。内部中断包括像除法运算中遇到需要除以 0 时所产生的中断,或者程序中为了作某些处理而设置的中断指令等。外部中断则主要用来处理 I/O 设备与 CPU 之间的通信。

当 CPU 响应一次中断时,也要和调用子程序时类似地把(IP)或(EIP)和(CS)保存入栈。除此之外,为了能全面地保存现场信息,以便在中断处理结束时返回现场,还需要把反映现场状态的(FLAGS)或(EFLAGS)保存入栈,然后才能转到中断例行程序去执行。当然从中断返回时,除要恢复(IP)或(EIP)和(CS)外,还需要恢复(FLAGS)或(EFLAGS)。

中断例行程序的入口地址称为中断向量。在 80x86 中,在实模式下工作时,存储器的最低地址区的 1024 个字节(地址从 00000H 到 003FFH)为中断向量区,其中存放着 256 种类型中断例行程序的入口地址(中断向量)。图 3.27 所示为基于 8086 微机的中断向量区。由于每个中断向量占有 4 个字节单元,所以中断指令中指定的类型号 N 需要乘以 4 才能取得所指定类型的中断向量。例如,如果类型号为 9,则与其相应的中断向量存放在 00024~00027H 单元中。80x86 为每个类型规定了一定的功能,例如类型 0 为除以 0 时的中断例行程序入口,类型 3 为设置断点时的中断例行程序入口,类型 4 为溢出处理的中断例行程序入口,类型 20 为程序结束的中断例行程序入口,类型 21 为系统功能调用的中断例行程序入口等。除非特别注明,类型号是以十六进制形式表示的。

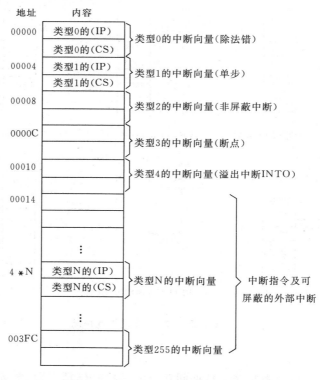

图 3.27 基于 8086 的微机存储器中的中断向量区

有关中断处理问题,还将在第 8 章中专门说明,这里只介绍有关中断的几条指令。

INT(interrupt)　　　　　　　　　　　　中断
INTO(interrupt if overflow)　　　　　　如溢出则中断
IRET/IRETD(return from interrupt)　　从中断返回

(1) INT 中断指令

格式： INT　 TYPE
　　　或 INT

执行的操作：Push (FLAGS)
　　　　　　IF←0
　　　　　　TF←0
　　　　　　AC←0
　　　　　　Push(CS)
　　　　　　Push (IP)
　　　　　　(IP)←(TYPE * 4)
　　　　　　(CS)←(TYPE * 4+2)

其中 TYPE 为类型号,它可以是常数或常数表达式,其值必须在 0～255 范围内。格式中的 INT 是一个字节的中断指令,它隐含的类型号为 3。INT 指令(包括下面的 INTO)不影响除 IF、TF 和 AC 以外的标志位。

(2) INTO　若溢出则中断指令

格式：INTO

执行的操作：若 OF=1,则： Push (FLAGS)
　　　　　　　　　　　　　IF←0
　　　　　　　　　　　　　TF←0
　　　　　　　　　　　　　AC←0
　　　　　　　　　　　　　Push (CS)
　　　　　　　　　　　　　Push (IP)
　　　　　　　　　　　　　(IP)←(10H)
　　　　　　　　　　　　　(CS)←(12H)

(3) IRET　从中断返回指令

格式：IRET

执行的操作：(IP)←Pop ( )
　　　　　　(CS)←Pop ( )
　　　　　　(FLAGS)←Pop ( )

(4) IRETD　从中断返回指令

格式：IRETD

执行的操作：(EIP)←Pop ( )
　　　　　　(CS)←Pop ( )
　　　　　　(EFLAGS)←Pop ( )

可见 IRET 适用于操作数长度为 16 位的情况,而 IRETD 则适用于操作数长度为 32 位的情况。

(3)和(4)两种指令的标志位由堆栈中取出的值来设置。

3.3.6 处理机控制与杂项操作指令

1. 标志处理指令

除有些指令影响标志位外,80x86 还提供了一组设置或清除标志位指令,它们只影响本指令指定的标志,而不影响其他标志位。这些指令是:

CLC(clear carry)	进位位置 0 指令	CF←0
CMC(complement carry)	进位位求反指令	CF←$\overline{CF}$
STC(set carry)	进位位置 1 指令	CF←1
CLD(clear direction)	方向标志位置 0 指令	DF←0
STD(set direction)	方向标志位置 1 指令	DF←1
CLI(clear interrupt)	中断标志置 0 指令	IF←0
STI(set interrupt)	中断标志置 1 指令	IF←1

2. 其他处理机控制与杂项操作指令

NOP(no operation)	无操作
HLT(halt)	停机
ESC(escape)	换码
WAIT(wait)	等待
LOCK(lock)	封锁
BOUND(bound)	界限
ENTER(enter)	建立堆栈帧
LEAVE(leave)	释放堆栈帧

这些指令可以控制处理机状态。它们都不影响条件码。

(1) NOP 无操作指令

该指令不执行任何操作,其机器码占有一个字节单元,在调试程序时往往用这条指令占有一定的存储单元,以便在正式运行时用其他指令取代。

(2) HLT 停机指令

该指令可使机器暂停工作,使处理机处于停机状态以等待一次外部中断的到来,中断结束后可继续执行下面的程序。

(3) ESC 换码指令

格式:ESC　　op,reg/mem

这条指令在使用协处理机(coprocessor)时,可以指定由协处理器执行的指令。指令的第一个操作数即指定其操作码,第二个操作数即指定其操作数。协处理机(如 8087、80287、80387 等。)是为提高速度而可以选配的硬件。自 486 起,浮点处理部件已装入 CPU 芯片,系统可直接支持协处理器指令,因此 ESC 指令已成为未定义指令,如遇到程序中的 ESC 指令,将引起一次异常处理。

(4) WAIT　等待指令

该指令使处理机处于空转状态,它也可以用来等待外部中断发生,但中断结束后仍返回 WAIT 指令继续等待。它也可以与 ESC 指令配合等待协处理机的执行结果。

(5) LOCK　封锁指令

该指令是一种前缀,它可与其他指令联合,用来维持总线的锁存信号直到与其联合的指令执行完为止。当 CPU 与其他处理机协同工作时,该指令可避免破坏有用信息。

LOCK 前缀可与下列指令联用:

BTS, BTR, BTC　　　　　　　mem, reg/imm
XCHG　　　　　　　　　　　reg, mem
XCHG　　　　　　　　　　　mem, reg
ADD, OR, ADC, SBB, AND, SUB, XOR　　mem, reg/imm
NOT, NEG, INC, DEC　　　　mem
CMPXCHG, XADD

(6) BOUND　界限指令(286 及其后继机型可用)

格式:BOUND　　reg, mem

执行的操作:BOUND 指令检查给出的数组下标是否在规定的上下界之内。如在上下界之内,则执行下一条指令;如超出了上下界范围,则产生中断 5。如发生中断,则中断返回时返回地址仍指向 BOUND 指令,而不是其下一条指令。

指令中的第一个操作数必须使用寄存器,用来存放当前数组下标。当操作数长度为 16 位时,使用 16 位寄存器;而当操作数长度为 32 位时,则使用 32 位寄存器。第二个操作数必须使用存储器寻址方式,该操作数用来存放数组的上下界。当操作数长度为 16 位时,使用相继的两个 16 位字存放下界和上界,下界存放在低地址所指定的字单元中,上界存放在高地址所指定的字单元中。当操作数长度为 32 位时,使用相继的两个双字来存放下界和上界,次序同上。

例 3.83　BOUND　　SI, DATA
　　　　　MOV　　　AX, ARRAY[SI]

在 DATA 单元中,存放数组 ARRAY 的起始下标(下界),DATA+2 单元中存放 ARRAY 的末尾下标(上界),SI 中存放当前要访问数组元素的下标。BOUND 指令用来检查当前下标是否在规定的上下界之内。如小于下界或大于上界则产生中断 5,否则执行下一条指令。

(7) ENTER　建立堆栈帧指令(286 及其后继机型可用)

ENTER 指令用于过程调用时为便于过程间传递参数而建立堆栈帧所用。堆栈帧是一个动态存储区,其工作原理将在第 6 章说明,这里只简单介绍 ENTER 指令所执行的操作。

格式:ENTER　　imm16, imm8

执行的操作:指令中所给出的两个操作数均为立即数。第一个操作数为 16 位立即数,由其指定堆栈帧的大小,即其所占据的字节数。第二个操作数为 8 位立即数,它给出

过程的嵌套层数,此数的范围应为0~31。该指令完成以下操作:

① Push(BP)或Push(EBP),以保存该寄存器的原始内容。

②(BP)←(SP)或(EBP)←(ESP),使BP或EBP寄存器保存当前堆栈指针SP或ESP的内容,以便在过程运行期间,以BP(或EBP)为基准访问堆栈帧中存放的变量。

③(BP)←(SP)-imm16或(EBP)←(ESP)-imm16,这样就建立了堆栈帧所占有的存储空间。

例3.84　ENTER　8,0

该指令所建立的堆栈帧空间如图3.28所示。图中原(SP)和原(BP)是指指令执行前的SP和BP的内容,新(SP)和新(BP)是指指令执行后的SP和BP的内容。该指令一般位于过程的入口点,为过程建立起如图3.28所示的堆栈帧,便于存放过程的局部变量,也便于过程间参数的传递。图3.28所示的新(BP)则可为对堆栈帧中变量的访问提供方便。

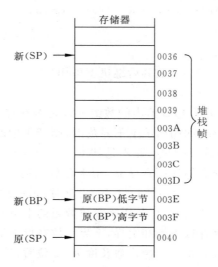

图3.28　ENTER　8,0指令所建立的堆栈帧

(8) LEAVE　释放堆栈帧指令(286及其后继机型可用)

LEAVE指令在程序中位于退出过程的RET指令之前,用来释放由ENTER指令建立的堆栈帧存储区。

格式:LEAVE

执行的操作:

①(SP)←(BP)或(ESP)←(EBP)

②(BP)←Pop( )或(EBP)←Pop( )

以上介绍了80x86为用户所提供的主要指令。对于一些常用指令并已作了比较完整的说明,但对于一些应用频度较小的指令只作了简单介绍。考虑到极少使用汇编语言编写浮点运算程序,因此在本书中并未介绍有关浮点指令的情况。此外,这里也未说明系统程序员所用的特权指令,必要时请读者自行查阅有关手册。

习 题

3.1 给定
(BX)=637DH,(SI)=2A9BH,位移量 D=7237H,试确定在以下各种寻址方式下的有效地址是什么?

(1) 立即寻址

(2) 直接寻址

(3) 使用 BX 的寄存器寻址

(4) 使用 BX 的间接寻址

(5) 使用 BX 的寄存器相对寻址

(6) 基址变址寻址

(7) 相对基址变址寻址

3.2 试根据以下要求写出相应的汇编语言指令。

(1) 把 BX 寄存器和 DX 寄存器的内容相加,结果存入 DX 寄存器中。

(2) 用寄存器 BX 和 SI 的基址变址寻址方式把存储器中的一个字节与 AL 寄存器的内容相加,并把结果送到 AL 寄存器中。

(3) 用寄存器 BX 和位移量 0B2H 的寄存器相对寻址方式把存储器中的一个字和(CX)相加,并把结果送回存储器中。

(4) 用位移量为 0524H 的直接寻址方式把存储器中的一个字与数 2A59H 相加,并把结果送回该存储单元中。

(5) 把数 0B5H 与(AL)相加,并把结果送回 AL 中。

3.3 写出把首地址为 BLOCK 的字数组的第 6 个字送到 DX 寄存器的指令。要求使用以下几种寻址方式:

(1) 寄存器间接寻址

(2) 寄存器相对寻址

(3) 基址变址寻址

3.4 现有(DS)=2000H,(BX)=0100H,(SI)=0002H,(20100)=12H,(20101)=34H,(20102)=56H,(20103)=78H,(21200)=2AH,(21201)=4CH,(21202)=B7H,(21203)=65H,试说明下列各条指令执行完后 AX 寄存器的内容。

(1) MOV　　AX,1200H

(2) MOV　　AX,BX

(3) MOV　　AX,[1200H]

(4) MOV　　AX,[BX]

(5) MOV　　AX,1100[BX]

(6) MOV　　AX,[BX][SI]

(7) MOV　　AX,1100[BX][SI]

3.5 给定
(IP)=2BC0H,(CS)=0200H,位移量 D=5119H,(BX)=1200H,(DS)=212AH,(224A0)=0600H,(275B9)=098AH,试为以下的转移指令找出转移的偏移地址。

(1) 段内直接寻址。
(2) 使用 BX 及寄存器间接寻址方式的段内间接寻址。
(3) 使用 BX 及寄存器相对寻址方式的段内间接寻址。

3.6 设当前数据段寄存器的内容为 1B00H,在数据段的偏移地址 2000H 单元内,含有一个内容为 0FF10H 和 8000H 的指针,它们是一个 16 位变量的偏移地址和段地址,试写出把该变量装入 AX 的指令序列,并画图表示出来。

3.7 在 0624 单元内有一条二字节 JMP　SHORT OBJ 指令,如其中位移量为 (1) 27H,(2) 6BH,(3) 0C6H,试问转向地址 OBJ 的值是多少?

3.8 假定(DS)=2000H,(ES)=2100H,(SS)=1500H,(SI)=00A0H,(BX)=0100H,(BP)=0010H,数据段中变量名 VAL 的偏移地址值为 0050H,试指出下列源操作数字段的寻址方式是什么? 其物理地址值是多少?

(1) MOV　　AX,0ABH　　　　　(2) MOV　　AX,BX
(3) MOV　　AX,[100H]　　　　(4) MOV　　AX,VAL
(5) MOV　　AX,[BX]　　　　　(6) MOV　　AX,ES:[BX]
(7) MOV　　AX,[BP]　　　　　(8) MOV　　AX,[SI]
(9) MOV　　AX,[BX+10]　　　(10) MOV　　AX,VAL[BX]
(11) MOV　　AX,[BX][SI]　　　(12) MOV　　AX,VAL[BX][SI]

3.9 在 ARRAY 数组中依次存储了七个字数据,紧接着是名为 ZERO 的字单元,表示如下:

ARRAY　　DW　　23,36,2,100,32000,54,0
ZERO　　　DW　　?

(1) 如果 BX 包含数组 ARRAY 的初始地址,请编写指令将数据 0 传送给 ZERO 单元。
(2) 如果 BX 包含数据 0 在数组中的位移量,请编写指令将数据 0 传送给 ZERO 单元。

3.10 如 TABLE 为数据段中 0032 单元的符号名,其中存放的内容为 1234H,试问以下两条指令有什么区别? 指令执行完后 AX 寄存器的内容是什么?

MOV　　AX,TABLE
LEA　　AX,TABLE

3.11 执行下列指令后,AX 寄存器中的内容是什么?

TABLE　　DW　　10,20,30,40,50
ENTRY　　DW　　3
　　　　⋮
　　　　MOV　　BX,OFFSET TABLE
　　　　ADD　　BX,ENTRY
　　　　MOV　　AX,[BX]

3.12 下列 ASCII 码串(包括空格符)依次存储在起始地址为 CSTRING 的字节单元中：

CSTRING　　DB　　′BASED ADDRESSING′

请编写指令将字符串中的第 1 个和第 7 个字符传送给 DX 寄存器。

3.13 已知堆栈段寄存器 SS 的内容是 0FFA0H，堆栈指针寄存器 SP 的内容是 00B0H，先执行两条把 8057H 和 0F79H 分别进栈的 PUSH 指令，再执行一条 POP 指令。试画出堆栈区和 SP 的内容变化过程示意图(标出存储单元的物理地址)。

3.14 设(DS)=1B00H，(ES)=2B00H，有关存储单元的内容如图 3.29 所示。请写出两条指令把字变量 X 装入 AX 寄存器。

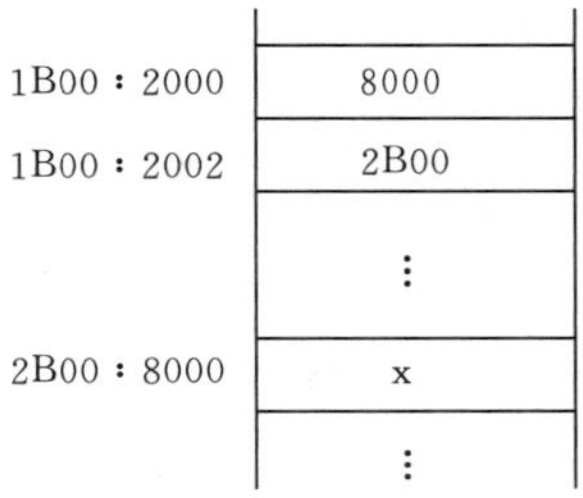

```
1B00:2000    8000
1B00:2002    2B00
              ⋮
2B00:8000     x
              ⋮
```

图 3.29　3.14 题的存储区情况

3.15 求出以下各十六进制数与十六进制数 62A0 之和，并根据结果设置标志位 SF，ZF，CF 和 OF 的值。

　　(1) 1234　　(2) 4321　　(3) CFA0　　(4) 9D60

3.16 求出以下各十六进制数与十六进制数 4AE0 的差值，并根据结果设置标志位 SF，ZF，CF 和 OF 的值。

　　(1) 1234　　(2) 5D90　　(3) 9090　　(4) EA04

3.17 写出执行以下计算的指令序列，其中 X，Y，Z，R 和 W 均为存放 16 位带符号数单元的地址。

　　(1) Z←W+(Z−X)

　　(2) Z←W−(X+6)−(R+9)

　　(3) Z←(W*X)/(Y+6)，R←余数

　　(4) Z←((W−X)/5*Y)*2

3.18 已知程序段如下：

```
MOV    AX,1234H
MOV    CL,4
ROL    AX,CL
DEC    AX
MOV    CX,4
MUL    CX
INT    20H
```

试问：

(1) 每条指令执行完后，AX 寄存器的内容是什么？

(2) 每条指令执行完后，进位、符号和零标志的值是什么？

(3) 程序结束时，AX 和 DX 的内容是什么？

3.19 下列程序段中的每条指令执行完后，AX 寄存器及 CF,SF,ZF 和 OF 的内容是什么？

```
MOV     AX, 0
DEC     AX
ADD     AX, 7FFFH
ADD     AX, 2
NOT     AX
SUB     AX, 0FFFFH
ADD     AX, 8000H
SUB     AX, 1
AND     AX, 58D1H
SAL     AX, 1
SAR     AX, 1
NEG     AX
ROR     AX, 1
```

3.20 变量 DATAX 和变量 DATAY 的定义如下：

```
DATAX   DW      0148H
        DW      2316H
DATAY   DW      0237H
        DW      4052H
```

请按下列要求写出指令序列：

(1) DATAX 和 DATAY 两个字数据相加，和存放在 DATAY 中。

(2) DATAX 和 DATAY 两个双字数据相加，和存放在从 DATAY 开始的字单元中。

(3) 解释下列指令的作用：

```
STC
MOV     BX, DATAX
ADC     BX, DATAY
```

(4) DATAX 和 DATAY 两个字数据相乘(用 MUL)。

(5) DATAX 和 DATAY 两个双字数据相乘(用 MUL)。

(6) DATAX 除以 23(用 DIV)。

(7) DATAX 双字除以字 DATAY(用 DIV)。

3.21 写出对存放在 DX 和 AX 中的双字长数求补的指令序列。

3.22 试编写一个程序求出双字长数的绝对值。双字长数在 A 和 A+2 单元中，结果存放在 B 和 B+2 单元中。

3.23 假设(BX)=0E3H,变量 VALUE 中存放的内容为 79H,确定下列各条指令单独执行后的结果。

    (1) XOR   BX,VALUE
    (2) AND   BX,VALUE
    (3) OR   BX,VALUE
    (4) XOR   BX,0FFH
    (5) AND   BX,0
    (6) TEST   BX,01H

3.24 试写出执行以下指令序列后 BX 寄存器的内容。执行前(BX)=6D16H。
    MOV   CL,7
    SHR   BX,CL

3.25 试用移位指令把十进制数+53 和-49 分别乘以 2。它们应该用什么指令? 得到的结果是什么? 如果要除以 2 呢?

3.26 试分析下面的程序段完成什么功能?

    MOV   CL,04
    SHL   DX,CL
    MOV   BL,AH
    SHL   AX,CL
    SHR   BL,CL
    OR   DL,BL

3.27 假定(DX)=0B9H,(CL)=3,(CF)=1,确定下列各条指令单独执行后 DX 中的值。

    (1) SHR   DX,1
    (2) SAR   DX,CL
    (3) SHL   DX,CL
    (4) SHL   DL,1
    (5) ROR   DX,CL
    (6) ROL   DL,CL
    (7) SAL   DH,1
    (8) RCL   DX,CL
    (9) RCR   DL,1

3.28 下列程序段执行完后,BX 寄存器中的内容是什么?

    MOV   CL,3
    MOV   BX,0B7H
    ROL   BX,1
    ROR   BX,CL

3.29 假设数据定义如下：

　　　CONAME　　DB　　'SPACE EXPLORERS INC.'
　　　PRLINE　　DB　　20 DUP(' ')

用串指令编写程序段分别完成以下功能：

（1）从左到右把 CONMAE 中的字符串传送到 PRLINE。
（2）从右到左把 CONMAE 中的字符串传送到 PRLINE。
（3）把 CONAME 中的第 3 和第 4 个字节装入 AX。
（4）把 AX 寄存器的内容存入从 PRLINE+5 开始的字节中。
（5）检查 CONAME 字符串中有无空格字符，如有则把它传送给 BH 寄存器。

3.30 编写程序段，把字符串 STRING 中的 '&' 字符用空格符代替。

　　　STRING　　DB　　'The date is FEB&03'

3.31 假设程序中数据定义如下：

　　　STUDENT_NAME　　DB　　30 DUP(?)
　　　STUDENT_ADDR　　DB　　9 DUP(?)
　　　PRINT_LINE　　　DB　　132 DUP(?)

分别编写下列程序段：

（1）用空格符清除 PRINT_LINE 域。
（2）在 STUDENT_ADDR 中查找第一个 '—'。
（3）在 STUDENT_ADDR 中查找最后一个 '—'。
（4）如果 STUDENT_NAME 域中全是空格符时，填入 '*'。
（5）把 STUDENT_NAME 移到 PRINT_LINE 的前 30 个字节中，把 STUDENT_ADDR 移到 PRINT_LINE 的后 9 个字节中。

3.32 编写一程序段，比较两个 5 字节的字符串 OLDS 和 NEWS，如果 OLDS 字符串不同于 NEWS 字符串则执行 NEW_LESS；否则顺序执行程序。

3.33 假定 AX 和 BX 中的内容为带符号数，CX 和 DX 中的内容为无符号数，请用比较指令和条件转移指令实现以下判断：

（1）若 DX 的内容超过 CX 的内容，则转去执行 EXCEED。
（2）若 BX 的内容大于 AX 的内容，则转去执行 EXCEED。
（3）若 CX 的内容等于零，则转去执行 ZERO。
（4）BX 与 AX 的内容相比较是否产生溢出？若溢出则转 OVERFLOW。
（5）若 BX 的内容小于等于 AX 的内容，则转 EQ_SMA。
（6）若 DX 的内容低于等于 CX 的内容，则转 EQ_SMA。

3.34 试分析下列程序段：

　　　ADD　　AX, BX
　　　JNO　　L1
　　　JNC　　L2
　　　SUB　　AX, BX

```
        JNC     L3
        JNO     L4
        JMP     SHORT L5
```

如果 AX 和 BX 的内容给定如下：

	AX	BX
(1)	147B	80DC
(2)	B568	54B7
(3)	42C8	608D
(4)	D023	9FD0
(5)	94B7	B568

问该程序执行完后，程序转向哪里？

3.35 指令 CMP AX,BX 后面跟着一条格式为 J…L1 的条件转移指令，其中…可以是 B,NB,BE,NBE,L,NL,LE 和 NLE 中的任一个。如果 AX 和 BX 的内容给定如下：

	AX	BX
(1)	1F52	1F52
(2)	88C9	88C9
(3)	FF82	007E
(4)	58BA	020E
(5)	FFC5	FF8B
(6)	09A0	1E97
(7)	8AEA	FC29
(8)	D367	32A6

问以上 8 条转移指令中的哪几条将引起转移到 L1？

3.36 假设 X 和 X+2 单元的内容为双精度数 p,Y 和 Y+2 单元的内容为双精度数 q,X 和 Y 为低位字，试说明下列程序段做什么工作？

```
        MOV     DX,X+2
        MOV     AX,X
        ADD     AX,X
        ADC     DX,X+2
        CMP     DX,Y+2
        JL      L2
        JG      L1
        CMP     AX,Y
        JBE     L2
L1:     MOV     AX,1
        JMP     SHORT EXIT
L2:     MOV     AX,2
EXIT:   INT     20H
```

3.37 要求测试在 STATUS 中的一个字节,如果第 1,3,5 位均为 1 则转移到 ROUTINE_1;如果此三位中有两位为 1 则转移到 ROUTINE_2;如果此三位中只有一位为 1 则转移到 ROUTINE_3;如果此三位全为 0 则转移到 ROUTINE_4。试画出流程图,并编制相应的程序段。

3.38 在下列程序的括号中分别填入如下指令:

(1) LOOP   L20
(2) LOOPE   L20
(3) LOOPNE   L20

试说明在三种情况下,当程序执行完后,AX,BX,CX 和 DX 四个寄存器的内容分别是什么?

```
TITLE         EXLOOP.COM
CODESG        SEGMENT
              ASSUME   CS:CODESG, DS:CODESG, SS:CODESG
              ORG      100H
BEGIN:        MOV      AX,01
              MOV      BX,02
              MOV      DX,03
              MOV      CX,04
L20:          INC      AX
              ADD      BX,AX
              SHR      DX,1
              (     )
              RET
CODESG        ENDS
              END      BEGIN
```

3.39 考虑以下的调用序列:

(1) MAIN 调用 NEAR 的 SUBA 过程(返回的偏移地址为 0400);
(2) SUBA 调用 NEAR 的 SUBB 过程(返回的偏移地址为 0A00);
(3) SUBB 调用 FAR 的 SUBC 过程(返回的段地址为 B200,偏移地址为 0100);
(4) 从 SUBC 返回 SUBB;
(5) SUBB 调用 NEAR 的 SUBD 过程(返回的偏移地址为 0C00);
(6) 从 SUBD 返回 SUBB;
(7) 从 SUBB 返回 SUBA;
(8) 从 SUBA 返回 MAIN;
(9) 从 MAIN 调用 SUBC(返回的段地址是 1000,偏移地址是 0600);

请画出每次调用及返回时的堆栈状态。

3.40 假设(EAX)=00001000H,(EBX)=00002000H,(DS)=0010H,试问下列

指令访问内存的物理地址是什么?

  (1) MOV ECX,[EAX+EBX]

  (2) MOV [EAX+2*EBX],CL

  (3) MOV DH,[EBX+4*EAX+1000H]

3.41 假设(EAX)=9823F456H,(ECX)=1F23491H,(BX)=348CH,(SI)=2000H,(DI)=4044H。在 DS 段中从偏移地址 4044H 单元开始的 4 个字节单元中,依次存放的内容为 92H,6DH,0A2H 和 4CH,试问下列各条指令执行完后目的地址及其中的内容是什么?

  (1) MOV [SI],EAX

  (2) MOV [BX],ECX

  (3) MOV EBX,[DI]

3.42 说明下列指令的操作

  (1) PUSH AX

  (2) POP ESI

  (3) PUSH [BX]

  (4) PUSHAD

  (5) POP DS

  (6) PUSH 4

3.43 请给出下列各指令序列执行完后目的寄存器的内容。

  (1) MOV EAX,299FF94H

    ADD EAX,34FFFFH

  (2) MOV EBX,40000000

    SUB EBX,1500000

  (3) MOV EAX,39393834H

    AND EAX,0F0F0F0FH

  (4) MOV EDX,9FE35DH

    XOR EDX,0F0F0F0H

3.44 请给出下列各指令序列执行完后目的寄存器的内容。

  (1) MOV   BX,-12

    MOVSX EBX,BX

  (2) MOV   CL,-8

    MOVSX EDX,CL

  (3) MOV   AH,7

    MOVZX ECX,AH

  (4) MOV   AX,99H

    MOVZX EBX,AX

3.45 请给出下列指令序列执行完后 EAX 和 EBX 的内容。

  MOV ECX,307F455H

```
BSF    EAX,ECX
BSR    EBX,ECX
```

3.46 请给出下列指令序列执行完后 AX 和 DX 的内容。

```
MOV    BX,98H
BSF    AX,BX
BSR    DX,BX
```

3.47 试编写一程序段,要求把 ECX、EDX 和 ESI 的内容相加,其和存入 EDI 寄存器中(不考虑溢出)。

3.48 请说明 IMUL   BX,DX,100H 指令的操作。

3.49 试编写一程序段,要求把 BL 中的数除以 CL 中的数,并把其商乘以 2,最后的结果存入 DX 寄存器中。

3.50 请说明 JMP   DI 和 JMP   [DI]指令的区别。

3.51 试编写一程序段,要求在长度为 100H 字节的数组中,找出大于 42H 的无符号数的个数并存入字节单元 UP 中;找出小于 42H 的无符号数的个数并存入字节单元 DOWN 中。

3.52 请用图表示 ENTER   16,0 所生成的堆栈帧的情况。

第4章 汇编语言程序格式

4.1 汇编程序功能

图 4.1 表示了汇编语言程序的建立及处理过程。首先用编辑程序(可用任一种文本编辑程序,如 EDIT 等)产生汇编语言的源程序(属性为 ASM 的源文件),源程序就是用汇编语言的语句编写的程序,它是不能为机器所识别的,所以要经过汇编程序加以翻译,因此汇编程序的作用就是把源文件转换成用二进制代码表示的目标文件(称为 OBJ 文件)。在转换的过程中,汇编程序将对源程序进行扫视,如果源程序中有语法错误,则汇编结束后,汇编程序将指出源程序中的错误,用户还可以用编辑程序来修改源程序中的错误,最后得到无语法错误的 OBJ 文件。

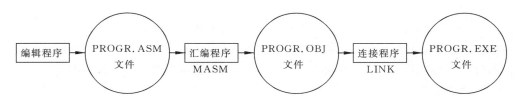

图 4.1 汇编语言程序的建立及汇编过程

OBJ 文件虽然已经是二进制文件,但它还不能直接上机运行,必须经过连接程序(link)把目标文件与库文件或其他目标文件连接在一起形成可执行文件(EXE 文件),这个文件可以由 DOS 装入存储器,并在机器上运行。

因此,在计算机上运行汇编语言程序的步骤是:
(1) 用编辑程序建立 ASM 源文件;
(2) 用 MASM 程序把 ASM 文件转换成 OBJ 文件;
(3) 用 LINK 程序把 OBJ 文件转换成 EXE 文件;
(4) 用 DOS 命令直接键入文件名就可执行该程序。

目前常用的汇编程序有 Microsoft 公司推出的宏汇编程序 MASM(Macro Assembler)和 Borland 公司推出的 TASM(Turbo Assembler)两种。本书采用 MASM5.0 版来说明汇编程序所提供的伪操作和操作符。如读者使用的是 MASM 的其他版本或是 TASM 等其他汇编程序,则由于它们之间在多数情况下是兼容的,而不会造成很大影响;如遇到细微差别,请读者查阅有关手册。

汇编程序的主要功能是:
(1) 检查源程序。
(2) 测出源程序中的语法错误,并给出出错信息。
(3) 产生源程序的目标程序,并可给出列表文件(同时列出汇编语言和机器语言的文件,称为 LST 文件)。

(4) 展开宏指令。

4.2 伪 操 作

　　汇编语言程序的语句除指令以外还可以由伪操作和宏指令组成。关于宏指令将在第 7 章加以说明,这一节只讨论伪操作。伪操作又称为伪指令,它们不像机器指令那样是在程序运行期间由计算机来执行的,而是在汇编程序对源程序汇编期间由汇编程序处理的操作,它们可以完成如处理器选择、定义程序模式、定义数据、分配存储区、指示程序结束等功能。在这一节里,只说明一些常用的伪操作。有些伪操作,如有关过程定义及外部过程所使用的伪操作将在第 6 章中介绍,有关宏汇编及条件汇编所使用的伪操作将在第 7 章中讨论。另外还有一些内容在本书中未涉及,若读者需要时请查阅相关手册。

4.2.1 处理器选择伪操作

　　由于 80x86 的所有处理器都支持 8086/8088 指令系统,但每一种高档的机型又都增加一些新的指令,因此在编写程序时要对所用处理器有一个确定的选择。也就是说,要告诉汇编程序应该选择哪一种指令系统。这一组伪操作的功能就是做这件事的。

　　此类伪操作主要有以下几种:

.8086　　　选择 8086 指令系统

.286　　　选择 80286 指令系统

.286 P　　选择保护方式下的 80286 指令系统

.386　　　选择 80386 指令系统

.386 P　　选择保护方式下的 80386 指令系统

.486　　　选择 80486 指令系统

.486 P　　选择保护方式下的 80486 指令系统

.586　　　选择 Pentium 指令系统

.586 P　　选择保护方式下的 Pentium 指令系统

有关"选择保护方式下的××××指令系统"的含义是指包括特权指令在内的指令系统。此外,上述伪操作均支持相应的协处理器指令。

　　这类伪操作一般放在整个程序的最前面。如不给出,则汇编程序认为其默认值为 .8086。它们可放在程序中,如程序中使用了一条 80486 所增加的指令,则可在该指令的上一行加上 .486。

4.2.2 段定义伪操作

1. 完整的段定义伪操作

　　存储器的物理地址是由段地址和偏移地址组合而成的,汇编程序在把源程序转换为目标程序时,必须确定标号和变量(代码段和数据段的符号地址)的偏移地址,并且需要把有关信息通过目标模块传送给连接程序,以便连接程序把不同的段和模块连接在一起,形成一个可执行程序。为此,需要用段定义伪操作,其格式如下:

  segment name   SEGMENT
      ⋮
  segment name   ENDS

其中删节号部分,对于数据段、附加段和堆栈段来说,一般是存储单元的定义、分配等伪操作;对于代码段则是指令及伪操作。

  此外,还必须明确段和段寄存器的关系,这可用 ASSUME 伪操作来实现,其格式为:
    ASSUME assignment,…, assignment
其中 assignment 说明分配情况,其格式为:
    segment register name : segment name
其中段寄存器名必须是 CS,DS,ES 和 SS(对于 386 及其后继机型还有 FS 和 GS)中的一个,而段名则必须是由 SEGMENT 定义的段中的段名。ASSUME NOTHING 则可取消前面由 ASSUME 所指定的段寄存器。

  举例说明如下:

例 4.1

```
;*************************************************************
data_seg1 segment                   ; define data segment
              ⋮
data_seg1 ends
;*************************************************************
data_seg2 segment                   ; define extra segment
              ⋮
data_seg2 ends
;*************************************************************
code_seg segment                    ; define code segment
        assume cs:code_seg, ds:data_seg1, es:data_seg2
start:                              ; starting execution address
; set DS register to current data segment
        mov   ax, data_seg1         ; data segment addr
        mov   ds, ax                ;   into DS register
; set ES register to current extra segment
        mov   ax, data_seg2         ; extra segment addr
        mov   es, ax                ;   into ES register
              ⋮
code_seg ends                       ; end of code segment
;*************************************************************
        end   start
```

由于 ASSUME 伪操作只是指定某个段分配给哪一个段寄存器，它并不能把段地址装入段寄存器中，所以在代码段中，还必须把段地址装入相应的段寄存器中。为此，例 4.1 的程序中，分别用两条 MOV 指令完成这一操作。如果程序中有堆栈段，也需要把段地址装入 SS 中。但是，代码段不需要这样做。代码段的这一操作是在程序初始化时完成的。

为了对段定义作进一步地控制，SEGMENT 伪操作还可以增加类型及属性的说明，其格式如下：

  segname SEGMENT [align_type] [combine_type] [use_type] ['class']
        ⋮
  segname ENDS

一般情况下，这些说明可以不用。但是，如果需要用连接程序把本程序与其他程序模块相连接时，就需要使用这些说明，分别叙述如下：

(1) 定位类型(align_type)说明段的起始地址应有怎样的边界值，它们可以是：

  PARA 指定段的起始地址必须从小段边界开始，即段起始地址的最低的 16 进制数位必须为 0。这样，偏移地址可以从 0 开始。

  BYTE 该段可以从任何地址开始。这样，起始偏移地址可能不是 0。

  WORD 该段必须从字的边界开始，即段起始地址必须为偶数。

  DWORD 该段必须从双字的边界开始，即段起始地址的最低十六进制数位必须为 4 的倍数。

  PAGE 该段必须从页的边界开始，即段起始地址的最低两个十六进制数位必须为 0（该地址能被 256 整除）。

定位类型的默认项是 PARA，即如未指定定位类型，则汇编程序默认为 PARA。

(2) 组合类型(combine_type)说明程序连接时的段合并方法，它们可以是：

  PRIVATE 该段为私有段，在连接时将不与其他模块中的同名分段合并。

  PUBLIC 该段连接时可以把不同模块中的同名段相连接而形成一个段，其连接次序由连接命令指定。每一分段都从小段的边界开始，因此各模块的原有段之间可能存在小于 16 个字节的间隙。

  COMMON 该段在连接时可以把不同模块中的同名段重叠而形成一个段，由于各同名分段有相同的起始地址，所以会产生覆盖。COMMON 的连接长度是各分段中的最大长度。重叠部分的内容取决于排列在最后一段的内容。

  AT expression 使段地址是表达式所计算出来的 16 位值。但它不能用来指定代码段。

  MEMORY 与 PUBLIC 同义。

  STACK 把不同模块中的同名段组合而形成一个堆栈段。该段的长度为各原有段的总和，各原有段之间并无 PUBLIC 所连接段中的间隙，而且栈顶可自动指向连接后形成的大堆栈段的栈顶。

组合类型的默认项是 PRIVATE。

(3) 使用类型(use_type)只适用于 386 及其后继机型,它用来说明使用 16 位寻址方式还是 32 位寻址方式。它们可以是:

USE16　　使用 16 位寻址方式。

USE32　　使用 32 位寻址方式。

当使用 16 位寻址方式时,段长不超过 64KB,地址的形式是 16 位段地址和 16 位偏移地址;当使用 32 位寻址方式时,段长可达 4GB,地址的形式是 16 位段地址和 32 位偏移地址。可以看出,在实模式下,应该使用 USE16。

使用类型的默认项是 USE16。

(4) 类别('class')在引号中给出连接时组成段组的类型名。类别说明并不能把相同类别的段合并起来,但在连接后形成的装入模块中,可以把它们的位置靠在一起。

2. 存储模型与简化段定义伪操作

较新版本的汇编程序(MASM5.0 与 MASM6.0)除支持"1. 完整的段定义伪操作"中所讨论的 segment 伪操作外,还提供了一种新的较简单的段定义方法。这种方法虽然不能像 segment 伪操作那样具有较完整的表达能力,但它确实比较简单易用,这就是下面要讨论的"存储模型与简化的段定义伪操作"。

(1) MODEL 伪操作

MODEL 伪操作的格式如下:

.MODEL memory_model [, model options]

它用来表示存储模型(memory_model),即用来说明在存储器中是如何安放各个段的。也就是说,它说明代码段在程序中如何安排,代码的寻址是近还是远;数据段在程序中又是如何安排,数据的寻址是近还是远。根据它们的不同组合,可以建立如下七种存储模型:

① Tiny　　所有数据和代码都放在一个段内,其数据和代码都是近访问。Tiny 程序可以写成.COM 文件的形式,COM 程序必须从 0100H 的存储单元开始。这种模型一般用于小程序。

② Small　　所有数据放在一个 64KB 的数据段内,所有代码放在另一个 64KB 的代码段内,数据和代码也都是近访问的。这是一般应用程序最常用的一种模型。

③ Medium　　代码使用多个段,一般一个模块一个段,而数据则合并成一个 64KB 的段组。这样,数据是近访问的,而代码则可远访问。

④ Compact　　所有代码都放在一个 64KB 的代码段内,数据则可放在多个段内,形成代码是近访问的,而数据则可为远访问的格式。

⑤ Large　　代码和数据都可用多个段,所以数据和代码都可以远访问。

⑥ Huge　　与 Large 模型相同,其差别是允许数据段的大小超过 64KB。

⑦ Flat　　允许用户用 32 位偏移量,但 DOS 下不允许使用这种模型,只能在 OS/2 下或其他保护模式的操作系统下使用。MASM 5 版本不支持这种模型,但 MASM 6 可以支持。

model options 允许用户指定三种选项:高级语言接口、操作系统和堆栈距离。

高级语言接口选项是指该汇编语言程序作为某一种高级语言程序的过程而为该高级语言程序调用时,应该用如 C,BASIC,FORTRAN,PASCAL 等来加以说明。

操作系统选项是要说明程序运行于哪个操作系统之下,可用 OS_DOS 或 OS_OS2 来说明,默认项是 OS_DOS。

堆栈距离选项可用 NEARSTACK 或 FARSTACK 来说明。其中 NEARSTACK 是指把堆栈段和数据段组合到一个 DGROUP 段中,DS 和 SS 均指向 DGROUP 段;FARSTACK 是指堆栈段和数据段并不合并。当存储模型为 TINY,SMALL,MEDIUM 和 FLAT 时,默认项为 NEARSTACK;当存储模型为 COMPACT、LARGE 和 HUGE 时,默认项为 FARSTACK。

例如： . MODEL　　SMALL, C
　　　　. MODEL　　LARGE, PASCAL, OS_DOS, FARSTACK

(2) 简化的段定义伪操作

汇编程序给出的标准段有下列几种：

① code　　　　　　　　　代码段
② initialized data　　　　初始化数据段
③ uninitialized data　　　未初始化数据段
④ far initialized data　　远初始化数据段
⑤ far uninitialized data　远未初始化数据段
⑥ constants　　　　　　　常数段
⑦ stack　　　　　　　　　堆栈段

可以看出,这种分段方法把数据段分得更细：一是把常数段和数据段分开;二是把初始化数据段和未初始化数据段分开(其中初始化数据是指程序中已指定初始值的数据);三是把近和远的数据段分开。这样做的结果可便于与高级语言兼容。读者可采取这些标准段来组织和编写程序,在连接时可把它们组成段组,以便提高程序的运行效率。实际上,如果你是为高级语言编写一个汇编语言过程,那你可以使用以上标准段模式;如果你编写一个独立的汇编语言程序,就不必非要这样细分了。一般采用下述的.CODE,.DATA 和.STACK 等定义三个标准段就可以了。

对应以上的标准段,可有如下简化段伪操作：

.CODE [name]　　对于一个代码段的模型,段名为可选项;
　　　　　　　　对于多个代码段的模型,则应为每个代码段指定段名。

.DATA

.DATA?

.FARDATA [name]　　可指定段名。如不指定,则将以 FAR_DATA 命名。

.FARDATA? [name]　可指定段名。如不指定,则将以 FAR_BSS 命名。

.CONST

.STACK [size]　　可指定堆栈段大小。如不指定,则默认值为 1KB。

必须注意：当使用简化段伪操作时,必须在这些简化段伪操作出现之前,即程序的一开始先用.MODEL 伪操作定义存储模型,然后再用简化段伪操作定义段。每一个新段的开始就是上一段的结束,而不必用 ENDS 作为段的结束符。

(3) 与简化段定义有关的预定义符号

汇编程序给出了与简化段定义有关的一组预定义符号,它们可在程序中出现,并由汇编程序识别使用。如在完整的段定义情况下,在程序的一开始,需要用段名装入数据段寄存器,如例 4.1 中的

  mov ax,data_seg1
  mov ds,ax

若用简化段定义,则数据段只用 .data 来定义,而并未给出段名,此时可用

  mov ax,@data
  mov ds,ax

这里预定义符号@data 就给出了数据段的段名。

另外,还有一些预定义符号,它们也可与条件汇编伪操作相配合,以帮助用户编写一些较为复杂的代码。

(4) 用 MODEL 定义存储模型时的段默认属性

表 4.1 给出了使用 MODEL 伪操作时的段默认情况。

表 4.1 用 MODEL 伪操作时的段默认属性

模型	伪操作	名字	定位	组合	类	组
Tiny	.CODE	_TEXT	Word	PUBLIC	'CODE'	DGROUP
	.FARDATA	FAR_DATA	Para	Private	'FAR_DATA'	
	.FARDATA?	FAR_BSS	Para	Private	'FAR_BSS'	
	.DATA	_DATA	Word	PUBLIC	'DATA'	DGROUP
	.CONST	CONST	Word	PUBLIC	'CONST'	DGROUP
	.DATA?	_BSS	Word	PUBLIC	'BSS'	DGROUP
Small	.CODE	_TEXT	Word	PUBLIC	'CODE'	
	.FARDATA	FAR_DATA	Para	Private	'FAR_DATA'	
	.FARDATA?	FAR_BSS	Para	Private	'FAR_BSS'	
	.DATA	_DATA	Word	PUBLIC	'DATA'	DGROUP
	.CONST	CONST	Word	PUBLIC	'CONST'	DGROUP
	.DATA?	_BSS	Word	PUBLIC	'BSS'	DGROUP
	.STACK	STACK	Para	STACK	'STACK'	DGROUP
Medium	.CODE	name_TEXT	Word	PUBLIC	'CODE'	
	.FARDATA	FAR_DATA	Para	Private	'FAR_DATA'	
	.FARDATA?	FAR_BSS	Para	Private	'FAR_BSS'	
	.DATA	_DATA	Word	PUBLIC	'DATA'	DGROUP

续表

模型	伪操作	名字	定位	组合	类	组
	.CONST	CONST	Word	PUBLIC	'CONST'	DGROUP
	.DATA?	_BSS	Word	PUBLIC	'BSS'	DGROUP
	.STACK	STACK	Para	STACK	'STACK'	DGROUP
Compact	.CODE	_TEXT	Word	PUBLIC	'CODE'	
	.FARDATA	FAR_DATA	Para	Private	'FAR_DATA'	
	.FARDATA?	FAR_BSS	Para	Private	'FAR_BSS'	
	.DATA	_DATA	Word	PUBLIC	'DATA'	DGROUP
	.CONST	CONST	Word	PUBLIC	'CONST'	DGROUP
	.DATA?	_BSS	Word	PUBLIC	'BSS'	DGROUP
	.STACK	STACK	Para	STACK	'STACK'	DGROUP
Large 或 Huge	.CODE	name_TEXT	Word	PUBLIC	'CODE'	
	.FARDATA	FAR_DATA	Para	Private	'FAR_DATA'	
	.FARDATA?	FAR_BSS	Para	Private	'FAR_BSS'	
	.DATA	_DATA	Word	PUBLIC	'DATA'	DGROUP
	.CONST	CONST	Word	PUBLIC	'CONST'	DGROUP
	.DATA?	_BSS	Word	PUBLIC	'BSS'	DGROUP
	.STACK	STACK	Para	STACK	'STACK'	DGROUP
Flat	.CODE	_TEXT	Dword	PUBLIC	'CODE'	
	.FARDATA	_DATA	Dword	PUBLIC	'DATA'	
	.FARDATA?	_BSS	Dword	PUBLIC	'BSS'	
	.DATA	_DATA	Dword	PUBLIC	'DATA'	
	.CONST	CONST	Dword	PUBLIC	'CONST'	
	.DATA?	_BSS	Dword	PUBLIC	'BSS'	
	.STACK	STACK	Dword	PUBLIC	'STACK'	

其中,模型列给出了可定义的 7 种模型。伪操作列给出了对应每一种模型可定义 7 种段的伪操作。名字列给出对应各段所用段名,其中可有多个代码的模型 medium、large 和 huge 中的段,可以在.CODE 伪操作中定义段名 name。此外,可以看到凡未初始化的数据段给出的段扩展名为 BSS。定位列给出段的起始地址边界的类型,组合列给出段的组合类型,类列给出各段所属类,组列则给出各种模型下所建立的段组。

(5) 简化段定义举例

例 4.2

```
         .MODEL      SMALL
         .STACK      100H               ; define stack segment
         .DATA                          ; define data segment
```

```
            ⋮
        .CODE                           ; define code segment
START：
        MOV         AX,@DATA            ; starting execution address
                                        ; data segment addr
        MOV         DS,AX               ;   into DS register
            ⋮
        MOV         AX,4C00H
        INT         21H
        END         START
```

可见其比完整的段定义简单得多。但由于完整的段定义可以全面地说明段的各种类型与属性,因此在很多情况下仍需使用它。

例 4.3

```
        .MODEL      SMALL
        .STACK      100H
        .CONST                          ; define constant segment
            ⋮
        .DATA                           ; define data segment
            ⋮
        .CODE                           ; define code segment
START：
        MOV         AX,DGROUP           ; data segment addr
        MOV         DS,AX               ;   into DS register
            ⋮
        MOV         AX,4C00H
        INT         21H
        END         START
```

此时,也可把段组名 DGROUP 作为段地址装入 DS 寄存器中。这样,在访问 CONST 段和 DATA 段中的变量时,都用 DS 作为段寄存器来访问,以提高运行效率。

例 4.4

```
        .MODEL      SMALL
        .FARDATA
            ⋮
        .CODE
START：
        MOV         AX,@DATA
        MOV         DS,AX
        MOV         AX,@FARDATA
        MOV         ES,AX
        ASSUME      ES：@FARDATA
            ⋮
```

```
            MOV         AX, 4C00H
            INT         21H
            END         START
```

.FARDATA 和 .FARDATA? 建立的是独立的段,所以必须为它们建立一个段寄存器(常用 ES),本例就说明了其定义方法。应当注意其中 ASSUME 伪操作的使用方法。

3. 段组定义伪操作

从"2. 存储模型与简化段定义伪操作"中,已经知道在各种存储模型中,汇编程序自动地把各数据段组成一个段组 DGROUP,以便程序在访问各数据段时使用一个数据段寄存器 DS。下面所给出的 GROUP 伪操作允许用户自行指定段组,其格式如下:

```
    grpname    GROUP    segname [, segname …]
```

其中 grpame 为段组名,segname 则为段名。

例 4.5

```
    DSEG1      SEGMENT    WORD     PUBLIC    'DATA'
                              ⋮
    DSEG1      ENDS
    DSEG2      SEGMENT    WORD     PUBLIC    'DATA'
                              ⋮
    DSEG2      ENDS
    DATAGROUP  GROUP      DSEG1, DSEG2
    CSEG       SEGMENT    PARA     PUBLIC    'CODE'
               ASSUME     CS : CSEG, DS : DATAGROUP
    START:     MOV        AX, DATAGROUP
               MOV        DS, AX
                              ⋮
               MOV        AX, 4C00H
               INT        21H
    CSEG       ENDS
               END        START
```

这样,程序中对定义在不同段中的变量,都可以用同一个段寄存器进行访问。

4.2.3 程序开始和结束伪操作

在程序的开始可以用 NAME 或 TITLE 作为模块的名字,NAME 的格式是:

```
    NAME    module_name
```

汇编程序将以给出的 module_name 作为模块的名字。如果程序中没有使用 NAME 伪操作,则也可使用 TITLE 伪操作,其格式为:

```
    TITLE   text
```

TITLE 伪操作可指定列表文件的每一页上打印的标题。同时,如果程序中没有使用 NAME 伪操作,则汇编程序将用 text 中的前六个字符作为模块名。text 最多可有 60 个字符。如果程序中既无 NAME 又无 TITLE 伪操作,则将用源文件名作为模块名。所以 NAME 及 TITLE 伪操作并不是必要的,但一般经常使用 TITLE,以便在列表文件中能打印出标题来。

表示源程序结束的伪操作的格式为:

END　[label]

其中标号(label)指示程序开始执行的起始地址。如果多个程序模块相连接,则只有主程序要使用标号,其他子程序模块则只用 END 而不必指定标号。例 4.1～4.5 已使用 END START 表示程序结束。汇编程序将在遇 END 时结束汇编,而程序则将从 START 开始执行。

MASM 6.0 版的汇编程序还增加了定义程序的入口点和出口点的伪操作。

.STARTUP 用来定义程序的初始入口点,并且产生设置 DS,SS 和 SP 的代码。如果程序中使用了.STARTUP,则结束程序的 END 伪操作中不必再指定程序的入口点标号。

.EXIT 用来产生退出程序并返回操作系统的代码,其格式为:

.EXIT　[return_value]

其中 return_value 为返回给操作系统的值。常用 0 作为返回值。

例 4.6

```
.MODEL      SMALL
.DATA
            ⋮
.CODE
.STARTUP
            ⋮
.EXIT       0
END
```

4.2.4　数据定义及存储器分配伪操作

这一类伪操作的格式是:

[Variable]　Mnemonic　Operand,…,Operand　[;Comments]

其中变量(Variable)字段是可有可无的,它用符号地址表示,其作用与指令语句前的标号相同,但它的后面不跟冒号。如果语句中有变量,则汇编程序使其记以第一个字节的偏移地址。

注释(Comments)字段用来说明该伪操作的功能,它也是可有可无的。

助记符(Mnemonic)字段说明所用伪操作的助记符。即伪操作,说明所定义的数据类型。常用的有以下几种：

DB 伪操作用来定义字节，其后的每个操作数都占有一个字节(8位)。

DW 伪操作用来定义字，其后的每个操作数占有一个字(16位，其低位字节在第一个字节地址中，高位字节在第二个字节地址中。)。

DD 伪操作用来定义双字，其后的每个操作数占有两个字(32位)。

DF 伪操作用来定义6个字节的字，其后的每个操作数占有48位，可用来存放远地址。这一伪操作只能用于386及其后继机型中。

DQ 伪操作用来定义4字，其后的每个操作数占有4个字(64位)，可用来存放双精度浮点数。

DT 伪操作用来定义10个字节，其后的每个操作数占有10个字节，形成压缩的BCD码。

这些伪操作可以把其后跟着的数据存入指定的存储单元，形成初始化数据；或者只分配存储空间而并不存入确定的数值，形成未初始化数据。DW和DD伪操作还可存储地址，DF伪操作则可存储由16位段地址及32位偏移地址组成的远地址指针。下面举例说明。

例4.7 操作数可以是常数，或者是表达式(根据该表达式可以求得一个常数)，如

```
DATA_BYTE    DB    10,4,10H
DATA_WORD    DW    100,100H,-5
DATA_DW      DD    3*20,0FFFDH
```

汇编程序可以在汇编期间在存储器中存入数据，如图4.2所示。

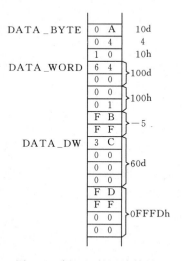

图4.2 例4.7的汇编结果

例4.8 操作数也可以是字符串，如

MESSAGE   DB   ′HELLO′

则存储器存储情况如图 4.3(1)所示,而 DB ′AB′ 和 DW ′AB′ 的存储情况则分别如图 4.3(2)和(3)所示。

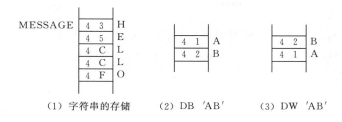

图 4.3 例 4.8 的汇编结果

例 4.9 操作数?可以保留存储空间,但不存入数据。如:

ABC   DB   0,?,?,?,0
DFF   DW   ?,52,?

经汇编后的存储情况如图 4.4 所示。

操作数字段还可以使用复制操作符(duplication operator)来复制某个(或某些)操作数。其格式为:

repeat_count   DUP(operand, …, operand)

其中 repeat_count 可以是一个表达式,它的值应该是一个正整数,用来指定括号中的操作数的重复次数。

例 4.10

ARRAY1   DB   2   DUP(0,1,2,?)
ARRAY2   DB   100 DUP(?)

汇编后的存储情况如图 4.5 所示。

由图可见,例 4.10 中的第一个语句和语句 ARRAY1   DB   0,1,2,?,0,1,2,? 是等价的。

图 4.4 例 4.9 的汇编结果   图 4.5 例 4.10 的汇编结果

例 4.11 DUP 操作可以嵌套,例如:

ARRAY3　DB　100　DUP(0,2　DUP(1,2),0,3)

则汇编结果如图 4.6 所示。

可以用 DW 或 DD 伪操作把变量或标号的偏移地址(DW)或由 16 位段地址和 16 位偏移地址组成的整个地址(DD)存入存储器。用 DD 伪操作存入地址时,第 1 个字为偏移地址,第 2 个字为段地址。

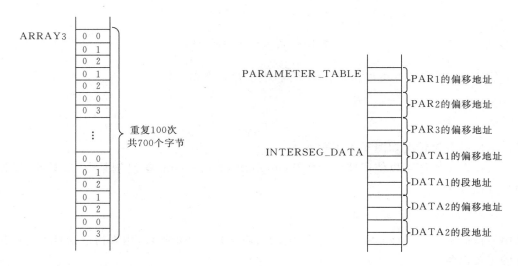

图 4.6　例 4.11 的汇编结果　　　　图 4.7　例 4.12 的汇编结果

例 4.12

```
PARAMETER_TABLE    DW    PAR1
                   DW    PAR2
                   DW    PAR3
INTERSEG_DATA      DD    DATA1
                   DD    DATA2
```

则汇编程序的存储情况如图 4.7 所示。其中偏移地址或段地址均占有一个字,其低位字节占有第 1 个字节,高位字节占有第 2 个字节。

386 及其后继机型具有 16 位段地址和 32 位偏移地址构成的 48 位远地址,这可用 DF 伪操作来定义。

例 4.13

```
                   .386
Dataseg            SEGMENT    PARA    USE32  'DATA'
Parse_table        DB    2048 DUP(0)
Tblptr             DF    Parse_table
Dataseg            ENDS
```

汇编后的存储情况如图 4.8 所示。

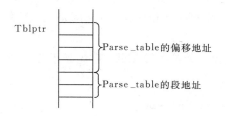

图 4.8　例 4.13 的汇编结果

几点说明

① 这里操作数字段中的变量或标号可以使用表达式,如:

variable±constant expression
label±constant expression

在这种情况下,汇编后,存储器中应该存入表达式的值。

② DB、DW、DD、DF、DQ、DT 等伪操作在 MASM6 中可用 BYTE、WORD、DWORD、FWORD、QWORD、TBYTE 来取代,其含义是等同的。

③ 变量的类型属性(type attribute)问题。在数据定义伪操作前面的变量的值,是该伪操作中的第一个数据项在当前段内的第一个字节的偏移地址。此外,它还有一个类型属性,用来表示该语句中的每一个数据项的长度(以字节为单位表示),因此 DB 伪操作的类型属性为 1,DW 为 2,DD 为 4,DF 为 6,DQ 为 8,DT 为 10。变量表达式的属性和变量是相同的。汇编程序可以用这种隐含的类型属性来确定某些指令是字指令还是字节指令。

例 4.14

```
OPER1    DB    ?,?
OPER2    DW    ?,?
         ⋮
         MOV   OPER1,0
         MOV   OPER2,0
```

则第 1 条指令应为字节指令,而第 2 条指令应为字指令。如果有下列指令序列:

例 4.15

```
OPER1    DB    1,2
OPER2    DW    1234H,5678H
         ⋮
         MOV   AX,OPER1+1
         MOV   AL,OPER2
```

汇编程序在汇编这一段程序时,能发现两条 MOV 指令的两个操作数的类型属性是不相同的:OPER1+1 为字节类型属性,而 AX 为字类型属性;OPER2 为字类型属性,而 AL 为字节类型属性。因此,汇编程序将指示出错:这两条 MOV 指令中的两个操作数的类型不匹配。

有一个办法可以指定操作数的类型属性,它优先于隐含的类型属性,即使用 PTR 属性操作符。其格式为:

    type PTR Variable±constant expression

其中类型可以是 BYTE,WORD,DWORD,FWORD,QWORD 或 TBYTE,这样一来变量的类型就可以指定了。例 4.15 的第一条指令可以写成:

    MOV   AX, WORD   PTR   OPER1+1

这样就把 OPER1+1 的类型属性指定为字,两个操作数的属性也就一致了,汇编时不会出错。而运行时应把 OPER1+1 的字内容送 AX,即把 OPER1+1 的内容送 AL,把 OPER2 的第一个字节的内容送 AH,所以指令执行完后,(AX)=3402H。

例 4.15 的第二条指令应写成:

    MOV   AL, BYTE   PTR   OPER2

汇编时不会出错,运行时应把 OPER2 的第一个字节的内容送 AL,即(AL)=34H。而

    MOV   AL, BYTE   PTR   OPER2 +1

则应把 OPER2 中的第一个字的高位字节送 AL,即(AL)=12H。

从例 4.15 可以看出:同一个变量可以具有不同的类型属性。除了用属性操作符给以定义外,还可以用 LABEL 伪操作来定义,其格式为:

    name   LABEL   type

对于数据项可表示为:

    variable_name   LABEL   type

其中类型可以是 BYTE,WORD,DWORD,FWORD,QWORD 或 TBYTE。

对于可执行的代码,则可表示为:

    label_name   LABEL   type

其中类型可以是 NEAR 或 FAR。对于 16 位段,NEAR 为 2 字节,FAR 为 4 字节;对于 32 位段,NEAR 为 4 字节,FAR 是 6 字节。

例 4.16

    BYTE_ARRAY   LABEL   BYTE
    WORD_ARRAY   DW   50 DUP(?)

这样,在 100 个字节数组中的第一个字节的地址赋予二个不同类型的变量名:字节类型的变量 BYTE_ARRAY 和字类型变量 WORD_ARRAY。

指令：

  MOV WORD_ARRAY＋2,0

把该数组的第 3 个和第 4 个字节置 0,而

  MOV BYTE_ARRAY＋2,0

则把该数组的第 3 个字节置 0。

4.2.5 表达式赋值伪操作 EQU

  有时程序中多次出现同一个表达式,为方便起见,可以用赋值伪操作给表达式赋予一个名字。其格式如下：

  Expression_name EQU Expression

此后,程序中凡需要用到该表达式之处,就可以用表达式名来代替了。可见,EQU 的引入提高了程序的可读性,也使其更加易于修改。上式中的表达式可以是任何有效的操作数格式,可以是任何可求出常数值的表达式,也可以是任何有效的助记符。举例如下：

CONSTANT	EQU	256	数赋以符号名
DATA	EQU	HEIGHT＋12	地址表达式赋以符号名
ALPHA	EQU	7	这是一组赋值伪操作,把 7－2＝5 赋以符号名 BETA,VAR＋5 赋以符号名 ADDR。
BETA	EQU	ALPHA－2	
ADDR	EQU	VAR＋BETA	
B	EQU	[BP＋8]	变址引用赋以符号名 B
P8	EQU	DS:[BP＋8]	加段前缀的变址引用赋以符号名 P8

  必须注意：在 EQU 语句的表达式中,如果有变量或标号的表达式,则在该语句前应该先给出它们的定义。例如,语句

  AB EQU DATA_ONE＋2

则必须放在 DATA_ONE 的定义之后才行,否则汇编程序将指示出错。

  另外还有一个与 EQU 相类似的＝伪操作也可以作为赋值伪操作使用。它们之间的区别是 EQU 伪操作中的表达式名是不允许重复定义的,而＝伪操作则允许重复定义。

  例如,EMP＝6 或 EMP EQU 6 都可以使数 6 赋以符号名 EMP,然而不允许两者同时使用。但是

    ⋮

  EMP＝7

    ⋮

  EMP＝EMP＋1

    ⋮

在程序中是允许使用的,因为＝伪操作允许重复定义。在这种情况下,在第一个语句后的指令中 EMP 的值为 7,而在第二个语句后的指令中 EMP 的值为 8。

4.2.6 地址计数器与对准伪操作

1. 地址计数器 $

在汇编程序对源程序汇编的过程中,使用地址计数器(location counter)来保存当前正在汇编的指令的偏移地址。当开始汇编或在每一段开始时,把地址计数器初始化为零,以后在汇编过程中,每处理一条指令,地址计数器就增加一个值,此值为该指令所需要的字节数。地址计数器的值可用 $ 来表示,汇编语言允许用户直接用 $ 来引用地址计数器的值,因此指令

JNE $＋6

的转向地址是 JNE 指令的首地址加上 6。当 $ 用在指令中时,它表示本条指令的第一个字节的地址。在这里,$＋6 必须是另一条指令的首地址。否则,汇编程序将指示出错信息。当 $ 用在伪操作的参数字段时,则和它用在指令中的情况不同,它所表示的是地址计数器的当前值。

例 4.17

ARRAY DW 1,2,$＋4,3,4,$＋4

如汇编时 ARRAY 分配的偏移地址为 0074,则汇编后的存储区将如图 4.9 所示。

注意:ARRAY 数组中的两个 $＋4 得到的结果是不同的,这是由于 $ 的值是在不断变化的缘故。当在指令中用到 $ 时,它只代表该指令的首地址,而与 $ 本身所在的字节无关。

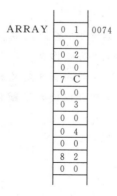

图 4.9 例 4.17 的汇编结果

2. ORG 伪操作

ORG 伪操作用来设置当前地址计数器的值,其格式为:

ORG　　constant expression

如常数表达式的值为 n,则 ORG 伪操作可以使下一个字节的地址成为常数表达式的值 n。例如:

```
VECTORS    SEGMENT
           ORG    10
VECT1      DW     47A5H
           ORG    20
VECT2      DW     0C596H
              ⋮
VECTORS    ENDS
```

则 VECT1 的偏移地址值为 0AH,而 VECT2 的偏移地址值为 14H。

常数表达式也可以表示从当前已定义过的符号开始的位移量,或表示从当前地址计数器值 $ 开始的位移量,如:

ORG　　$+8

可以表示跳过 8 个字节的存储区,亦即建立了一个 8 字节的未初始化的数据缓冲区。如程序中需要访问该缓冲区,则可用 label 伪操作来定义该缓冲区的如下变量名:

```
BUFFER    LABEL    BYTE
          ORG      $+8
```

当然,其完成的功能和

BUFFER   DB   8 DUP(?)

是一样的。

3. EVEN 伪操作

EVEN 伪操作使下一个变量或指令开始于偶数字节地址。一个字的地址最好从偶地址开始,所以对于字数组为保证其从偶地址开始,可以在其前用 EVEN 伪操作来达到这一目的。例如:

```
DATA_SEG       SEGMENT
                  ⋮
               EVEN
WORD_ARRAY     DW     100 DUP(?)
                  ⋮
DATA_SEG       ENDS
```

4. ALIGN 伪操作

ALIGN 伪操作为保证双字数组边界从 4 的倍数开始创造了条件,其格式为:

```
ALIGN    boundary
```

其中 boundary 必须是 2 的幂。例如：

```
            .DATA
              ⋮
            ALIGN    4
ARRAY       DB       100 DUP（?）
              ⋮
```

就可保证 ARRAY 的值为 4 的倍数。

当然，ALIGN   2 和 EVEN 是等价的。

4.2.7 基数控制伪操作

汇编程序默认的数为十进制数，因而除非专门指定，汇编程序把程序中出现的数均看作十进制数。为此，当使用其他基数表示的常数时，需要专门给以标记如下：

(1) 二进制数：由一串 0 和 1 组成其后跟以字母 B，如 00101100B。

(2) 十进制数：由 0～9 的数字组成的数。一般情况下，后面不必加上标记，在指定其他基数的情况下，后面可跟字母 D，例如 178D。

(3) 十六进制数：由 0～9 及 A～F 组成的数，后面跟字母 H。这个数的第一个字符必须是 0～9，所以如果第一个字符是 A～F 时，应在其前面加上数字 0，如 0FFFFH。

(4) 八进制数：由数字 0～7 组成的数，后面可跟字母 O 或 Q，如 1777O。

(5) .RADIX 伪操作可以把默认的基数改变为 2～16 范围内的任何基数。其格式如下：

```
.RADIX    expression
```

其中表达式用来表示基数值（用十进制数表示）。

例如：

```
MOV    BX，0FFH
MOV    BX，178
```

与

```
.RADIX    16
MOV       BX，0FF
MOV       BX，178D
```

是等价的。

注意：在用 .RADIX16 把基数定为十六进制后，十进制数后面都应跟字母 D。在这种情况下，如果某个十六进制数的末字符为 D，则应在其后跟字母 H，以免与十进制数发生混淆。

（6）字符串可以看成串常数，可以用单引号或双引号把字符串放在其中，得到的是字符串的 ASCII 码值，例如′ABCD′。

4.3　汇编语言程序格式

汇编语言源程序中的每个语句可以由 4 项组成，格式如下：

[name]　operation　operand　[；comment]

其中名字(name)项是一个符号。

操作(operation)项是一个操作码的助记符，它可以是指令、伪操作或宏指令名。

操作数(operand)项由一个或多个表达式组成，它提供为执行所要求的操作而需要的信息。

注释(comment)项用来说明程序或语句的功能。；为识别注释项的开始。；也可以从一行的第一个字符开始，此时整行都是注释，常用来说明下面一段程序的功能。

上面 4 项中带方括号的两项是可有可无的。各项之间必须用"空格"(space)或"水平制表"(TAB)符隔开。下面分别说明各项的表示方法。

4.3.1　名字项

源程序中用下列字符来表示名字：

字母 A～Z

数字 0～9

专用字符 ？，·，@ ，—，$

除数字外，所有字符都可以放在源语句的第一个位置。名字中如果用到·，则必须是第一个字符。可以用很多字符来说明名字，但只有前面的 31 个字符能被汇编程序所识别。

一般说来，名字项可以是标号或变量。它们都用来表示本语句的符号地址，都是可有可无的，只有当需要用符号地址来访问该语句时它才需要出现。

（1）标号，标号在代码段中定义，后面跟着冒号：，它也可以用 LABEL 或 EQU 伪操作来定义。此外，它还可以作为过程名定义，这将在第 6 章中加以说明。标号经常在转移指令或 CALL 指令的操作数字段出现，用以表示转向地址。

标号有 3 种属性：段、偏移及类型。

段属性定义标号的段起始地址，此值必须在一个段寄存器中，而标号的段则总是在 CS 寄存器中。

偏移属性，标号的偏移地址是从段起始地址到定义标号的位置之间的字节数。对于 16 位段是 16 位无符号数；对于 32 位段则是 32 位无符号数。

类型属性，用来指出该标号是在本段内引用还是在其他段中引用的。如是在段内引用的，则称为 NEAR。对于 16 位段，指针长度为 2 字节；对于 32 位段，指针长度为 4 字节。如在段外引用,则称为 FAR。对于 16 位段,指针长度为 4 字节(段地址 2 字节,偏移地址 2 字节);对于 32 位段,指针长度为 6 字节(段地址 2 字节,偏移地址 4 字节)。

(2) 变量,变量在数据段或附加数据段中定义,后面不跟冒号。它也可以用 LABEL 或 EQU 伪操作来定义。变量经常在操作数字段出现。它也有段、偏移及类型三种属性。

段属性:定义变量的段起始地址,此值必须在一个段寄存器中。

偏移属性:变量的偏移地址是从段的起始地址到定义变量的位置之间的字节数。对于 16 位段,是 16 位无符号数;对于 32 位段,则是 32 位无符号数。在当前段内给出变量的偏移值等于当前地址计数器的值,当前地址计数器的值可以用 $ 来表示。

类型属性:变量的类型属性定义该变量所保留的字节数。如 BYTE(1 个字节长)、WORD(2 个字节长)、DWORD(4 个字节长)、FWORD(6 个字节长)、QWORD(8 个字节长)、TBYTE(10 个字节长),这一点在数据定义伪操作中已作了说明。

在同一个程序中,同样的标号或变量的定义只允许出现一次,否则汇编程序会指示出错。

4.3.2 操作项

操作项可以是指令、伪操作或宏指令的助记符。对于指令,汇编程序将其翻译为机器语言指令。对于伪操作,汇编程序将根据其所要求的功能进行处理。对于宏指令,则将根据其定义展开。在第 7 章中将会专门论述。

4.3.3 操作数项

操作数项由一个或多个表达式组成,多个操作数项之间一般用逗号分开。对于指令,操作数项一般给出操作数地址,它们可能有一个或两个,或三个,或一个也没有。对于伪操作或宏指令,则给出它们所要求的参数。

操作数项可以是常数、寄存器、标号、变量或由表达式组成。在这里,将专门对表达式加以说明。表达式是常数、寄存器、标号、变量与一些操作符相组合的序列,可以有数字表达式和地址表达式两种。在汇编期间,汇编程序按照一定的优先规则对表达式进行计算后可得到一个数值或一个地址。为了能了解表达式的组成,下面先介绍一些常用的操作符。

1. 算术操作符

算术操作符有 +,-,*,/ 和 MOD。其中 MOD 是指除法运算后得到的余数,如 19/7 的商是 2,而 19 MOD 7 则为 5(余数)。

算术操作符可以用于数字表达式或地址表达式中,但当它用于地址表达式时,只有当其结果有明确的物理意义时才是有效的结果,例如,两个地址相乘或相除是无意义的。在地址表达式中,可以使用 + 或 - ,但也必须注意其物理意义,例如把两个不同段的地址相加也是无意义的。经常使用的是地址±数字量,它是有意义的,例如,SUM+1 是指 SUM 字节单元的下一个字节单元的地址(注意:不是指 SUM 单元的内容加 1),而 SUM-1 则是指 SUM 字节单元的前一个字节单元的地址。

例 4.18 如果要求把首地址为 BLOCK 的字数组的第 6 个字传送到 DX 寄存器,可用指令如下:

　　MOV　DX,BLOCK+(6-1)*2

例 4.19 如数组 ARRAY 定义如下,试写出把数组长度(字数)存入 CX 寄存器的指令。

　　ARRAY　　　DW　　1,2,3,4,5,6,7
　　ARYEND　　 DW　　?

其中 ARYEND 是为计算数组长度而建立的符号地址,所需指令如下:

　　MOV　　CX,(ARYEND-ARRAY)/2

汇编程序在汇编期间将计算出表达式的值而形成指令:

　　MOV　　CX,7

2. 逻辑与移位操作符

逻辑操作符有 AND,OR,XOR 和 NOT;移位操作符有 SHL 和 SHR。它们都是按位操作的,只能用于数字表达式中。逻辑操作符要求汇编程序对其前后两个操作数(或表达式)作指定的逻辑操作。移位操作符的格式是:

　　expression　SHL(或 SHR)numshift

汇编程序将 expression 左移或右移 numshift 位,如移位数大于 15,则结果为 0。

例 4.20

　　IN　　　AL,PORT_VAL
　　OUT　　PORT_VAL　AND　0FEH,AL

其中 PORT_VAL 为端口号,OUT 指令中的表达式说明当 PORT_VAL 为偶数时,输出端口号与输入端口号相同;而当 PORT_VAL 为奇数时,则输出端口号比输入端口号小 1。

例 4.21

　　AND　　DX,PORT_VAL　AND　0FEH

该指令在汇编时由汇编程序对指令中的表达式进行计算得到一个端口号,而在程序运行时,该指令的执行是把 DX 寄存器的内容与汇编程序计算得到的端口号相"与",结果保存在 DX 寄存器中。

3. 关系操作符

它有 EQ(相等)、NE(不等)、LT(小于)、GT(大于)、LE(小于或等于)、GE(大于或等于)6 种。

关系操作符的两个操作数必须都是数字,或是同一段内的两个存储器地址。计算的结果应为逻辑值,结果为真,表示为 0FFFFH;结果为假,则表示为 0。

例 4.22

  MOV BX,((PORT_VAL LT 5) AND 20) OR ((PORT_VAL GE 5) AND 30)

则当 PORT_VAL < 5 时,汇编结果应该是,

  MOV BX,20

否则,汇编结果应该是,

  MOV BX,30

4. 数值回送(Value_returning)操作符

  它主要有 TYPE,LENGTH,SIZE,OFFSET,SEG 等。这些操作符把一些特征或存储器地址的一部分作为数值回送。下面分别说明各个操作符的功能。

  (1) TYPE

  格式为,TYPE expression

如果该表达式是变量,则汇编程序将回送该变量的以字节数表示的类型:DB 为 1,DW 为 2,DD 为 4,DF 为 6,DQ 为 8,DT 为 10。如果表达式是标号,则汇编程序将回送代表该标号类型的数值:NEAR 为 −1,FAR 为 −2。如果表达式为常数,则应回送 0。

例 4.23

  ARRAY DW 1,2,3

则对于指令 ADD SI,TYPE ARRAY

汇编程序将其形成为:

  ADD SI,2

  (2) LENGTH

  格式为,LENGTH Variable

对于变量中使用 DUP 的情况,汇编程序将回送分配给该变量的单元数,而对于其他情况则送 1。

例 4.24

  FEES DW 100 DUP(0)

对于指令 MOV CX,LENGTH FEES

汇编程序将使其形成为:

  MOV CX,100

例 4.25

  ARRAY DW 1,2,3

对于指令 MOV CX,LENGTH ARRAY

汇编程序将使其形成为：

  MOV CX, 1

 例 4.26

  TABLE DB 'ABCD'

对于指令 MOV CX, LENGTH TABLE

汇编程序将使其形成为：

  MOV CX, 1

 （3）SIZE

  格式为：SIZE Variable

汇编程序应回送分配给该变量的字节数。但是，此值是 LENGTH 值和 TYPE 值的乘积，所以对于例 4.24 的情况：

  MOV CX, SIZE FEES

将形成为 MOV CX, 200

对于例 4.25 的情况：

  MOV CX, SIZE ARRAY

将形成为 MOV CX, 2

而对于例 4.26 的情况：

  MOV CX, SIZE TABLE

将形成为 MOV CX, 1

 （4）OFFSET

  格式为：OFFSET Variable 或 label

汇编程序将回送变量或标号的偏移地址值。

 例 4.27

  MOV BX, OFFSET OPER_ONE

则汇编程序将 OPER_ONE 的偏移地址作为立即数回送给指令，而在执行时则将该偏移地址装入 BX 寄存器中。所以这条指令与指令

  LEA BX, OPER_ONE

是等价的。

 （5）SEG

  格式为：SEG Variable 或 label

汇编程序将回送变量或标号的段地址值。

 例 4.28 如果 DATA_SEG 是从存储器的 05000H 地址开始的一个数据段的段名，OPER1 是该段中的一个变量名，则

  MOV BX, SEG OPER1

将把 0500H 作为立即数插入指令。实际上,由于段地址是由连接程序分配的,所以该立即数是连接时插入的。执行期间则使 BX 寄存器的内容成为 0500H。

  5. 属性操作符

  它主要有 PTR、段操作符、SHORT、THIS、HIGH、LOW、HIGHWORD 和 LOWWORD 等。

  (1) PTR

  格式为:type PTR expression

PTR 用来建立一个符号地址,但它本身并不分配存储器,只是用来给已分配的存储地址赋予另一种属性,使该地址具有另一种类型。格式中的类型字段表示所赋予的新的类型属性,而表达式字段则是被取代类型的符号地址。

  例 4.29 已有数据定义如下:

  TWO_BYTE DW ?

可以用以下语句对这两个字节赋予另一种类型定义:

  ONE_BYTE EQU BYTE PTR TWO_BYTE
  OTHER_BYTE EQU BYTE PTR (TWO_BYTE + 1)

后者也可以写成:

  OTHER_BYTE EQU ONE_BYTE + 1

这里 ONE_BYTE 和 TWO_BYTE 两个符号地址具有相同的段地址和偏移地址,但是它们的类型属性不同,前者为 1,后者为 2。

  前面已经说明类型可有 BYTE、WORD、DWORD、FWORD、QWORD、TBYTE、NEAR 和 FAR 等几种,所以 PTR 也可以用来建立字、双字、四字或段内及段间的指令单元等。

  此外,有时指令要求使用 PTR 操作符。例如用

  MOV [BX],5

指令把立即数存入 BX 寄存器内容指定的存储单元中,但汇编程序不能分清是存入字单元还是字节单元,此时必须用 PTR 操作符来说明属性,应该写明:

    MOV BYTE PTR [BX],5
  或 MOV WORD PTR [BX],5

  (2) 段操作符:用来表示一个标量、变量或地址表达式的段属性。例如,用段前缀指定某段的地址操作数

  MOV AX,ES:[BX + SI]

可见它是用段寄存器:地址表达式来表示的。此外,也可以用段名:地址表达式或组名:地址表达式来表示其段属性。

(3) SHORT

用来修饰 JMP 指令中转向地址的属性,指出转向地址是在下一条指令地址的±127 个字节范围之内。

(4) THIS

格式为, THIS attribute 或 type

它可以像 PTR 一样建立一个指定类型(BYTE、WORD、DWORD、FWORD、QWORD 或 TBYTE)的或指定距离(NEAR 或 FAR)的地址操作数。该操作数的段地址和偏移地址与下一个存储单元地址相同。例如：

```
FIRST_TYPE    EQU   THIS  BYTE
WORD_TABLE          100   DUP（?）
```

此时 FIRST_TYPE 的偏移地址和 WORD_TABLE 完全相同,但它是字节类型的;而 WORD_TABLE 则是字类型的。又如：

```
START    EQU   THIS  FAR
         MOV   CX,100
```

这样,MOV 指令有一个 FAR 属性的地址 START,这就允许其他段的 JMP 指令直接跳转到 START 来。

(5) HIGH 和 LOW

称为字节分离操作符,它接收一个数或地址表达式,HIGH 取其高位字节,LOW 取其低位字节。例如：

```
          CONST  EQU  0ABCDH
则        MOV    AH,  HIGH  CONST
将汇编成   MOV    AH,  0ABH
```

(6) HIGHWORD 和 LOWWORD

称为字分离操作符,它接收一个数或地址表达式,HIGHWORD 取其高位字,LOWWORD 取其低位字。这两个操作符是 MASM6 版支持的。

以上说明了 6 种类型的常用操作符。若知道表达式是常数、寄存器、标号、变量和操作符的组合,在计算表达式时,应该首先计算优先级高的操作符,然后从左到右地对优先级相同的操作符进行计算。括号也可以改变计算次序,括号内的表达式应优先计算。下面给出操作符的优先级别(其中有些操作符并未提到过,需要时可从手册中查到。),从高到低排列如下：

(1) 在圆括号中的项,方括号中的项,结构变量(变量,字段。),然后是 LENGTH、SIZE、WIDTH 和 MASK。

(2) 名:(段取代)。

(3) PTR,OFFSET,SEG,TYPE,THIS 及段操作符。

(4) HIGH 和 LOW。

(5) 乘法和除法：*,/,MOD,SHL,SHR。

(6) 加法和减法：＋，－。

(7) 关系操作：EQ,NE,LT,LE,GT,GE。

(8) 逻辑：NOT。

(9) 逻辑：AND。

(10) 逻辑：OR,XOR。

(11) SHORT。

现在,可以计算出表达式的值了。实际上表达式的值是由汇编程序计算的,而程序员应该正确掌握书写表达式的方法,以减少出错的可能性。

4.3.4 注释项

注释项用来说明一段程序、一条或几条指令的功能,它是可有可无的。但是,对于汇编语言程序来说,注释项的作用是很明显的,它可以使程序容易被读懂,因此汇编语言程序必须写好注释。注释应该写出本条(或本段)指令在程序中的功能和作用,而不应该只写指令的动作。读者在有机会阅读程序例子时,应注意学习注释的写法,在编制程序时,更应学会写好注释。

根据以上说明的汇编语言程序格式及有关的伪操作定义,给出源文件(ASM)格式。

汇编语言源程序格式举例(1)

```
;PROGRAM TITLE GOES HERE—
;Followed by descriptive phrases

;EQU STATEMENTS GO HERE
;**************************************************************
datarea segment                         ;define data segment

;DATA GOES HERE
datarea ends
;**************************************************************
prognam segment                         ;define code segment
;--------------------------------------------------------------
main    proc    far                     ;main part of program
        assume  cs:prognam,ds:datarea
start:                                  ;starting execution address
;set up stack for return
        push    ds                      ;save old data segment
        sub     ax,ax                   ;put zero in AX
        push    ax                      ;save it on stack
;set DS register to current data segment
        mov     ax,datarea              ;datarea segment addr
```

```
            mov     ds,ax                    ; into DS register
;MAIN PART OF PROGRAM GOES HERE
            ret                              ;return to DOS
main        endp                             ;end of main part of program
;------------------------------------------------------------------
sub1        proc    near                     ;define subprocedure
;SUBROUTINE GOES HERE
sub1        endp                             ;end of subprocedure
;------------------------------------------------------------------
prognam     ends                             ;end of code segment
;******************************************************************
            end     start                    ;end of assembly
```

几点说明

(1) 其中关于建立过程的 PROC 和 ENDP 伪操作对将在第 6 章中说明。这里只要知道利用这一对伪操作把程序段分为若干个过程，使程序的结构更加清晰就可以了。

(2) 这里只定义了最基本的代码段和数据段，如果程序中还需定义附加段和堆栈段，则定义的方式及建立段寄存器的方法是相同的，读者可自行设计。

(3) 这里把主程序建立为过程，由 DOS 调用该过程，进入程序后，首先把 DS 的内容和 0 作为段地址和偏移地址入栈，以便在程序结束时用 RET 指令返回 DOS，这是一种工作方式。如果在主程序开始时没有用上面三条指令在堆栈中建立返回信息，则在程序结束时就不能直接用 RET 返回指令，而应该使用编号为 4C 的功能调用返回 DOS，如下所示：

```
MOV     AX,4C00H
INT     21H
```

这种方式用得更加普遍。

汇编语言源程序格式举例(1)适用于 MASM 的各种版本，对于 MASM 5.0、6.0 版可采用汇编语言源程序格式举例(2)。

汇编语言源程序格式举例(2)

```
            .model   small                   ;define memory model
            .stack   100h                    ;define stack segment
            .data                            ;define data segment
;DATA GOES HERE
            .code                            ;define code segment
main        proc     far
```

```
start:
        mov     ax,@data        ;data segment addr
        mov     ds,ax           ;  into DS register
;MAIN PART OF PROGRAM GOES HERE
        mov     ax,4c00h
        int     21h             ;return to DOS
main    endp                    ;end of main program
        end     start           ;end of assembly
```

4.4 汇编语言程序的上机过程

在 4.1 节里,已经简单地说明了汇编语言程序从建立到执行的过程,这一节将说明这一过程的具体操作方法。在下面的叙述中,将以 Microsoft 提供的 MASM 5.0 版为基础,如果读者所用的是其他版本,或是 Borland 公司的 TASM,其基本使用方法均相类似,如有问题请查阅有关手册。

4.4.1 建立汇编语言的工作环境

为运行汇编语言程序至少要在磁盘上建立以下文件:
(1) 编辑程序,如 EDIT.EXE
(2) 汇编程序,如 MASM.EXE
(3) 连接程序,如 LINK.EXE
(4) 调试程序,如 DEBUG.COM
必要时,还可建立如 CREF.EXE,EXE2BIN.EXE 等文件。

4.4.2 建立 ASM 文件

为了说明汇编语言程序上机运行的过程,现举例如下:

例 4.30 请把 40 个字母 a 的字符串从源缓冲区传送到目的缓冲区。

可以用编辑程序 EDIT 在磁盘上建立如下的源程序 EX_MOVS.ASM。

例 4.30 的源程序 EX_MOVS.ASM

```
;PROGRAM TITLE GOES HERE—ex_movs
;*************************************************
data        segment                 ;define data segment
    source_buffer       db      40 dup('a')
data        ends
;*************************************************
```

```
extra      segment                   ;define extra segment
    dest_buffer              db        40 dup(?)
extra      ends
;**************************************************************
code       segment                   ;define code segment
;--------------------------------------------------------------
main       proc       far            ;main part of program
        assume cs:code,ds:data,es:extra
start:                                ;starting execution address
;set up stack for return
           push       ds             ;save old data segment
           sub        ax,ax          ;put zero in AX
           push       ax             ;save it on stack
;set DS register to current data segment
           mov        ax,data        ;data segment addr
           mov        ds,ax          ;   into DS register
;set ES register to current extra segment
           mov        ax,extra       ;extra segment addr
           mov        es,ax          ;   into ES register
;MAIN PART OF PROGRAM GOES HERE
           lea        si,source_buffer   ;put offset addr of source
                                     ;   buffer in SI
           lea        di,dest_buffer ;put offset addr of dest
                                     ;   buffer in DI
           cld                       ;set DF flag to forward
           mov        cx,40          ;put count in CX
           rep        movsb          ;move entire string
           ret                       ;return to DOS
main       endp                      ;end of main part of program
;--------------------------------------------------------------
code       ends                      ;end of code segment
;**************************************************************
           end        start          ;end of assembly
```

4.4.3 用 MASM 程序产生 OBJ 文件

源文件建立后,就要用汇编程序对源文件汇编,汇编后产生二进制的目标文件(OBJ文件),其操作与汇编程序回答如下:

C>masm ex_movs ↵
Microsoft (R) Macro Assembler Version 5.00
Copyright (C) Microsoft Corp 1981-1985,1987. All rights reserved.

Object filename [ex_movs. OBJ]：
Source listing[NUL. LST]：ex_movs
Cross-reference [NUL. CRF]：ex_movs
　　51646 ＋ 447778 Bytes symbol space free
　　　　0 Warning Errors
　　　　0 Severe　Errors

汇编程序的输入文件是 ASM 文件，其输出文件可以有三个，表示于上列汇编程序回答的第 3～5 行。第一个是 OBJ 文件，这是汇编的主要目的，所以这个文件我们是需要的，对于[ex_movs.OBJ]后的：应回答↙，这样就在磁盘上建立了这一目标文件。第二个是 LIST 文件，称为列表文件。这个文件同时列出源程序和机器语言程序清单，并给出符号表，因而可使程序调试更加方便。这个文件是可有可无的，如果不需要则可对[NUL.LST]：回答↙；如果需要这个文件，则可回答文件名，这里是 ex_movs ↙，这样例 4.30 的列表文件 EX_MOVS.LST 就建立起来了。LST 清单的最后部分为段名表和符号表，表中分别给出段名，段的大小及有关属性，以及用户定义的符号名、类型及属性。

例 4.30 的列表文件 EX_MOVS.LST

Microsoft（R）Macro Assembler Version 5.00　　　　　　　　3/4/98 00：27：43
　　　　　　　　　　　　　　　　　　　　　　　　　　　　　　Page　1－1

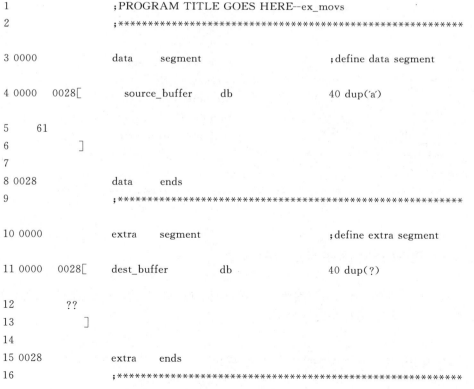

```
                1                    ;PROGRAM TITLE GOES HERE--ex_movs
                2                    ;****************************************************************

                3 0000               data    segment                  ;define data segment

                4 0000    0028[      source_buffer    db              40 dup('a')

                5    61
                6         ]
                7
                8 0028               data    ends
                9                    ;****************************************************************

               10 0000               extra   segment                  ;define extra segment

               11 0000    0028[      dest_buffer      db              40 dup(?)

               12         ??
               13         ]
               14
               15 0028               extra   ends
               16                    ;****************************************************************
```

```
17 0000               code    segment                    ;define code segment

18                    ;--------------------------------------------------------------------

19 0000               main    proc    far                ;main part of program

20
21                    assume cs:code,ds:data,es:extra

22
23 0000               start:                              ;starting execution address

24
25                    ;set up stack for return
26 0000  1E           push    ds                          ;save old data segment

27 0001  2B C0        sub     ax,ax                       ;put zero in AX

28 0003  50           push    ax                          ;save it on stack

29
30                    ;set DS register to current data segment

31 0004  B8----R      mov     ax,data                     ;data segment addr

32 0007  8E D8        mov     ds,ax                       ;   into DS register

33
34                    ;set ES register to current extra segment

35 0009  B8----R      mov     ax,extra                    ;extra segment addr
```

Microsoft (R) Macro Assembler Version 5.00   3/4/98 00:27:43
                                                                 Page   1-2

```
36 000C  8E C0        mov     es,ax                       ;   into ES register

37
```

```
38                              ;MAIN PART OF PROGRAM GOES HERE
39 000E  8D 36 0000 R    lea     si,source_buffer    ;put of fset addr of source
40                                                   ; buffer in  SI

41 0012  8D 3E 0000 R    ea      di,dest_buffer      ;put of fset addr of dest
42                                                   ; buffer in  DI

43 0016  FC              cld                         ;set DF flag to forward

44 0017  B9 0028         mov     cx,40               ;put count in CX

45 001A  F3/ A4          rep     movsb               ;move entire string

46 001C  CB              ret                         ;return to DOS

47
48 001D                  main    endp                ;end of main part of program
49                       ;--------------------------------------------------------

50 001D                  code    ends                ;end of code segment

51                       ;**************************************************************

52
53                               end     start       ;end of assembly
```

Microsoft (R) Macro Assembler Version 5.00   3/4/98 00:27:43
                                                          Symbols—1

Segments and Groups:

Name	Length	Align	Combine	Class
CODE	001D	PARA	NONE	
DATA	0028	PARA	NONE	
EXTRA	0028	PARA	NONE	

Symbols:

Name	Type	Value	Attr	
DEST_BUFFER.........	L BYTE	0000	EXTRA	Length = 0028
MAIN.............	F PROC	0000	CODE	Length = 001D
SOURCE_BUFFER........	L BYTE	0000	DATA	Length = 0028
START............	L NEAR	0000	CODE	
@FILENAME..........	TEXT	ex_movs		

```
      47 Source   Lines
      47 Total    Lines
       9 Symbols
   50552 + 448872 Bytes symbol space free
       0 Warning Errors
       0 Severe  Errors
```

汇编程序能提供的第三个文件是 CRF 文件,这个文件用来产生交叉引用表 REF。对于一般程序不需要建立此文件,所以对于第 5 行的[NUL.CRF]:可以用↙来回答,这样就完成了汇编过程。如果希望建立交叉引用表,则应该用文件名来回答,这里是 ex_movs ↙,这就产生了 EX_MOVS.CRF 文件。为了建立交叉引用表,必须调用 CREF 程序,所以如果需要就应该从系统中把 CREF.EXE 文件 COPY 到你的磁盘上,此后就键入

C>cref ex_movs ↙

机器回答

Microsoft (R) Cross-Reference Utility   Version 5.00
Copyright (C) Microsoft Corp 1981-1985, 1987.   All rights reserved.

Listing [ex_movs.REF]:
7 Symbols

用 ↙ 回答就建立了 EX_MOVS.REF 文件。

例 4.30 的交叉引用表 EX_MOVS.REF 文件

Microsoft Cross-Reference   Version 5.00   Wed Mar 04 00:34:07 1998

Symbol Cross-Reference   (# definition, + modification)   Cref-1

CODE	17#	21	50
DATA	3#	8 21	31
DEST_BUFFER.	11#	41	
EXTRA.	10#	15 21	35
MAIN	19#	48	
SOURCE_BUFFER.	4#	39	
START.	23#	53	

7 Symbols

交叉引用表给出了用户定义的所有符号,对于每个符号列出了其定义所在行号(加上#)及引用的行号。可以看出它为大程序的修改提供了方便,而一般较小的程序则可不使用。

到此为止,汇编过程已经完成了。但是,汇编程序还有另一个重要功能:可以给出源程序中的错误信息。警告错误(warning errors)指出汇编程序所认为的一般性错误;严重错误(severe errors)则指出汇编程序认为已使汇编程序无法进行正确汇编的错误。除给出错误的个数外,汇编程序还能指出错误信息。如果程序有错,则应重新调用编辑程序修改错误,并重新汇编直到汇编正确通过为止。当然汇编程序只能指出程序中的语法错误,至于程序的算法或编制程序中的其他错误则应在程序调试时去解决。

4.4.4 用 LINK 程序产生 EXE 文件

汇编程序已产生出二进制的目标文件(OBJ),但 OBJ 文件并不是可执行文件,还必须使用连接程序(LINK)把 OBJ 文件转换为可执行的 EXE 文件。当然,如果一个程序是由多个模块组成时,也应该通过 LINK 把它们连接在一起,操作方法及机器回答如下:

C>link ex_moves ↙
Microsoft (R) Overlay Linker   Version 3.60
Copyright (C) Microsoft Corp 1983－1987.   All rights reserved.

Run File [EX_MOVS.EXE]:
List File [NUL.MAP]:ex_movs
Libraries [.LIB]:
LINK : warning L4021: no stack segment

LINK 程序有两个输入文件 OBJ 和 LIB。OBJ 是我们需要连接的目标文件,LIB 则是程序中需要用到的库文件,如无特殊需要,则应对[LIB]:回答 ↙。LINK 程序有两个输出文件,一个是EXE文件,这当然是我们所需要的,应对[EX_MOVS.EXE]：回答↓,这样就在磁盘上建立了该可执行文件。LINK的另一个输出文件为MAP文件,它是连接程序的列表文件,又称为连接映像(Link map),它给出每个段在存储器中的分配情况。下面给出了例4.30的连接映像EX_MOVS.MAP文件。

LINK : warning L4021: no stack segment

Start   Stop   Length Name   Class
00000H 00027H 00028H DATA
00030H 00057H 00028H EXTRA
00060H 0007CH 0001DH CODE

Program entry point at 0006:0000

连接程序给出的无堆栈段的警告性错误并不影响程序的运行。所以,到此为止,连接过程已经结束,可以执行EX_MOVS程序了。

4.4.5 程序的执行

在建立了EXE文件后,就可以直接从DOS执行程序,如下所示：

C＞ex_movs ↓
C＞

程序运行结束并返回DOS。如果用户程序已直接把结果在终端上显示出来,那么程序已经运行结束,结果也已经得到了。但是,EX_MOVS程序并未显示出结果,那么怎么知道程序执行的结果是正确的呢？此外,大部分程序必须经过调试阶段才能纠正程序执行中的错误,得到正确的结果,那么又怎样来调试程序呢？这里要使用调试程序。有关调试程序的使用,请读者参看本书的配套教材《IBM PC汇编语言程序设计实验教程》或在本例中可使用DEBUG的D命令来检查附加段中是否已得到了从源缓冲区传来的40个字母a。

4.4.6 COM文件

COM文件也是一种可执行文件,由程序本身的二进制代码组成,它没有EXE文件所具有的包括有关文件信息的标题区(header),所以它占有的存储空间比EXE文件要小。COM文件不允许分段,它所占有的空间不允许超过64KB,因而只能用来编制较小的程序。由于其小而简单,装入速度比EXE文件要快。

使用COM文件时,程序不分段,其入口点(开始运行的起始点)必须是100H(其前的256个字节为程序段前缀所在地),且不必设置堆栈段。在程序装入时,由系统自动把SP建立在该段之末。对于所有的过程则应定义为NEAR。下面给出COM文件的源程序格式举例(1)。对于 MASM 的 5.0、6.0 版也可采用 COM 文件的源程序格式举例(2)。

COM 文件的源程序格式举例(1)

```
;PROGRAM   TITLE   GOES   HERE--
;   Followed by descriptive phrases
;EQU statements go here
;***************************************************************
prognam segment
        org         100h
        assume cs:prognam,ds:prognam,es:prognam,ss:prognam
;--------------------------------------------------------------------
main    proc        near
;Program goes here
        mov         ax,4c00h                      ;return to DOS
        int         21h
;(or    int         20h)
;--------------------------------------------------------------------
;DATA goes here
prognam ends
;***************************************************************
        end         main
```

COM 文件的源程序格式举例(2)

```
;PROGRAM TITLE goes here--

        .model      tiny
        .code
        org         100h
start: jmp          begin

;DATA goes here

begin:

;PROGRAM goes here

        mov         ax,4c00h
        int         21h

        end         start
```

用户在建立源文件以后,同样经过汇编、连接形成 EXE 文件,然后可以通过 EXE2BIN 程序来建立 COM 文件,操作方法如下:

C＞exe2bin filename filename.com ↵

请读者注意上行中的第一个 filename 给出已形成的 EXE 文件的文件名,但不必给出文件扩展名。第二个 filename 即为所要求的 COM 文件的文件名,它必须带有文件扩展名 COM,这样就形成了所要的 COM 文件。在 DOS 系统下,可直接在机器上用文件名执行。如果第二个 filename 后不跟扩展名,则将形成 BIN 文件,在 DOS 系统下运行该程序时,必须先用 rename 命令把它改名为 COM 文件才能直接运行。

此外,COM 文件还可以直接在调试程序 DEBUG 中用 A 或 E 命令建立,对于一些短小的程序,这也是一种相当方便的方法。

习　　题

4.1　指出下列指令的错误:
　　(1) MOV　　AH,BX
　　(2) MOV　　[BX],[SI]
　　(3) MOV　　AX,[SI][DI]
　　(4) MOV　　MYDAT[BX][SI],ES：AX
　　(5) MOV　　BYTE　PTR[BX],1000
　　(6) MOV　　BX,OFFSET　MYDAT[SI]
　　(7) MOV　　CS,AX
　　(8) MOV　　ECX,AX

4.2　下面哪些指令是非法的?(假设 OP1,OP2 是已经用 DB 定义的变量)
　　(1) CMP　　15,BX
　　(2) CMP　　OP1,25
　　(3) CMP　　OP1,OP2
　　(4) CMP　　AX,OP1

4.3　假设下列指令中的所有标识符均为类型属性为字的变量,请指出下列指令中哪些是非法的?它们的错误是什么?
　　(1) MOV　　BP,AL
　　(2) MOV　　WORD_OP[BX+4*3][DI],SP
　　(3) MOV　　WORD_OP1,WORD_OP2
　　(4) MOV　　AX,WORD_OP1[DX]
　　(5) MOV　　SAVE_WORD,DS
　　(6) MOV　　SP,SS：DATA_WORD[BX][SI]
　　(7) MOV　　[BX][SI],2
　　(8) MOV　　AX,WORD_OP1＋WORD_OP2

(9) MOV　AX,WORD_OP1－WORD_OP2＋100

(10) MOV　WORD_OP1,WORD_OP1－WORD_OP2

4.4　假设 VAR1 和 VAR2 为字变量,LAB 为标号,试指出下列指令的错误之处：

(1) ADD　VAR1,VAR2

(2) SUB　AL,VAR1

(3) JMP　LAB[SI]

(4) JNZ　VAR1

(5) JMP　NEAR　LAB

4.5　画图说明下列语句所分配的存储空间及初始化的数据值。

(1) BYTE_VAR　DB　'BYTE',12,－12H,3 DUP(0,?,2 DUP(1,2),?)

(2) WORD_VAR　DW　5 DUP(0,1,2),?,－5,'BY','TE',256H

4.6　试列出各种方法,使汇编程序把 5150H 存入一个存储器字中(例如：DW 5150H)。

4.7　请设置一个数据段 DATASG,其中定义以下字符变量或数据变量。

(1) FLD1B 为字符串变量：'personal computer'；

(2) FLD2B 为十进制数字节变量：32；

(3) FLD3B 为十六进制数字节变量：20；

(4) FLD4B 为二进制数字节变量：01011001；

(5) FLD5B 为数字的 ASCII 字符字节变量：32654；

(6) FLD6B 为 10 个零的字节变量；

(7) FLD7B 为零件名(ASCII 码)及其数量(十进制数)的表格：

　　PART1　20

　　PART2　50

　　PART3　14

(8) FLD1W 为十六进制数字变量：FFF0；

(9) FLD2W 为二进制数字变量：01011001；

(10) FLD3W 为(7)中零件表的地址变量；

(11) FLD4W 为包括 5 个十进制数的字变量：5,6,7,8,9；

(12) FLD5W 为 5 个零的字变量；

(13) FLD6W 为本段中字数据变量和字节数据变量之间的地址差。

4.8　假设程序中的数据定义如下：

```
PARTNO    DW    ?
PNAME     DB    16    DUP(?)
COUNT     DD    ?
PLENTH    EQU   $－PARTNO
```

问 PLENTH 的值为多少？它表示什么意义？

4.9　有符号定义语句如下：

```
BUFF        DB    1,2,3,'123'
EBUFF       DB    0
L           EQU   EBUFF-BUFF
```

问 L 的值是多少?

4.10 假设程序中的数据定义如下：

```
LNAME       DB    30  DUP(?)
ADDRESS     DB    30  DUP(?)
CITY        DB    15  DUP(?)
CODE_LIST   DB    1,7,8,3,2
```

(1) 用一条 MOV 指令将 LNAME 的偏移地址放入 AX。

(2) 用一条指令将 CODE_LIST 的头两个字节的内容放入 SI。

(3) 写一条伪操作使 CODE_LENGHT 的值等于 CODE_LIST 域的实际长度。

4.11 试写出一个完整的数据段 DATA_SEG,它把整数 5 赋予一个字节,并把整数 -1,0,2,5 和 4 放在 10 字数组 DATA_LIST 的头 5 个单元中。然后,写出完整的代码段,其功能为:把 DATA_LIST 中头 5 个数中的最大值和最小值分别存入 MAX 和 MIN 单元中。

4.12 给出等值语句如下：

```
ALPHA       EQU   100
BETA        EQU   25
GAMMA       EQU   2
```

下列表达式的值是多少?

(1) ALPHA * 100+BETA

(2) ALPHA MOD GAMMA+BETA

(3) (ALPHA+2)*BETA-2

(4) (BETA/3)MOD 5

(5) (ALPHA+3)*(BETA MOD GAMMA)

(6) ALPHA GE GAMMA

(7) BETA AND 7

(8) GAMMA OR 3

4.13 对于下面的数据定义,三条 MOV 指令分别汇编成什么?(可用立即数方式表示)

```
TABLEA      DW    10 DUP(?)
TABLEB      DB    10 DUP(?)
TABLEC      DB    '1234'
            ⋮
MOV         AX,LENGTH  TABLEA
```

```
            MOV        BL,LENGTH   TABLEB
            MOV        CL,LENGTH   TABLEC
```

4.14 对于下面的数据定义,各条 MOV 指令单独执行后,有关寄存器的内容是什么?

```
    FLDB       DB    ?
    TABLEA     DW    20 DUP(?)
    TABLEB     DB    'ABCD'
```

  (1) MOV   AX,TYPE   FLDB
  (2) MOV   AX,TYPE   TABLEA
  (3) MOV   CX,LENGTH TABLEA
  (4) MOV   DX,SIZE   TABLEA
  (5) MOV   CX,LENGTH TABLEB

4.15 指出下列伪操作表达方式的错误,并改正之。

  (1) DATA_SEG   SEG
  (2) SEGMENT   'CODE'
  (3) MYDATA   SEGMENT/DATA
     ⋮
      ENDS
  (4) MAIN_PROC   PROC   FAR
     ⋮
      END   MAIN_PROC
  MAIN_PROC   ENDP

4.16 按下面的要求写出程序的框架。

  (1) 数据段的位置从 0E000H 开始,数据段中定义一个 100 字节的数组,其类型属性既是字又是字节;
  (2) 堆栈段从小段开始,段组名为 STACK;
  (3) 代码段中指定段寄存器,指定主程序从 1000H 开始,给有关段寄存器赋值;
  (4) 程序结束。

4.17 写一个完整的程序放在代码段 C_SEG 中,要求把数据段 D_SEG 中 AUGEND 和附加段 E_SEG 中的 ADDEND 相加,并把结果存放在 D_SEG 中的 SUM 中。其中 AUGEND,ADDEND 和 SUM 均为双精度数,AUGEND 赋值为 99251,ADDEND 赋值为 −15962。

4.18 请说明表示程序结束的伪操作和结束程序执行的语句之间的差别。它们在源程序中应如何表示?

4.19 试说明下述指令中哪些需要加上 PTR 伪操作:

```
    BVAL    DB    10H,20H
```

WVAL   DW   1000H

(1) MOV   AL, BVAL
(2) MOV   DL, [BX]
(3) SUB   [BX], 2
(4) MOV   CL, WVAL
(5) ADD   AL, BVAL+1

第 5 章　循环与分支程序设计

一般说来，编制一个汇编语言程序的步骤如下：

(1) 分析题意，确定算法。这一步是能否编制出高质量程序的关键，因此不应该一拿到题目就急于写程序，而是应该仔细地分析和理解题意，找出合理的算法和适当的数据结构。

(2) 根据算法，画出程序框图。这一点对初学者特别重要。这样做可以减少出错的可能性。画框图时，可以从粗到细把算法逐步地具体化。

(3) 根据框图编写程序。

(4) 上机调试程序。任何程序必须经过调试，才能检查出设计思想是否正确，以及程序是否符合设计思想。在调试程序的过程中，应该善于利用机器提供的调试工具（如 DEBUG）来进行工作，你会发现它将给你提供很大的帮助。

第 5 章和第 6 章将说明汇编语言程序设计方法，将举例说明上述程序设计步骤，但上机调试就只能由读者自己去实践了。

在转入汇编语言程序设计方法的讨论之前，需要说明的是，本书所讨论的编程环境将只限于在 DOS 操作系统下的实模式，前面已经提到，从 80286 开始，已经为用户提供了实模式和保护模式；从 80386 起，又增加了虚 86 模式。因此一共提供了三种工作模式。虚 86 模式是指在一台计算机上可以同时运行多个 8086 程序，所以从编程角度看，应该说存在着实模式和保护模式两种工作模式。在第 2 章里，已经对保护模式的引入目的以及这两种模式的寻址机制分别做了介绍。就编程而言，这两种工作模式并无实质上的区别，但它们所用环境和某些实现方法还是有些差异的。考虑到使用实模式已可解决应用程序所面向的大量问题，本书又面向汇编语言的基本程序设计方法，而且在 DOS 环境下目前还未能提供保护模式的良好编程环境。因此本书有关程序设计方法的说明将只限于实模式，但有了实模式的编程基础，在了解保护模式的编程环境后，转向保护模式编程也将不会很困难。

程序有顺序、循环、分支和子程序四种结构形式。顺序程序结构是指完全按顺序逐条执行的指令序列，这在程序段中是大量存在的，但作为完整的程序则很少见，这里不对它们作专门讨论。本章说明循环与分支程序结构，第 6 章则专门讨论子程序结构。

5.1　循环程序设计

5.1.1　循环程序的结构形式

在 3.3.5 小节"4. 循环指令"的讨论中，已经给出了两个简单的循环程序的例子，现在来分析一下循环结构。循环程序可以有两种结构形式,如图 5.1 所示。一种是 DO_WHILE 结构形式;另一种是 DO_UNTIL 结构形式。DO_WHILE 结构把对循环控制条件的判断放在循环的入口,先判断条件,满足条件就执行循环体,否则就退出循环。DO_UNTIL 结构则先执行循环体,然后再判断控制条件,不满足条件则继续执行循环操作,一旦满足条件则退出循环。这两种结构可以根据具体情况选择使用。一般说来,如果有循环次数等于 0 的可能,则应选择 DO_WHILE 结构,否则则使用 DO_UNTIL 结构。不论哪一种结构形式,循环程序都可由如下三部分组成:

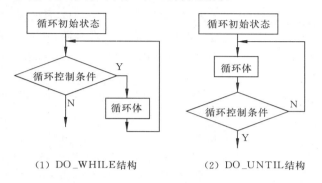

图 5.1 循环程序的结构形式

(1) 设置循环的初始状态。如设置循环次数的计数值,以及为循环体正常工作而建立的初始状态等。

(2) 循环体。这是循环工作的主体,它由循环的工作部分及修改部分组成。循环的工作部分是为完成程序功能而设计的主要程序段;循环的修改部分则是为保证每一次重复(循环)时,参加执行的信息能发生有规律的变化而建立的程序段。

(3) 循环控制部分。循环控制本来应该属于循环体的一部分,由于它是循环程序设计的关键,所以要对它作专门的讨论。每个循环程序必须选择一个循环控制条件来控制循环的运行和结束,而合理地选择该控制条件就成为循环程序设计的关键问题。有时,循环次数是已知的,此时可以用循环次数作为循环的控制条件,LOOP 指令使这种循环程序设计能很容易地实现。有时循环次数是已知的,但有可能使用其他特征或条件来使循环提前结束,LOOPZ 和 LOOPNZ 指令则又是设计这种循环程序的工具。在 3.3.5 小节"4. 循环指令"中,已经给出的例子就说明了这两种情况。然而,有时循环次数是未知的,那就需要根据具体情况找出控制循环结束的条件。循环控制条件的选择是很灵活的,有时可供选择的方案不止一种,此时就应分析比较,选择一种效率最高的方案来实现。下面,用例子来说明。

5.1.2 循环程序设计方法

例 5.1 试编制一个程序把 BX 寄存器内的二进制数用十六进制数的形式在屏幕上显示出来。

根据题意应该把 BX 的内容从左到右每四位为一组在屏幕上显示出来,显然这可以用循环结构来完成,每次循环显示一个十六进制数位,因而循环次数是已知的,计数值为4。循环体中则应包括从二进制到所显示字符的 ASCII 之间的转换,以及每个字符的显示,后者可以使用 DOS 功能调用来实现。画出程序框图如图 5.2 所示。这里采用了循环移位的方法把所要显示的 4 位二进制数移到最右面,以便作数字到字符的转换工作。另外,由于数字 0~9 的 ASCII 为 30~39H,而字母 A~F 的 ASCII 为 41~46H,所以在把 4 位二进制数加上 30H 后还需作一次判断,如果为字符 A~F,则还应加上 7 才能显示出正确的十六进制数。

以 BINIHEX.ASM 为文件名建立"二进制到十六进制数转换程序"源文件。在程序中没有使用 LOOP 指令,这是因为循环移位指令要使用 CL 寄存器,而 LOOP 指令要使用 CX 寄存器,为了解决 CX 寄存器的冲突问题,这里用 CH 寄存器存放循环计数值,而用 DEC 及 JNZ 两条指令完成 LOOP 指令的功能,说明即使用计数值控制循环结束也不是非用 LOOP 指令不可,这里只是提供了另一种方法而已。当然也可以把计数值初始化为 0,用每循环一次加 1 然后比较次数是否达到要求的方法来实现;或者仍用 LOOP 指令,而用堆栈保存其中的一个信息(例如计数值)来解决 CX 寄存器的冲突问题等。总之,程序设计是很灵活的,只要算法和指令的使用没有错误,都可以达到目的。但是,怎样做才能使效率最高,那是需要仔细斟酌的。

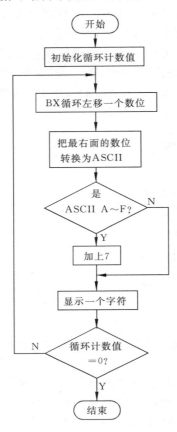

图 5.2 二进制到十六进制数转换的程序框图

二进制到十六进制转换程序

```
prognam segment                 ;define code segment
main proc far
        assume cs:prognam
start:                          ;starting execution addr
;set up stack for return
        push    ds              ;save old data segment
        sub     ax,ax           ;put zero in AX
        push    ax              ;save zero on stack
;main part of program
        mov     ch,4            ;number of digits
rotate: mov     cl,4            ;set count to 4 bits
        rol     bx,cl           ;left digit to right
        mov     al,bl           ;move to AL
```

```
            and     al,0fh              ;mask off left digit
            add     al,30h              ;convert hex to ASCII
            cmp     al,3ah              ;is it > 9 ?
            jl      printit             ;jump if digit = 0 to 9
            add     al,7h               ;digit is A to F
printit:
            mov     dl,al               ;put ASCII char in DL
            mov     ah,2                ;display output funct
            int     21h                 ;call DOS
            dec     ch                  ;done 4 digits?
            jnz     rotate              ;not yet
            ret                         ;return to DOS
main        endp                        ;end of main part of prog.

prognam     ends                        ;end of segment

            end                         ;end of assembly
```

例 5.2 在 ADDR 单元中存放着数 Y 的地址,试编制一程序把 Y 中 1 的个数存入 COUNT 单元中。

要测出 Y 中 1 的个数就应逐位测试。一个比较简单的办法是可以根据最高有效位是否为 1 来计数,然后用移位的方法把各位数逐次移到最高位去。循环的结束可以用计数值为 16 来控制,但更好的办法是结合上述方法可以用测试数是否为 0 来作为结束条件,这样可以在很多情况下缩短程序的执行时间。此外,考虑到 Y 本身为 0 的可能性,应该采用 DO_WHILE 的结构形式。根据以上考虑,可以画出图 5.3 的程序框图,编制了"数 1 的程序"。

这个例子说明算法和循环控制条件的选择对程序的工作效率有很大的影响,而循环控制条件的选择又是很灵活的,应该根据具体情况来确定。

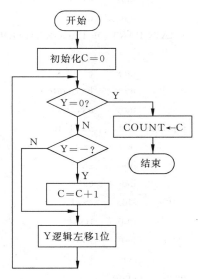

图 5.3 数 1 的程序框图

数 1 的程序

```
;**************************************************
datarea segment                         ;define data segment
        addr    dw      number
```

```
        number   dw      Y
        count    dw      ?
        datarea  ends
;******************************************************************
        prognam segment                 ;define code segment
;------------------------------------------------------------------
        main    proc    far             ;main part of program

               assume cs:prognam,ds:datarea

        start:                          ;starting execution address

;set up stack for return
                push    ds              ;save old data segment
                sub     ax,ax           ;put zero in AX
                push    ax              ;save it on stack

;set DS register to current data segment
                mov     ax,datarea      ;datarea segment addr
                mov     ds,ax           ;   into DS register
;MAIN PART OF PROGRAM GOES HERE

                mov     cx,0            ;initialize C to 0
                mov     bx, addr
                mov     ax,[bx]         ;put Y in AX
        repeat: test    ax,0ffffh       ;test Y
                jz      exit            ;if Y=0,get exit
                jns     shift           ;if MSB=0,C unchanged
                inc     cx              ;else,C=C+1
        shift:  shl     ax,1            ;shift Y one bit left
                jmp     repeat          ;repeat
        exit:   mov     count,cx
                ret                     ;return to DOS
        main    endp                    ;end of main part of prog.
;------------------------------------------------------------------
        prognam ends                    ;end of code segment
;******************************************************************
                end     start           ;end of assembly
```

例 5.3 在附加段中,有一个首地址为 LIST 和未经排序的字数组。在数组的第一个字中,存放着该数组的长度,数组的首地址已存放在 DI 寄存器中,AX 寄存器中存放着一个数。要求编制一程序:在数组中查找该数,如果找到此数,则把它从数组中删除。

这一程序应该首先查找数组中是否有(AX),如果没有则不对数组作任何处理就结束程序。如果找到这一元素,则应把数组中位于其前(指地址比该元素高)的元素后移一个字(即向低地址方向移动),并修改数组长度值。如果找到的元素正好位于数组末尾,则不必移动任何元素,只要修改数组长度值就可以了。这里,第一部分的查找元素可以使用串处理指令,第二部分的删除元素则可使用循环结构。由于查找结束时就可以知道该元素的位置,因此可以作为循环次数已知的情况来设计。画出程序框图如图 5.4 所示。程序主体部分见"删除数组中一元素程序"。

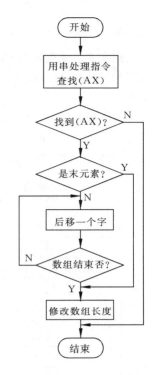

图 5.4 删除数组中一元素的程序框图

删除数组中一元素程序

```
;Delete the value in AX from an unordered list in the extra
;    segment, if that value is in the list.
;Inputs:(DI)=starting address of the list.
;        First location of the list=Length of list(words)

del_ul   proc    near
         cld                     ;make DF=0,to scan forward
         push    di              ;save starting address
         mov     cx,es:[di]      ;fetch element count
         add     di,2            ;make DI point to 1st data el.
         repne   scasw           ;value in the list?
         je      delete          ;if so,go delete it
         pop     di              ;otherwise,exit
         jmp     short exit
;The following instructions delete an element from the
;  list,as follows:
;   (1) If the element lies at the end of the list,delete
;       it by decreasing the element count by 1.
;   (2) Otherwise,delete the element by moving all
;       subsequent elements up by one position.
delete:  jcxz    dec_cnt         ;if (CX)=0,delete last el.
next_el: mov     bx,es:[di]      ;move one el. up in list
         mov     es:[di-2],bx
         add     di,2            ;point to next el.
         loop    next_el         ;repeat until all el. moved
dec_cnt: pop     di              ;decrease el. count by 1
```

```
                dec     word ptr es:[di]
exit:           ret                             ;exit
del_ul          endp                            ;end of main part of prog.
```

例 5.4 将正数 N 插入一个已整序的字数组的正确位置。该数组的首地址和末地址分别为 ARRAY_HEAD 和 ARRAY_END,其中所有数均为正数且已按递增的次序排列。

由于数组的首地址和末地址都是已知的,因此数组长度是可以确定的。但是,这里只要求插入一个数,并不一定要扫描整个数组,所以可以用找到应插入数的位置作为循环的结束条件。此外,为空出要插入数的位置,其前的全部元素都应前移一个字(即向地址增

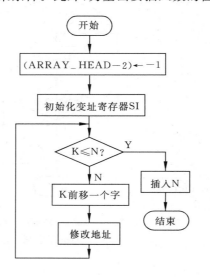

图 5.5 数组中插入一元素的程序框图

大的方向移动一个字,这里的前后是指程序运行的方向为前,反之则为后)所以算法上应该从数组的尾部向头部查找,可逐字取出数组中的一个数 K 与 N 作比较,如 K > N,则把 K 前移一个字,然后继续往后查找;如 K≤N,则把 N 插在 K 之前就可以结束程序了。

在考虑算法时,必须把可能出现的边界情况考虑在内,如例 5.3 中对有可能出现要删除的元素正好在数组末尾的考虑。这个问题是初学者易于忽略的,应该引起注意。在例 5.4 中,应该考虑 N 大于或小于数组中所有数的两种可能性。如果 N 大于数组中所有数,则第一次比较就可以结束循环,也就是说循环次数有可能等于 0,所以应该选用 DO_WHILE 结构形式。如果 N 小于数组中所有数,则必须使循环及时结束,也就是说不允许查找的范围超过数组的首地址,这当然可以把数组的首地址也同时作为结束条件来考虑,或者同时用循环次数作为结束条件之一来考虑。本例的更好办法是:可以利用所有数均为正数的条件,在 ARRAY_HEAD−2 单元中存放'−1'这个数,这样可以保证如果数 N 小于数组中所有数,那它必然大于 −1,这样就可以正确地把 N 放在数组之首了,循环的结束仍然可以使用 K>N 这一条件。根据上述有关算法的考虑可画出程序框图如图 5.5 所示。编制了"数组中插入一元素程序"。

数组中插入一元素程序

```
;****************************************************
datarea segment                         ;define data segment
    x               dw      ?
    array_head      dw      3,5,15,23,37,49,52,65,78,99
    array_end       dw      105
    n               dw      32
```

```
datarea ends
;************************************************************
prognam  segment                    ;define code segment
;--------------------------------------------------------------------------
main     proc    far                ;main part of program

    assume cs:prognam,ds:datarea

start:                              ;starting execution address

;set up stack for return
        push    ds                  ;save old data segment
        sub     ax,ax               ;put zero in AX
        push    ax                  ;save it on stack

;set DS register to current data segment
        mov     ax,datarea          ;datarea segment addr
        mov     ds,ax               ;   into DS register
;MAIN PART OF PROGRAM GOES HERE
        mov     ax,n
        mov     array_head-2,0ffffh
        mov     si,0
compare:cmp     array_end[si],ax
        jle     insert
        mov     bx,array_end[si]
        mov     array_end[si+2],bx
        sub     si,2
        jmp     short compare
insert: mov     array_end[si+2],ax
        ret                         ;return to DOS

main    endp                        ;end of main part of program
;--------------------------------------------------------------------------
prognam ends                        ;end of code segment
;************************************************************
        end     start               ;end of assembly
```

上述例子说明,循环控制条件是循环程序设计的关键,必须结合对算法的分析与考虑合理地选择。同时,必须仔细地考虑边界情况出现的可能性,以免在特殊情况下造成错误。

例 5.5 设有数组 X 和 Y。X 数组中有 $X_1,\cdots,X_{10}$;Y 数组中有 $Y_1,\cdots,Y_{10}$。试编制程序计算

$$Z_1 = X_1 + Y_1 \qquad Z_5 = X_5 - Y_5 \qquad Z_8 = X_8 - Y_8$$
$$Z_2 = X_2 + Y_2 \qquad Z_6 = X_6 + Y_6 \qquad Z_9 = X_9 + Y_9$$
$$Z_3 = X_3 - Y_3 \qquad Z_7 = X_7 - Y_7 \qquad Z_{10} = X_{10} + Y_{10}$$
$$Z_4 = X_4 - Y_4$$

结果存入 Z 数组。

对于这种问题,也可用循环程序结构来完成。已知循环计数值为 10,每次循环的操作数是可以顺序取出的,但所作的操作却有所不同,这里有加法和减法两种操作。为了区别每次应该做哪一种操作,可以设立标志位,如标志位为 0 做加法;为 1 则做减法。这样,进入循环后只要判别标志位就可确定应该做的操作了。显然,这里要做 10 次操作就应该设立 10 个标志位,我们把它放在一个存储单元 LOGIC_RULE 中,这种存储单元一般称为逻辑尺,本例设定的逻辑尺为:

0000000011011100

从低位开始所设的标志位反映了每次要做的操作顺序,最高的 6 位没有意义,把它们设为 0。可以画出程序框图如图 5.6 所示,编制了"例 5.5 的程序"。

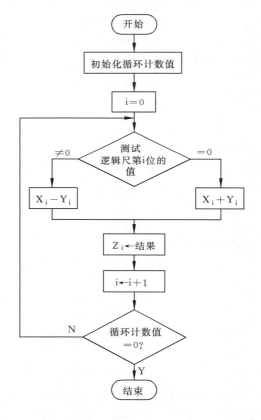

图 5.6 例 5.5 的程序框图

例 5.5 的程序实现

```
;***************************************************
datarea segment                    ;define data segment
    x         dw    x1,x2,x3,x4,x5,x6.x7,x8,x9,x10
    y         dw    y1,y2,y3,y4,y5,y6,y7,y8,y9,y10
    z         dw    z1,z2,z3,z4,z5,z6,z7,z8,z9,z10
    logic_rule dw   00dch
datarea ends
;***************************************************
prognam segment                    ;define code segment
;-----------------------------------------------------------------
main    proc    far                ;main part of program

    assume cs:prognam,ds:datarea

start:                             ;starting execution address

;set up stack for return
        push    ds                 ;save old data segment
        sub     ax,ax              ;put zero in AX
        push    ax                 ;save it on stack

;set DS register to current data segment
        mov     ax,datarea         ;datarea segment addr
        mov     ds,ax              ;   into DS register

;MAIN PART OF PROGRAM GOES HERE
        mov     bx,0
        mov     cx,10
        mov     dx,logic_rule
next:   mov     ax,x[bx]
        shr     dx,1
        jc      subtract
        add     ax,y[bx]
        jmp     short result
subtract:
        sub     ax,y[bx]
result: mov     z[bx],ax
        add     bx,2
        loop    next
```

```
               ret                              ;return to DOS
main    endp                                    ;end of main part of program
;------------------------------------------------------------------------
        prognam ends                            ;end of code segment
;**********************************************************************

        end     start                           ;end of assembly
```

这种设置逻辑尺的方法是很常用的。例如，在矩阵运算中，为了跳过操作数为 0 的计算，经常采用这种方法。又如，把一组数据存入存储器时，如果其中数值为 0 的元素很多，也可用这种方法设立一个每位表示一个下标的逻辑尺（这样的逻辑尺可能占有几个字，由数组的长度确定），0 元素就可不占有存储单元了。在例 5.5 中，每个标志只占一位，如果要表示的特征数更多，则每个标志可占有几位，而在处理方法上是完全相同的。

设立标志位的方法除了如逻辑尺那样可静态地预置外，还可以在程序中动态地修改标志位的值，以达到控制的目的，下例将说明这种方法。

例 5.6 试编制一程序：从键盘输入一行字符，要求第一个键入的字符必须是空格符，如不是，则退出程序；如是，则开始接收键入的字符并顺序存放在首地址为 BUFFER 的缓冲区中（空格符不存入），直到接收到第二个空格符时退出程序。

这一程序要求接收的字符从空格符开始又以空格符结束，因此程序中必须区分所接收的字符是否是第一个字符。为此，设立作为标志的存储单元 FLAG。一开始将其置为 0，接收第一个字符后可将其置 1。整个程序的框图如图 5.7 所示，并将例 5.6 的程序实现。

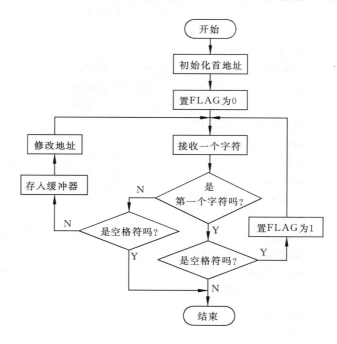

图 5.7 例 5.6 的程序框图

例 5.6 的程序实现

```
;**************************************************
datarea segment                     ;define data segment
    buffer      db      80 dup(?)
    flag        db      ?
datarea ends
;**************************************************
prognam segment                     ;define code segment
;----------------------------------------------------------------------
main        proc        far         ;main part of program

        assume cs:prognam,ds:datarea

start:                              ;starting execution address

;set up stack for return
        push        ds              ;save old data segment
        sub         ax,ax           ;put zero in AX
        push        ax              ;save it on stack

;set DS register to current data segment
        mov         ax,datarea      ;datarea segment addr
        mov         ds,ax           ;   into DS register

;MAIN PART OF PROGRAM GOES HERE
        lea         bx,buffer
        mov         flag,0
next:   mov         ah,01
        int         21h
        test        flag,01h
        jnz         follow
        cmp         al,20h
        jnz         exit
        mov         flag,1
        jmp         next
follow: cmp         al,20h
        jz          exit
        mov         [bx],al
        inc         bx
        jmp         next
exit:   ret                         ;return to DOS
```

```
            main    endp                              ;end of main part of program
;------------------------------------------------------------------
            prognam ends                              ;end of code segment
;*****************************************************************

            end     start                             ;end of assembly
```

5.1.3 多重循环程序设计

循环可以有多重结构。多重循环程序设计的基本方法和单重循环程序设计是一致的,应分别考虑各重循环的控制条件及其程序实现,相互之间不能混淆。另外,应该注意在每次通过外层循环再次进入内层循环时,初始条件必须重新设置。下面,举例加以说明。

例 5.7 有一个首地址为 A 的 N 字数组,编制程序使该数组中的数按照从大到小的次序整序。

这里采用起泡排序算法,从第一个数开始依次对相邻两个数进行比较。如次序对,则不做任何操作;如次序不对,则使两个数交换位置,表 5.1 表示了这种算法的例子。从中可以看出,在做了第一遍的(N−1)次比较后,最小的数已经放到了最后,所以第二遍比较只需要考虑(N−1)个数,即只需要比较(N−2)次。第三遍则只需要做(N−3)次比较……总共最多(N−1)遍比较就可以完成排序。图 5.8 表示了起泡排序算法的程序框图,并编制了"例5.7 起泡排序算法"的程序。

表 5.1 起泡排序算法举例

序号	数	比较遍数		
		1	2	3
1	8	8	16	84
2	5	16	84	32
3	16	84	32	16
4	84	32	8	8
5	32	5	5	5

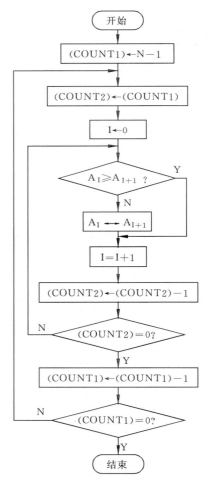

图 5.8 例 5.7 起泡排序算法的程序框图

例5.7 起泡排序算法的程序实现之一

```
;***********************************************************
datarea segment                     ;define data segment
    a       dw      n dup(?)
datarea ends
;***********************************************************
prognam segment                     ;define code segment
;-----------------------------------------------------------
main    proc    far                 ;main part of program

        assume cs:prognam,ds:datarea

start:                              ;starting execution address

;set up stack for return
        push    ds                  ;save old data segment
        sub     ax,ax               ;put zero in AX
        push    ax                  ;save it on stack

;set DS register to current data segment
        mov     ax,datarea          ;datarea segment addr
        mov     ds,ax               ;  into DS register

;MAIN PART OF PROGRAM GOES HERE
        mov     cx,n                ;set count1
        dec     cx                  ;  to n-1
loop1:  mov     di,cx               ;save count1 in DI
        mov bx,0                    ;clear  BX
loop2:  mov     ax,a[bx]            ;load a(i) into AX and
        cmp     ax,a[bx+2]          ;  compare with a(i+1)
        jge     cotinue             ;swap if
        xchg    ax,a[bx+2]          ;  a(i) < a(i+1) and
        mov     a[bx],ax            ;store greater number
cotinue: add    bx,2                ;increment index
        loop    loop2               ;if not the end of a pass,
                                    ;repeat
        mov     cx,di               ;restore count1
                                    ;  for the either loop
        loop    loop1               ;if not the final pass
                                    ;repeat
```

```
            ret                                    ;return to DOS
main        endp                                   ;end of main part of program
;----------------------------------------------------------------
prognam     ends                                   ;end of code segment
;****************************************************************
            end     start                          ;end of assembly
```

例 5.8 在附加段中有一个字数组,其首地址已存放在 DI 寄存器中,在数组的第一个字中存放着该数组的长度。要求编制一个程序使该数组中的数按照从小到大的次序排列整齐。

这里,当然也可以采用起泡排序算法。但是,在例 5.7 中,内、外循环的次数都是已知的,在整个程序运行过程中,内循环次数虽然是在不断变化着,但它是按每次减 1 的规律变化的;而外循环次数则是一个由数组长度确定的数(其值为 N−1)。也就是说,不管数组的原始排列情况如何,算法保证只要做(N−1)遍比较,总可以达到排序的目的。显然,在很多情况下,数组的比较遍数并未达到(N−1)就已经整序完毕,但程序必须继续运行到(N−1)遍才能结束。为了提高效率,可以采用另一种结束外循环的办法:设立一个交换标志位,每次进入外循环就将交换标志位置 1;在内循环中每做一次交换操作就将该标志位置 0。在每次内循环结束后,可以测试交换标志位,如该位为 0 则再一次进入外循环;如该位为 1,则说明上一遍比较已经未引起交换操作,数组已整序完毕,这样就可以立即结束外循环了。当然,这种方法在数组已整序完毕后会多做一遍比较。但在多数情况下,其比较遍数会少于(N−1),因而算法效率较高。这种算法的程序框图如图 5.9 所示,并编制了"例 5.3 起泡排序算法的程序"。

例 5.7 和例 5.8 的情况再次说明算法以及循环控制条件的选择对程序效率的影响。另外,例 5.8 中关于设立测试标志的方法和例 5.6 类似是循环程序设计中常用的一种方法。总之,循环控制条件的选择是很灵活的,读者应该根据具

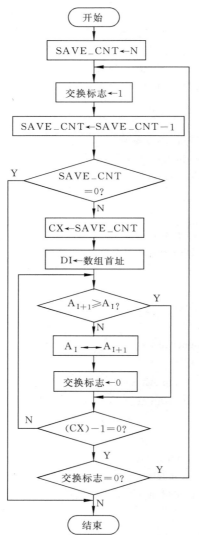

图 5.9 例 5.8 起泡排序算法的程序框图

体情况合理选择。

例 5.8 起泡排序算法的程序实现之三

```
;Arranges the 16-bit elements of a list in the extra segment
;   in ascending order,using bubble sort.
;Inputs: ES:DI=Starting address of the list.
;   First location of the list=Length of the list (words)
;**************************************************************
data    segment                     ;define data segment
    save_cnt    dw      ?
    start_addr  dw      ?
    data ends
;**************************************************************
prognam segment                     ;define code segment
;----------------------------------------------------------------
bubble  proc    far
    assume  cs:prognam,ds:data
        push    ds              ;save caller's registers
        push    cx
        push    ax
        push    bx
        mov     ax,data         ;initialize DS
        mov     ds,ax
;
        mov     start_addr,di   ;save starting address
                                ;   in memory
        mov     cx,es:[di]      ;fetch element count N
        mov     save_cnt,cx     ;   save this value in memory
init:   mov     bx,1            ;exchange flag (BX)=1
        dec     save_cnt        ;get ready for
                                ;   count-1 compares
        jz      sorted          ;exit if save_cnt is 0
        mov     cx,save_cnt     ;and load this count
                                ;   into CX
        mov     di,start_addr   ;load start address
                                ;   into DI
next:   add     di,2            ;address a data el.
        mov     ax,es:[di]      ;   and load it into AX
        cmp     es:[di+2],ax    ;is next el. < this el. ?
        jae     cont            ;no,go check next pair
        xchg    es:[di+2],ax    ;yes,exchange
        mov     es:[di],ax      ;   these elements
        sub     bx,bx           ;and make exchange flag 0
```

```
cont:       loop    next                    ;process entire list
            cmp     bx,0                    ;any exchanges made?
            je      init                    ;if so,process list again
sorted:     mov     di,start_addr           ;if not so,
            pop     bx                      ;   restore registers
            pop     ax
            pop     cx
            pop     ds
            ret                             ;exit

bubble      endp
;--------------------------------------------------------------------
prognam     ends                            ;end of code segment
;****************************************************************
            end                             ;end of assembly
```

5.2　分支程序设计

5.2.1　分支程序的结构形式

分支程序结构可以有两种形式,如图 5.10 所示。它们分别相当于高级语言中的 IF_THEN_ELSE 语句和 CASE 语句,适用于要求根据不同条件作不同处理的情况。IF_THEN_ELSE 语句可以引出两个分支。CASE 语句则可以引出多个分支。不论哪一种形式,它们的共同特点是:运行方向是向前的,在某一种特定条件下,只能执行多个分支中的一个分支。

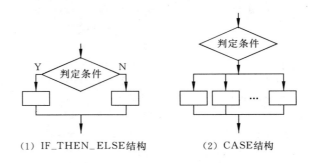

(1) IF_THEN_ELSE结构　　(2) CASE结构

图 5.10　分支程序的结构形式

5.2.2　分支程序设计方法

程序的分支一般用条件转移指令来产生,在 3.3 节的例 3.78 中,区分正数、0 和负数的程序就是分支程序。可以看出,这里是利用转移指令不影响条件码的特性,连续地使用条件转移指令使程序产生了三个不同的分支(尽管负数的分支并未执行任何操作),而对于数组中的每一个数,它只能是三个分支中的某一个。下面我们再来看一个例子。

例 5.9 在附加段中,有一个按从小到大顺序排列的无符号数数组,其首地址存放在 DI 寄存器中,数组中的第一个单元存放着数组长度。在 AX 中有一个无符号数,要求在数组中查找(AX),如找到,则使 CF=0,并在 SI 中给出该元素在数组中的偏移地址;如未找到,则使 CF=1。

我们已经遇到过多个表格查找的例子,都是使用顺序查找的方法。本例是一个已经排序的数组,可以采用折半查找法以提高查找效率。折半查找法先取有序数组的中间元素与查找值相比较,如相等则查找成功;如查找值大于中间元素,则再取高半部的中间元素与查找值相比较;如查找值小于中间元素,则再取低半部的中间元素与查找值相比较;如此重复直到查找成功或最终未找到该数(查找不成功)为止。折半查找法的效率高于顺序查找法,对于长度为 N 的表格,顺序查找法平均要作 N/2 次比较,而折半查找法的平均比较次数为 $\log_2 N$。所以,如果数组长度为 100,则顺序查找法平均要作 50 次比较,而折半查找法平均作 7 次比较就可以了。

在一个长度为 n 的有序数组 r 中,查找元素 k 的折半查找算法可描述如下:

(1) 初始化被查找数组的首尾下标,low ← 1,high ← n;

(2) 若 low>high,则查找失败,置 CF=1,退出程序。

否则,计算中点:mid ← (low + high)/2;

(3) k 与中点元素 r[mid]比较。若 k=r[mid],则查找成功,程序结束;若 k <

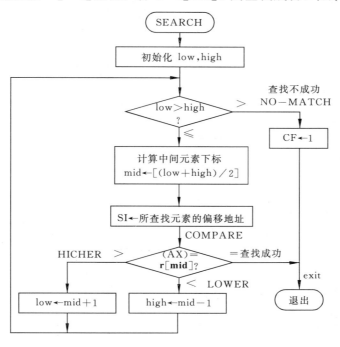

图 5.11 折半查找算法的程序框图

r[mid],则转步骤(4);若 k>r[mid],则转步骤(5);

(4) 低半部分查找 (lower),high ← mid－1,返回步骤(2),继续查找;

(5) 高半部分查找 (higher),low ← mid＋1,返回步骤(2),继续查找。

图 5.11 表示了折半查找算法的程序框图。给出的程序首先把查找值与数组的第一个元素和最后一个元素相比较,如果找到或者该数小于第一个元素或大于最后一个元素则结束查找;否则从 SEARCH 开始折半查找。SEARCH 以前的工作在图 5.11 中未表示出来。折半查找算法的程序实现如后所示。

例 5.9 折半查找算法的程序实现

```
;SEARCH_HALF--EX5_9
;Search an ordered list in the extra segment for the word
;    value contained in AX
;Inputs: ES:DI=starting address of the list
;         First location=Length of list(words)
;Results:If the value is in the list,
;         CF=0
;         SI=Offset of matching element.
;        If thr value is not in the list,
;         CF=1
;         SI=Offset of last element compared.
;****************************************************************
dseg      segment                          ;define data segment

   low_idx    dw       ?
   high_idx   dw       ?

dseg      ends
;****************************************************************
cseg      segment                          ;define code segment
;----------------------------------------------------------------
b_search           proc       near
    assume   cs:cseg,ds:dseg,es:dseg

             push     ds                   ;save caller's DS
             push     ax
             mov      ax,dseg              ;initialize DS
             mov      ds,ax
             mov      es,ax
             pop      ax
;
;Find out if AX lies beyond the boundaries of the list.
```

```
            cmp     ax,es:[di+2]            ;search value<or=first el. ?
            ja      chk_last                ;no,go check last el.
            lea     si,es:[di+2]            ;yes,fetch addr of first el.
            je      exit                    ;if value =1st el. ,exit
            stc                             ;if value < 1st el. ,set CF
            jmp     exit                    ;   and then exit
chk_last:
            mov     si,es:[di]              ;point to last el.
            shl     si,1
            add     si,di
            cmp     ax,es:[si]              ;search value>or=last el. ?
            jb      search                  ;no, go search list
            je      exit                    ;yes, exit if value =last el.
            stc                             ;if value >last el. ,set CF
            jmp     exit                    ;   and then exit
; Search for value within the list.
search:
            mov     low_idx,1               ;fetch index
            mov     bx,es:[di]
            mov     high_idx,bx
            mov     bx,di
mid:        mov     cx,low_idx              ;calculate middle point
            mov     dx,high_idx
            cmp     cx,dx
            ja      no_match
            add     cx,dx
            shr     cx,1
            mov     si,cx
            shl     si,1                    ;calculate next search addr
compare:    cmp     ax,es:[bx+si]           ;search value found?
            je      exit                    ;if so,exit
            ja      higher                  ;otherwise, find correct half
; These instructions are executed if the search value is
;    lower in the list.
            dec     cx
            mov     high_idx,cx
            jmp     mid
higher:     inc     cx
            mov     low_idx,cx
            jmp     mid
no_match:
```

```
            stc                         ;if so, set CF
; Following are exit instructions.
exit:       pop     ds
            ret                         ; and exit

b_search    endp
;------------------------------------------------------------
cseg        ends                        ;end of code segment
;************************************************************
            end                         ;end of assembly
```

如果数组如下：

LIST DW 12,11,22,33,44,55,66,77,88,99,111,222,333 要求查找的数为(AX)=55。数组长度为12,第一次比较的是数组的第六个元素；因55＜66,所以第二次用低半部折半查找,比较的是第三个元素33；因55＞33,所以第三次用高半部折半查找,比较的是第四个元素44。因55＞44,所以第四次用高半部折半查找,比较的是第五个元素55。这样经过四次比较后,因查找成功而退出程序。

如果要查找的数是(AX)=57,则第一次比较的仍是第六个元素66；因57＜66,所以第二次用低半部折半查找,比较的是第三个元素33；因57＞33,所以第三次用高半部折半查找,比较的是第四个元素44；因57＞44,所以第四次用高半部折半查找,比较的是第五个元素55,因57＞55,在用高半部折半查找时,因low＞high而以查找失败退出程序。

可以看出,在这个例子中,同样用CMP指令以及条件转移指令产生二个或多个程序分支。当然,由于多数运算型指令置条件码,所以在条件转移指令之前并不一定要使用CMP或TEST指令,只要保证使用条件转移指令时的条件码符合要求就可以了。在循环结构的例5.5中建立了逻辑尺,根据对逻辑尺中每一位不同值的判断,产生不同的程序分支,也是分支程序的一种实现方法。以上多个例子都是既有分支结构而又包括循环结构,实际上,多数程序都是各种程序结构的组合。而且,循环结构可以看作分支结构的一种特例,它只是多次走一个分支,只在满足循环结束条件时,走另一个分支罢了。

5.2.3 跳跃表法

分支程序的两种结构形式都可以用上面所述的方法来实现。此外,在实现CASE结构时,还可以使用跳跃表法,使程序能根据不同的条件转移到多个程序分支中去,下面举例说明。

例5.10 试根据AL寄存器中哪一位为1(从低位到高位)把程序转移到8个不同的程序分支中去。

下面列出了用变址寻址方式实现跳跃表法的程序,还可以使用寄存器间接和基址变址寻址方式来达到同一目的,"例5.10用寄存器间接寻址方式实现跳跃表法的程序"和"例5.10用基址变址寻址方式实现跳跃表法的程序"分别列出了修改后的主程序段,程序的其他部分与"例 5.10 用变址寻址方式实现跳跃表法的程序"相同。其实,这三种方法并无实质的区别,只是其中关键的 JMP 指令所用的寻址方式不同而已。

跳跃表法是一种很有用的分支程序设计方法,希望读者能通过所举例子掌握要领,灵活使用。

例 5.10 用变址寻址方式实现跳跃表法的程序

```
;******************************************************************
branch_addresses segment                ;define data segment
branch_table    dw      routine_1
                dw      routine_2
                dw      routine_3
                dw      routine_4
                dw      routine_5
                dw      routine_6
                dw      routine_7
                dw      routine_8
branch_addresses ends
;******************************************************************
procedure_select segment                ;define code segment
;--------------------------------------------------------------------
main    proc    far                     ;main part of program

assume cs:procedure_select,ds:branch_addresses

start:                                  ;starting execution address

;set up stack for return
        push    ds                      ;save old data segment
        sub     bx,bx                   ;put zero in BX
        push    bx                      ;   save it on stack

;set DS register to current data segment
        mov     bx,branch_addresses     ;data segment addr
        mov     ds,bx                   ;   into DS register

;MAIN PART OF PROGRAM GOES HERE
        cmp     al,0                    ;this test assures that some
                                        ;   bit of AL has been set by
        je      continue_main_line      ;   earlier instruction to
                                        ;       specify a routine
        mov     si,0
```

```
1:          shr     al,1                    ;puts least_significant bit
                                            ;   of AL into the CF
            jnb     not_yet                 ;if CF=0,the on bit in AL
                                            ;   has not yet been found
            jmp     branch_table[si]        ;if CF=1,then control is
                                            ;   transferred
not_yet:    add     si,type branch_table    ;if no transfer,then
                                            ;the bit that is on has not
                                            ;   yet been found,so SI is
                                            ;   set to point to the
                                            ;       next entry in the
                                            ;           address table by
                                            ;               adding 2
            jmp     1                       ;jump to l to shift and
                                            ;   retest
continue_main_line:                         ;we reach here only if no
            ⋮                               ;   bit was set to indicate
                                            ;       a desired routine
routine_1:  ⋮
routine_2:  ⋮

            ret                             ;exit

main        endp                            ;end of main part of prog.
;------------------------------------------------------------------------
procedure_select ends                       ;end of code segment
;************************************************************************

            end     start                   ;end of assembly
;MAIN PART OF PROGRAM GOES HERE
            cmp     al,0                    ;this test assures that some
                                            ;   bit of AL has been set by
            je      continue_main_line      ;   earlier instruction to
                                            ;       specify a routine
            lea     bx,branch_table         ;BX set to location holding
                                            ;   address of first routine
1:          shr     al,1                    ;puts least_significant bit
                                            ;   of AL into the CF
            jnb     not_yet                 ;if CF=0,the on bit in AL
                                            ;   has not yet been found
            jmp     word ptr[bx]            ;if CF=1,then control is
```

```
not_yet:   add     bx,type branch_table    ; transferred
                                           ;if no transfer,then
                                           ;the bit that is on has not
                                           ;   yet been found,so BX is
                                           ;   set to point to the
                                           ;   next entry in the
                                           ;     address table by
                                           ;       adding 2
           jmp     l                       ;jump to l to shift and
                                           ;   retest
continue_main_line:
```

例 5.10 用基址变址寻址方式实现跳跃表法的程序

```
;MAIN PART OF PROGRAM GOES HERE
           cmp     al,0                    ;this test assures that some
                                           ;   bit of AL has been set by
           je      continue_main_line      ;     earlier instruction to
                                           ;       specify a routine
           lea     bx,branch_table         ;BX set to location holding
                                           ;   address of first routine
           mov     si,7 * type branch_table ;points initially to
                                           ;   last such entry in list
           mov     cx,8                    ;loop counter allowing
                                           ;   shift maximum
l:         shl     al,1                    ;shifts high_order AL bit
                                           ;   into CF
           jnb     not_yet                 ;if CF=0,routine represented
                                           ;   by that bit not desired
           jmp     word ptr[bx][si]        ;if CF=1,transfer to
                                           ;   procedure represented by
                                           ;     most recent bit tested
not_yet:   sub     si,type branch_table    ;adjust index register
                                           ;   to point to "next"
                                           ;     branch address
           loop    l
continue_main_line:
```

5.3　如何在实模式下发挥 80386 及其后继机型的优势

在本章的开始，已经说明本书将以实模式为基础来阐述程序设计方法。5.1 节和 5.2 节所给出的都是基于 8086 的程序，由于 80x86 的兼容性，这些程序都可以在任何一种

80x86 机型的实模式下运行,而且它们在高档机运行可比在低档机上运行获得更高的性能。但是,386 及其后继机型除提供更大容量、更高的速度和保护模式的支持外,还提供了一些良好的特性,如能在程序设计中充分利用这些优势,将有利于提高编程质量。在这一节里,将就这方面的问题加以讨论。

5.3.1 充分利用高档机的 32 位字长特性

80x86 系列从 80386 起就把机器字长从 16 位增加到 32 位。字长的增加除有利于提高运算精度外,也能提高编程效率。例如,在 3.3.2 小节中,例 3.46 和例 3.50 所讨论的双字长数的加法和减法,在 8086 中必须用 ADD 和 ADC 序列或 SUB 和 SBB 序列来完成;而在 386 及其后继机型中可只用一条 ADD 或 SUB 指令完成就可以了。因此,不论在空间方面还是时间方面都有利于程序效率的提高。为了更进一步说明问题,看一看下面的程序举例。

例 5.11 如有两个 4 字长(64 位)数分别存放在 DATA1 和 DATA2 中,请用 8086 指令编写一程序求出它们的和,并把结果存放于 DATA3 中。

为了得到 4 字长数的和,在 8086 中需要分 4 段计算,每段一个字长(16 位),用 4 次循环可得到 4 字长数的和。考虑到每次求和可能有进位值,要用 ADC(而不是 ADD)指令求和,而且在进入循环前应先清除 CF 位。在循环中修改地址指针时用 INC 指令而不用 ADD 指令,以免影响求和时得到的进位值(INC 指令不影响 CF 位)。所编程序见"例 5.11 的程序实现 1"。

例 5.11 的程序实现 1

```
        .model   small
        .data                      ;define data segment
data1   dq       123456789abcdefh
data2   dq       0fedcba987654321h
data3   dq       ?
        .code                      ;define code segment

start:                             ;starting execution addr
        mov      ax,@data          ;data segment addr
        mov      ds,ax             ;   into DS register
;
        clc
        lea      si,data1          ;SI is pointer for operand1
        lea      di,data2          ;DI is pointer for operand2
        lea      bx,data3          ;BX is pointer for the sum
        mov      cx,4              ;CX is the loop counter
back:   mov      ax,[si]           ;move the first operand to AX
```

```
        adc      ax,[di]              ;add the second operand to AX
        mov      [bx],ax              ;store the sum
        inc      si                   ;point to next word of operand1
        inc      si
        inc      di                   ;point to next word of operand2
        inc      di
        inc      bx                   ;point to next word of sum
        inc      bx
        loop     back                 ;if not finished,continue adding
        mov      ax,4c00h             ;go back to DOS
        int      21h
;
        end      start
```

例 5.12 编制 386 及其后继机型的程序,实现例 5.11 的要求。

在 386 及其后继机型中可以充分利用其 32 字长的特性,每次可对双字求和,这样循环 2 次就可得到 4 字长数之和,程序见"例 5.11 的程序实现 2"。此程序可得到正确的结果,且由于循环次数的减少,速度上要优于"例 5.11 的程序实现 1"。

例 5.11 的程序实现 2

```
            .model    small
            .386
            .data                     ;define data segment
data1       dq        123456789abcdefh
data2       dq        0fedcba987654321h
data3       dq        ?
            .code                     ;define code segment

start:                                ;starting execution addr
            mov       ax,@data        ;data segment addr
            mov       ds,ax           ;   into DS register
;
            clc
            lea       si,data1        ;SI is pointer for operand1
            lea       di,data2        ;DI is pointer for operand2
            lea       bx,data3        ;BX is pointer for the sum
            mov       cx,2            ;CX is the loop counter
back:       mov       eax,[si]        ;move the first operand to EAX
            adc       eax,[di]        ;add the second operand to EAX
            mov       [bx],eax        ;store the sum
            inc       si              ;point to next dword of operand1
            inc       si
```

```
        inc     si
        inc     si
        inc     di              ;point to next dword of operand2
        inc     di
        inc     di
        inc     di
        inc     bx              ;point to next dword of sum
        inc     bx
        inc     bx
        inc     bx
        loop    back            ;if not finished,continue adding
        mov     ax,4c00h        ;go back to DOS
        int     21h
;
        end     start
```

但是,由于循环中的地址修改使用了 INC 指令,使程序变得冗长,从而影响了其运行效率。其实,使用 INC 指令的原因,是为了避免破坏 CF 值。我们可以在加法后先把标志寄存器的内容保持在堆栈中,在修改指针后再进行恢复。这样,指针的修改可用 ADD 指令来实现,因而使程序的效率得到提高,其程序主体部分修改如下:

例 5.11 的程序实现 3

```
          mov     cx,2            ;CX is the loop counter
back:     mov     eax,[si]        ;move the first operand to EAX
          adc     eax,[di]        ;add the second operand to EAX
          mov     [bx],eax        ;store the sum
          pushf                   ;save CF
          add     si,4            ;point to next dword of operand1
          add     di,4            ;point to next dword of operand2
          add     bx,4            ;point to next dword of sum
          popf                    ;restore CF
          loop    back            ;if not finished,continue adding
```

实际上,对于这样的 2 个 4 字长数求和的例子,可直接用 3.3.2 小节中使用 16 位指令对双字长数相加的方法,能得到"例 5.11 的程序实现 4"效率更高的程序。

例 5.11 的程序实现 4

```
          .model   small
          .386
          .stack   200h           ;define stack segment
          .data                   ;define data segment
data1     dq       123456789abcdefh
```

```
        data2    dq       0fedcba987654321h
        data3    dq       ?
                 .code                                ;define code segment

        start:                                        ;starting execution addr
                 mov      ax,@data                    ;data segment addr
                 mov      ds,ax                       ;   into DS register
        ;
                 mov      eax,dword ptr data1         ;move lower dword of data1 into EAX
                 add      eax,dword ptr data2         ;add lower dword of data2 to EAX
                 mov      edx,dword ptr data1+4       ;move upper dword of data1 into EDX
                 adc      edx,dword ptr data2+4       ;add upper dword of data2 to EDX
                 mov      dword ptr data3,eax         ;store lower dword of sum
                 mov      dword ptr data3+4,edx       ;store upper dword of sum
        ;
                 mov      ax,4c00h                    ;go back to DOS
                 int      21h
        ;
                 end      start
```

参考文献[2]上详细分析了例 5.11 各种程序实现所需要的时钟周期数,得出的结论是:用 386 或 486 运行上述程序,用 32 位字长计算可获得比用 16 位字长计算快 5～7 倍的效果。可见,尽可能利用高档机的 32 位字长特性是很有意义的。

这里只是以加法运算为例,说明利用高档机 32 位字长特性的重要性。实际上,对于其他指令也有类似的效果,对于如乘/除法等更复杂的指令,收到的效果会更好。

上面所说的充分利用高档机的 32 位字长的特性,当然也包括对其提供的 32 位寄存器在内。386 及其后继机型除可访问 8086,80286 所提供的 8 位和 16 位寄存器外,还提供了 8 个 32 位通用寄存器,所有这些寄存器在实模式下都可以被访问。此外,除 8086,80286 提供的 4 个段寄存器外,还增加 2 个附加数据段寄存器 FS 和 GS,在实模式下也都可以使用。只有指令指针寄存器 EIP 和标志寄存器 EFLAGS 在实模式下只有低 16 位可用。在实模式下,段的大小最大可到 64KB,EIP 的高 16 位应为 0。

5.3.2 通用寄存器可作为指针寄存器

在 3.1.1 小节中,已经说明 386 及其后继机型除提供 16 位寻址外,还提供了 32 位寻址。在实模式下,这两种寻址方式可同时使用。在表 3.1 中,已列出了 16 位寻址和 32 位寻址所允许使用的寄存器情况。可以看出,在使用 32 位寻址时,32 位通用寄存器可以作为基址或变址寄存器使用。也就是说,允许 32 位通用寄存器作指针寄存器用。在实模式下,段的大小被限制于 64KB,这样段内的偏移地址范围应为 0000～FFFFH,所以在把 32 位通用寄存器用作指针寄存器时,应该注意它们的高 16 位应为 0。

提示:32位通用寄存器可用作指针寄存器,但16位通用寄存器中仍然只有BX,BP和SI,DI可用作指针寄存器。所以,下列指令是合法的:

    MOV EAX,[BX]
    MOV EAX,[EDX]
    MOV AX,WORD PTR [ECX]

而下列指令是非法的:

    MOV AX,[DX]
    MOV EAX,[CX]

在386及其后继机型中,允许同一寄存器既用于基址寄存器,也用于变址寄存器。因此,下列指令也是合法的:

    MOV AX,[EBX][EBX]

5.3.3 与比例因子有关的寻址方式

在3.1.1小节中,已经给出了80386及其后继机型所提供的与比例因子有关的三种寻址方式:比例变址寻址方式,基址比例变址寻址方式和相对基址比例变址寻址方式。这些寻址方式为表格处理和多维数组处理提供了有力的工具。

例 5.13 用比例变址寻址方式编写一程序,要求把5个双字相加并保存其结果。下面给出了所编"例5.13的程序实现"。从程序中可以清楚地看出,采用比例变址寻址方式可以直接把数组的元素下标存入变址寄存器中,而比例因子1,2,4和8,正好对应于数组元素为字节、字、双字和4字的不同情况。因此,这类寻址方式为数组处理提供了极大的方便。

例 5.13 的程序实现

```
        .model   small
        .386
        .stack   200h                    ;define stack segment
        .data                            ;define data segment
array   dd       234556h,0f983f5h,6754ae2h,0c5231239h,0af34acb4h
result  dq       ?
        .code                            ;define code segment

start:                                   ;starting execution addr
        mov      ax,@data                ;data segment addr
        mov      ds,ax                   ;   into DS register
```

```
        ;
                sub     ebx,ebx                 ;EBX=0
                mov     edx,ebx                 ;clear EDX
                mov     eax,ebx                 ;clear EAX
                mov     cx,5                    ;set the counter to 5
        back:   add     eax,array[ebx*4]        ;add the 32-bit operand
                adc     edx,0                   ;save the carry
                inc     ebx                     ;point to next 32-bit data
                dec     cx                      ;decrement the counter
                jnz     back                    ;repeat until counter is zero
                mov     dword ptr result,eax    ;save the lower 32 bits
                mov     dword ptr result+4,edx  ;save the upper 32 bits
        ;
                mov     ax,4c00h                ;go back to DOS
                int     21h
        ;
                end     start
```

例 5.14 数据段中有一个 20 个字节的字节表,还有一个 20 个字的字表。试编制一程序,把字节表中的每个数据作为字表中的高位字节部分移入字表中。

如果用一般寻址方式编制程序,则如"例 5.14 的程序实现"。用比例变址寻址方式编制程序,如"例 5.14 的比例变址寻址方式程序实现"。可见比例变址寻址方式为表格处理也提供了很大的便利。

例 5.14 的程序实现

```
                .model    small
                .386
                .stack    200h                  ;define stack segment
                .data                           ;define data segment
        byte_table   db    20 dup(?)
        word_table   dw    20 dup(?)
                .code                           ;define code segment
        start:                                  ;starting execution addr
                mov     ax,@data                ;data segment addr
                mov     ds,ax                   ;   into DS register
        ;
                xor     si,si                   ;put zero in SI
                xor     di,di                   ;   and DI
        next:
                mov     ax,word_table[di]       ;load word into AX
                mov     ah,byte_table[si]       ;load byte into AH
```

```
            mov     word_table[di],ax      ;save it to word_table
            inc     si                     ;increment
            add     di,2                   ;  index
            cmp     si,20                  ;done?
            jl      next                   ;not yet
;
            mov     ax,4c00h               ;go back to DOS
            int     21h
;
            end     start
```

例 5.14 的比例变址寻址方式程序实现

```
            .model   small
            .386
            .stack   200h                  ;define stack segment
            .data                          ;define data segment
byte_table  db       20 dup(?)
word_table  dw       20 dup(?)
            .code                          ;define code segment
start:                                     ;starting execution addr
            mov     ax,@data               ;data segment addr
            mov     ds,ax                  ;  into DS register
;
            xor     esi,esi                ;put zero in ESI
next:
            mov     ax,word_table[esi*2]   ;load word into AX
            mov     ah,byte_table[esi]     ;load byte into AH
            mov     word_table[esi*2],ax   ;save it to word_table
            inc     esi                    ;increment index
            cmp     esi,20                 ;done?
            jl      next                   ;not yet
;
            mov     ax,4c00h               ;go back to DOS
            int     21h
;
            end     start
```

例 5.15 用比例变址寻址方式编写一程序,要求把两个长度为 10 的双字长数组的对应元素分别相乘,并保存其乘积。其程序实现如下。

例 5.15 的程序实现

```
                .model    small
                .386
                .stack    200h                    ;define stack segment
                .data                             ;define data segment
data1           dd        10 dup(?)
data2           dd        10 dup(?)
result          dq        10 dup(?)
                .code                             ;define code segment
start:                                            ;starting execution addr
                mov       ax,@data                ;data segment addr
                mov       ds,ax                   ;   into DS register
;
                sub       ebx,ebx                 ;EBX=0
                mov       cx,10                   ;set the counter to 10
back:           mov       eax,data1[ebx*4]        ;multiply data1(i)
                mul       data2[ebx*4]            ;   by data2(i)
                mov       dword ptr result[ebx*8],eax     ;store product
                mov       dword ptr result+4[ebx*8],edx   ;   in result(i)
                inc       ebx                     ;point to next data
                dec       cx                      ;decrement the counter
                jnz       back                    ;repeat until counter is zero
;
                mov       ax,4c00h                ;go back to DOS
                int       21h
;
                end       start
```

例 5.15 的程序和例 5.14 的程序很类似,它们都利用两个表格或两个数组的下标相同的特点,用一个变址寄存器保存其下标而实现了程序的要求。从以上三个例子可以看出,与比例因子有关的寻址方式为数组和表格处理提供了极大的方便,在使用 386 及其后继机型时,应尽量利用这一特性。

5.3.4 各种机型提供的新指令

在 80x86 系列中,新机型的产生往往为用户提供一些新指令或对原有指令的扩充,因此在编程时应尽量利用新机型所提供的这些有利条件。有关指令在第 3 章已经作了说明,这里只归纳一下各种机型所提供的主要新指令或指令扩充,以便于读者利用。

1. 286 提供的新指令

PUSHA/POPA

```
PUSH      IMM
IMUL      REG，SRC
IMUL      REG，SRC，IMM
INS
OUTS
BOUND
ENTER
LEAVE
```

2. 386 提供的新指令

```
MOVSX，MOVZX
PUSHAD/POPAD
PUSHFD/POPFD
CDQ，CWDE
BT，BTS，BTR，BTC
BSF，BSR
SHLD，SHRD
```

条件转移的段内近转移格式

```
JECXZ
SET  cc
IRETD
```

3. 486 提供的新指令

```
BSWAP
XADD
CMPXCHG
```

4. Pentium 提供的新指令

CMPXCHG8B

本节已经给出了 386 及其后继机型在实模式下的程序举例。可以看出，它既不是纯 16 位模块，也不是纯 32 位模块。

纯 16 位模块应该只有以下特性：

(1) 所有段长都小于或等于 64KB；

(2) 数据项主要是 8 位或 16 位的；

(3) 指向代码或数据的指针只有 16 位偏移地址；

(4) 只在 16 位段之间传送控制。

8086/80286 的程序具有以上特性。

纯 32 位模块应该只有以下特性：

(1) 段长可大于 64KB(0～4GB)；

(2) 数据项主要是 8 位或 32 位的；

(3) 指向代码或数据的指针具有 32 位偏移地址；

(4) 只在 32 位段之间传送控制。

386 及其后继机型的保护模式程序具有以上特性。

实模式下的程序是一种混合的 16 位和 32 位代码。也就是说,在同一模块中,允许同时使用 16 位和 32 位操作数和寻址方式。实模式规定必须使用 16 位段,而段中的指令则可以是混合的 16 位和 32 位代码。在例 5.11 的程序实现 2、例 5.11 的程序实现 4、例 5.13 的程序实现、例 5.14 的比例变址寻址方式程序实现和例 5.15 的程序实现的程序中,均可找到有关指令。表 5.2 列出了其中某些指令。

表 5.2 实模式下,混合的 16 位与 32 位指令举例

程 序 名 称	汇 编 指 令	
例 5.11 的程序实现 1	adc	eax, [di]
例 5.11 的程序实现 2	mov	dword ptr data3, eax
例 5.13 的程序实现	sub	ebx, edx
例 5.14 的比例变址寻址方式程序实现	mov	ah, byte_table[esi]
例 5.14 的比例变址寻址方式程序实现	mov	word_table[esi * 2], ax
例 5.13 的程序实现	add	eax, array[ebx * 4]
例 5.15 的程序实现	mov	dword ptr result+4[ebx * 8], edx

习 题

5.1 试编写一个汇编语言程序,要求对键盘输入的小写字母用大写字母显示出来。

5.2 编写程序,从键盘接收一个小写字母,然后找出它的前导字符和后续字符,再按顺序显示这三个字符。

5.3 将 AX 寄存器中的 16 位数分成 4 组,每组 4 位,然后把这四组数分别放在 AL、BL、CL 和 DL 中。

5.4 试编写一程序,要求比较两个字符串 STRING1 和 STRING2 所含字符是否相同,若相同则显示′MATCH′,若不相同则显示′NO MATCH′。

5.5 试编写一程序,要求能从键盘接收一个个位数 N,然后响铃 N 次(响铃的 ASCII 码为 07)。

5.6 编写程序,将一个包含有 20 个数据的数组 M 分成两个数组:正数数组 P 和负数数组 N,并分别把这两个数组中数据的个数显示出来。

5.7 试编制一个汇编语言程序,求出首地址为 DATA 的 100D 字数组中的最小偶数,并把它存放在 AX 中。

5.8 把 AX 中存放的 16 位二进制数 K 看作是 8 个二进制的"四分之一字节"。试编写一程序,要求数一下值为 3(即 11B)的四分之一字节数,并将该数在终端上显示出来。

5.9 试编写一汇编语言程序,要求从键盘接收一个四位的十六进制数,并在终端上显示与它等值的二进制数。

5.10 设有一段英文,其字符变量名为 ENG,并以＄字符结束。试编写一程序,查对单词 SUN 在该文中的出现次数,并以格式"SUN××××"显示出次数。

5.11 从键盘输入一系列以＄为结束符的字符串,然后对其中的非数字字符计数,并显示出计数结果。

5.12 有一个首地址为 MEM 的 100D 字数组,试编制程序删除数组中所有为零的项,并将后续项向前压缩,最后将数组的剩余部分补上零。

5.13 在 STRING 到 STRING+99 单元中存放着一个字符串,试编制一程序测试该字符串中是否存在数字。如有,则把 CL 的第 5 位置 1,否则将该位置 0。

5.14 在首地址为 TABLE 的数组中按递增次序存放着 100H 个 16 位补码数,试编写一个程序把出现次数最多的数及其出现次数分别存放于 AX 和 CX 中。

5.15 数据段中已定义了一个有 n 个字数据的数组 M,试编写一程序求出 M 中绝对值最大的数,把它放在数据段的 M+2n 单元中,并将该数的偏移地址存放在 M+2(n+1)单元中。

5.16 在首地址为 DATA 的字数组中,存放了 100H 个 16 位补码数,试编写一程序,求出它们的平均值放在 AX 寄存器中;并求出数组中有多少个数小于此平均值,将结果放在 BX 寄存器中。

5.17 试编制一个程序,把 AX 中的十六进制数转换为 ASCII 码,并将对应的 ASCII 码依次存放到 MEM 数组中的四个字节中。例如,当(AX)=2A49H 时,程序执行完后,MEM 中的 4 个字节内容为 39H,34H,41H 和 32H。

5.18 把 0～100D 之间的 30 个数存入以 GRADE 为首地址的 30 个字数组中,GRADE+i 表示学号为 i+1 的学生的成绩。另一个数组 RANK 为 30 个学生的名次表,其中 RANK+i 的内容是学号为 i+1 的学生的名次。编写一程序,根据 GRADE 中的学生成绩,将学生名次填入 RANK 数组中。(提示:一个学生的名次等于成绩高于这个学生的人数加 1)。

5.19 已知数组 A 包含 15 个互不相等的整数,数组 B 包含 20 个互不相等的整数。试编制一程序,把既在 A 中又在 B 中出现的整数存放于数组 C 中。

5.20 设在 A,B 和 C 单元中分别存放着三个数。若三个数都不是 0,则求出三数之和并存放于 D 单元中;若其中有一个数为 0,则把其他两个单元也清零。请编写此程序。

5.21 试编写一程序,要求比较数组 ARRAY 中的三个 16 位补码数,并根据比较结果在终端上显示如下信息:
  (1) 如果三个数都不相等则显示 0;
  (2) 如果三个数有两个相等则显示 1;
  (3) 如果三个数都相等则显示 2。

5.22 从键盘输入一系列字符(以回车符结束),并按字母、数字及其他字符分类计数,最后显示出这三类的计数结果。

5.23 已定义了两个整数变量 A 和 B,试编写程序完成下列功能:
  (1) 若两个数中有一个是奇数,则将奇数存入 A 中,偶数存入 B 中;
  (2) 若两个数均为奇数,则将两数均加 1 后存回原变量;

（3）若两个数均为偶数，则两个变量均不改变。

5.24 假设已编制好 5 个歌曲程序，它们的段地址和偏移地址存放在数据段的跳跃表 SINGLIST 中。试编制一程序，根据从键盘输入的歌曲编号 1～5，转去执行五个歌曲程序中的某一个。

5.25 试用 8086 的乘法指令编制一个 32 位数和 16 位数相乘的程序；再用 80386 的乘法指令编制一个 32 位数和 16 位数相乘的程序，并定性比较两个程序的效率。

5.26 如数据段中在首地址为 mess1 的数据区内存放着一个长度为 35 的字符串，要求把它们传送到附加段中的缓冲区 mess2 中去。为提高程序执行效率，希望主要采用 movsd 指令来实现。试编写这一程序。

5.27 试用比例变址寻址方式编写一 386 程序，要求把两个 64 位整数相加并保存结果。

第 6 章 子程序结构

子程序又称为过程,它相当于高级语言中的过程和函数。在一个程序的不同部分,往往要用到类似的程序段,这些程序段的功能和结构形式都相同,只是某些变量的赋值不同,此时就可以把这些程序段写成子程序形式,以便需要时可以调用它。有些程序段对于某个用户可能只用到一次,但它是一般用户经常用到的。例如,十进制数转换成二进制数,二进制数转换为十六进制数并显示输出等。对于这些常用的特定功能的程序段,也经常编制成子程序的形式供用户使用。

模块化程序设计方法是按照各部分程序所实现的不同功能把程序划分成多个模块,各个模块在明确各自的功能和相互间的连接约定后,就可以分别编制和调试程序,最后再把它们连接起来,形成一个大程序。这是一种很好的程序设计方法,而子程序结构就是模块化程序设计的基础。

6.1 子程序的设计方法

6.1.1 过程定义伪操作

在这一小节里,我们只给出 MASM 提供的基本的过程定义伪操作。从 MASM 5.1 开始,又为用户提供了功能更强的过程定义伪操作,这将在 6.1.5 小节加以说明。

过程定义伪操作用在过程(子程序)的前后,使整个过程形成清晰的、具有特定功能的代码块。其格式为:

```
procedure name      PROC        Attribute
                     ⋮
procedure name      ENDP
```

其中过程名为标识符,它又是子程序入口的符号地址。它的写法和标号的写法相同。属性(attribute)是指类型属性,它可以是 NEAR 或 FAR。

在第 3 章中,已经介绍过 CALL 和 RET 指令都有 NEAR 和 FAR 的属性,段内调用使用 NEAR 属性,段间调用使用 FAR 属性。为了使用户的工作更加方便,80x86 的汇编程序用 PROC 伪操作的类型属性来确定 CALL 和 RET 指令的属性。也就是说,如果所定义的过程是 FAR 属性的,那么对它的调用和返回一定都是 FAR 属性的;如果所定义的过程是 NEAR 属性的,那么对它的调用和返回也一定是 NEAR 属性的。这样,用户只需在定义过程时考虑它的属性,而 CALL 和 RET 的属性可以由汇编程序来确定。用户对过程属性的确定原则很简单,即:

(1) 如调用程序和过程在同一个代码段中,则使用 NEAR 属性;

（2）如调用程序和过程不在同一个代码段中，则使用 FAR 属性。

现举例说明如下：

例 6.1　调用程序和子程序在同一代码段中。

```
MAIN      PROC      FAR
            ⋮
          CALL      SUBR1
            ⋮
          RET
MAIN      ENDP
;
SUBR1     PROC      NEAR
            ⋮
          RET
SUBR1     ENDP
```

由于调用程序 MAIN 和子程序 SUBR1 是在同一代码段中的，所以 SUBR1 定义为 NEAR 属性。这样，MAIN 中对 SUBR1 的调用和 SUBR1 中的 RET 就都是 NEAR 属性的。但是一般说来，主过程 MAIN 应定义为 FAR 属性，这是由于把程序的主过程看作 DOS 调用的一个子过程，因而 DOS 对 MAIN 的调用以及 MAIN 中的 RET 就是 FRA 属性的。当然，CALL 和 RET 的属性是汇编程序确定的，用户只需正确选择 PROC 的属性就可以了。

例 6.1 的情况也可以写成如下的程序：

```
MAIN      PROC      FAR
            ⋮
          CALL      SUBR1
            ⋮
          RET
SUBR1     PROC      NEAR
            ⋮
          RET
SUBR1     ENDP
MAIN      ENDP
```

也就是说，过程定义也可以嵌套，一个过程定义中可以包括多个过程定义。

例 6.2　调用程序和子程序不在同一个代码段内。

```
SEGX      SEGMENT
            ⋮
SUBT      PROC      FAR
            ⋮
          RET
```

```
SUBT        ENDP
              ⋮
            CALL      SUBT
              ⋮
SEGX        ENDS
;
SEGY        SEGMENT
              ⋮
            CALL      SUBT
              ⋮
SEGY        ENDS
```

SUBT 为一过程,它有两处被调用,一处是与它同在 SEGX 段内,另一处是在 SEGY 段内。为此,SUBT 必须具有 FAR 属性以适应 SEGY 段调用的需要。SUBT 既然有 FAR 属性,则不论在 SEGX 段或 SEGY 段对 SUBT 的调用就都具有 FAR 属性了,这样不会发生什么错误;反之,如果这里的 SUBT 使用了 NEAR 属性,则在 SEGY 段内对它的调用就要出错了。

6.1.2 子程序的调用和返回

过程的正确执行是由子程序的正确调用和正确返回保证的。80x86 的 CALL 和 RET 指令完成的就是调用和返回的功能。为保证其正确性,除 PROC 的属性要正确选择外,还应该注意子程序运行期间的堆栈状态。由于 CALL 时已使返回地址入栈,所以 RET 时应该使返回地址出栈,如果子程序中不能正确使用堆栈而造成执行 RET 前 SP 并未指向进入子程序时的返回地址,则必然会导致运行出错,因此子程序中对堆栈的使用应该特别小心,以免发生错误。

6.1.3 保存与恢复寄存器

由于调用程序(又称主程序)和子程序经常是分别编制的,所以它们所使用的寄存器往往会发生冲突。如果主程序在调用子程序之前的某个寄存器内容在从子程序返回后还有用,而子程序又恰好使用了同一个寄存器,这就破坏了该寄存器的原有内容,因而会造成程序运行错误,这是不允许的。为避免这种错误的发生,在一进入子程序后,就应该把子程序所需要使用的寄存器内容保存在堆栈中,而在退出子程序前把寄存器内容恢复原状。例如:

```
SUBT        PROC      NEAR
            PUSH      AX
            PUSH      BX
            PUSH      CX
            PUSH      DX
```

```
              ⋮
         POP       DX
         POP       CX
         POP       BX
         POP       AX
         RET
SUBT               ENDP
```

在子程序设计时,应仔细考虑哪些寄存器是必须保存的,哪些寄存器是不必要或不应该保存的。一般说来,子程序中用到的寄存器是应该保存的。但是,如果使用寄存器在主程序和子程序之间传送参数的话,则这种寄存器就不一定需要保存,特别是用来向主程序回送结果的寄存器,就更不应该因保存和恢复寄存器而破坏了应该向主程序传送的信息。

从286开始可用的PUSHA/POPA指令以及从386开始可用的PUSHAD/POPAD指令为子程序中保存和恢复寄存器提供了有力的支持。在6.1.5小节中,还会看到汇编程序MASM也在过程定义伪操作中对寄存器的保存与恢复给予了支持。

6.1.4 子程序的参数传送

调用程序在调用子程序时,经常需要传送一些参数给子程序;子程序运行完后也经常要回送一些信息给调用程序。这种调用程序和子程序之间的信息传送称为参数传送(或称变量传送或过程通信)。参数传送方式可以有以下几种:

1. 通过寄存器传送参数

这是最常用的一种方式,使用方便,但参数很多时不能使用这种方法。下面举例说明。

例6.3 十进制到十六进制数转换程序。程序要求从键盘取得一个十进制数,然后把该数以十六进制形式在屏幕上显示出来。

这里采用子程序结构,用一个子程序DECIBIN实现从键盘取得十进制数并把它转

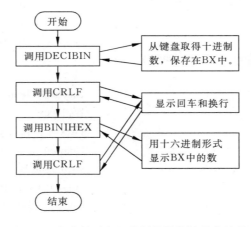

图6.1 十进制到十六进制数转换的程序结构

换为二进制数;另一个子程序 BINIHEX 把此二进制数以十六进制数的形式在屏幕上显示出来。为避免屏幕上的重叠,另外用 CRLF 子程序取得回车和换行效果。整个程序结构如图 6.1 所示。在这里,各个子程序之间用 BX 寄存器来传送信息。在过程 DECIBIN 中取得的输入数据转换为二进制数后保存在 BX 寄存器中,而过程 BINIHEX 需要把 BX 寄存器中的数用十六进制形式显示出来。也就是说,BX 寄存器用来在过程间传递要转换的数。该程序见"例 6.3 十进制到十六进制数转换的程序实现"。

例 6.3 十进制到十六进制数转换的程序实现

```
;DECIHEX--main Program--EX6_3
;   Convert decimal on keybd to hex on screen
;****************************************************************
decihex segment
        assume cs:decihex

;MAIN PART OF PROGRAM. Connects procedure together.
main        proc        far
repeat:     call        decibin             ;keyboard to binary
            call        crlf                ;print cr and lf
            call        binihex             ;binary to screen
            call        crlf                ;print cr and lf
            jmp         repeat              ;do it again

main        endp
;--------------------------------------------------------------------------------
;PROCEDURE TO CONVERT DEC ON KEYBD TO BINARY,
;    RESULT IS LEFT IN BX REGISTER

decibin     proc        near

            mov         bx,0                ;clear BX for number
;Get digit from keyboard, convert to binary
newchar:
            mov         ah,1                ;keyboard input
            int         21h                 ;call DOS
            sub         al,30h              ;ASCII to binary
            jl          exit                ;jump if < 0
            cmp         al,9d               ;is it >9d ?
            jg          exit                ;yes, not dec digit
            cbw                             ;byte in AL to word in AX
;(digit is now in AX)
;Multiply number in BX by 10 decimal
```

```
                xchg    ax,bx               ;trade digit & number
                mov     cx,10d              ;put 10 dec in CX
                mul     cx                  ;number times 10
                xchg    ax,bx               ;trade number & digit
;Add digit in AX to number in BX
                add     bx,ax               ;add digit to number
                jmp     newchar             ;get next digit
exit:
                ret                         ;return from decibin

decibin         endp                        ;end of decibin proc
;--------------------------------------------------------------------
;PROCEDURE TO CONVERT BINARY NUMBER IN BX TO HEX ON CONSOLE
;   SCREEN
binihex         proc    near

                mov     ch,4                ;number of digits
rotate:         mov     cl,4                ;set count to 4 bits
                rol     bx,cl               ;left digit to right
                mov     al,bl               ;move to AL
                and     al,0fh              ;mask off left digit
                add     al,30h              ;convert hex to ASCII
                cmp     al,3ah              ;is it >9?
                jl      printit             ;jump if digit =0 to 9
                add     al,7h               ;digit is A to F
printit:
                mov     dl,al               ;put ASCII char in DL
                mov     ah,2                ;Display Output funct
                int     21h                 ;call DOS
                dec     ch                  ;done 4 digits?
                jnz     rotate              ;not yet
                ret                         ;return from binihex

binihex         endp
;--------------------------------------------------------------------
;PROCEDURE TO PRINT CARRIAGE RETURN AND LINEFEED
crlf            proc    near

                mov     dl,0dh              ;carriage return
                mov     ah,2                ;display function
                int     21h                 ;call DOS
```

```
            mov      dl,0ah              ;linefeed
            mov      ah,2                ;display function
            int      21h                 ;call DOS
            ret                          ;return from crlf

crlf        endp
;------------------------------------------------------------------
decihex     ends
;******************************************************************
            end      main
```

2. 如过程和调用程序在同一源文件(同一程序模块)**中,则过程可直接访问模块中的变量**

例 6.4 主程序 MAIN 和过程 PROADD 在同一源文件中,要求用过程 PROADD 累加数组中的所有元素,并把和(不考虑溢出的可能性)送到指定的存储单元中去。该程序见"例 6.4 的程序实现"。在这里,过程 PROADD 直接访问模块的数据区。

例 6.4 的程序实现

```
;PROADD_1--EX6_4_1
;******************************************************************
data        segment                      ;define data segment
   ary              dw       100 dup(?)
   count            dw       100
   sum              dw       ?
data        ends
;******************************************************************
code        segment                      ;define code segment
;------------------------------------------------------------------
main        proc     far                 ;main part of program

            assume cs:code,ds:data

start:                                   ;starting execution address
            push     ds
            sub      ax,ax
            push     ax
;
            mov      ax,data
            mov      ds,ax
;           .
;           .
;           .
```

```
                call        near ptr proadd
;                .
;                .
;                .
                ret
main            endp                                    ;end of main part of prog.
;------------------------------------------------------------------------
proadd          proc        near                        ;define subprocedure
                push        ax                          ;save the registers
                push        cx
                push        si
                lea         si,ary                      ;put addr of ary in SI
                mov         cx,count                    ;put count in CX
                xor         ax,ax                       ;clear AX
next:           add         ax,[si]                     ;add the elements of the
                add         si,2                        ;   array to AX
                loop        next
                mov         sum,ax                      ;store the result in sum
                pop         si                          ;restore the registers
                pop         cx
                pop         ax
                ret                                     ;return
proadd          endp                                    ;end of subprocedure
;------------------------------------------------------------------------
code            ends                                    ;end of code segment
;************************************************************
                end         start                       ;end of assembly
```

如果数据段中有如下两个数组：

```
data            segment
    ary         dw          100     dup(?)
    count       dw          100
    sum         dw          ?
;
    num         dw          100     dup(?)
    n           dw          100
    total       dw          ?
data            ends
```

主程序两次调用 PROADD 要求分别累加 ARY 和 NUM 数组的内容。在这种情况下，如果还是采用例 6.4 中所述的参数传送方式的话，那就只有在调用 PROADD 累加 NUM 数组前，先把 NUM 数组及 N 的内容全部传送到 ARY 数组和 COUNT 单元去，

PROADD 运行完后,再把 SUM 的结果送到 TOTAL 去。这显然不是一种好办法。为解决这一类问题,下面再分别介绍"通过地址表传送参数地址"和"通过堆栈传送参数或参数地址"两种参数传送的方法。

3. 通过地址表传送参数地址

从下面"例 6.4 采用通过地址表传送参数法的程序实现"中可以看出,这种方法是在主程序中建立一个地址表,把要传送给子程序的参数都存放在地址表中,然后把地址表的首地址通过寄存器 BX 传送到子程序中去。子程序通过地址表取得所需参数,并把结果存入指定的存储单元中去。这样做的结果,对于上述程序中要累加 NUM 数组的内容时,只需在程序中增加下述指令,再调用 PROADD 后就能在 TOTAL 中得到 NUM 的累加和。

```
    MOV        TABLE, OFFSET NUM
    MOV        TABLE+2, OFFSET N
    MOV        TABLE+4, OFFSET TOTAL
    MOV        BX, OFFSET TABLE
    CALL       PROADD
```

此外,该程序采用 COM 文件形式,代码、数据和堆栈都设在一个段中。当然,这并不说明这里必须使用 COM 文件,只是给出另一种程序格式而已。

例 6.4 采用通过地址表传送参数法的程序实现

```
;PROADD_2--EX6_4_2
;**************************************************************
prog_seg    segment

            org         100h

            assume cs:prog_seg,ds:prog_seg,ss:prog_seg

main        proc        near
;
            mov         ax,prog_seg
            mov         ds,ax
;           .
;           .
;           .
            mov         table,offset ary        ;put the addresses
            mov         table+2,offset count    ;  of ary,count and sum
            mov         table+4,offset sum      ;  in parameter table
            mov         bx,offset table         ;put addr of table in BX
            call        proadd                  ;call proadd
```

```
        ;                       .
        ;                       .
        ;                       .
                    mov         ax,4c00h
                    int         21h
main                endp
;--------------------------------------------------------------------
proadd              proc        near                    ;define subprocedure
                    push        ax                      ;save registers
                    push        cx
                    push        si
                    push        di
                    mov         si,[bx]                 ;get the addr of array,
                    mov         di,[bx+2]               ;   the value of count,
                    mov         cx,[di]                 ;   and the addr of
                    mov         di,[bx+4]               ;   result
                    xor         ax,ax                   ;clear AX
next:               add         ax,[si]                 ;add the elements
                    add         si,2                    ;   of the array
                    loop        next                    ;   to AX
                    mov         [di],ax                 ;return result
                    pop         di                      ;restore registers
                    pop         si
                    pop         cx
                    pop         ax
                    ret                                 ;return
proadd              endp                                ;end of subprocedure
;--------------------------------------------------------------------
ary                 dw          100 dup(?)              ;reserve 100 words for ary
count               dw          100                     ;   and 1 word for count
sum                 dw          ?                       ;   and 1 word for sum
table               dw          3 dup(?)                ;reserve 3 words for
                                                        ;   parameter addresses
;--------------------------------------------------------------------
prog_seg            ends
;****************************************************************
                    end         main
```

4. 通过堆栈传送参数或参数地址

"例6.4采用通过堆栈传送参数地址法的程序实现"的方法,是在主程序里把参数地址保存到堆栈中,在子程序里从堆栈中取出参数以达到传送参数的目的。如本例中堆栈最满时的状态如图6.2所示。必须注意,子程序结束时的RET指令应使用带常数的返回指令，以便返回主程序后，堆栈能恢复原始状态不变。

例 6.4 采用通过堆栈传送参数地址法的程序实现

```
;PROADD_3--EX6_4_3
;**************************************************************
parm_seg        segment                 ;define data segment
        ary     dw      100 dup(?)
        count   dw      100
        sum     dw      ?
parm_seg        ends
;**************************************************************
stack_seg       segment
                dw      100 dup(?)
        tos     label   word
stack_seg       ends
;**************************************************************
code1           segment                 ;define code segment
;----------------------------------------------------------------
main            proc    far             ;main part of program
        assume cs:code1,ds:parm_seg,ss:stack_seg

start:                                  ;starting execution address

;set up SS and SP register
        mov     ax,stack_seg
        mov     ss,ax
        mov     sp,offset tos

;set up stack for return
        push    ds                      ;save old data segment
        sub     ax,ax                   ;put zero in AX
        push    ax                      ;save it on stack

;set DS register to current data segment
        mov     ax,parm_seg             ;data segment addr
        mov     ds,ax                   ;   into DS register

;       .
;       .
;       .
```

```
                mov         bx,offset ary       ;push addr of ary
                push        bx                  ;  onto stack
                mov         bx,offset count     ;push addr of count
                push        bx                  ;  onto stack
                mov         bx,offset sum       ;push addr of sum
                push        bx                  ;  onto stack
                call        far ptr proadd
;                           .
;                           .
;                           .
                ret
main            endp
;----------------------------------------------------------------------
code1           ends
;**************************************************************
code2           segment

                assume      cs:code2

;----------------------------------------------------------------------
proadd          proc        far
                push        bp                  ;save BP
                mov         bp,sp               ;use BP to access parameter
                push        ax                  ;save other registers
                push        cx
                push        si
                push        di
                mov         si,[bp+0ah]         ;get addr of ary
                mov         di,[bp+8]           ;get addr of count
                mov         cx,[di]             ; CX ← (count)
                mov         di,[bp+6]           ;get addr of sum
                xor         ax,ax               ;clear AX
next:           add         ax,[si]             ;compute sum
                add         si,2
                loop        next
                mov         [di],ax             ;store result
                pop         di                  ;restore registers
                pop         si
                pop         cx
                pop         ax
                pop         bp
                ret         6                   ;adjust stack and return
```

```
proadd      endp
;------------------------------------------------------------
code2       ends
;************************************************************

         end    start              ;end of assembly
```

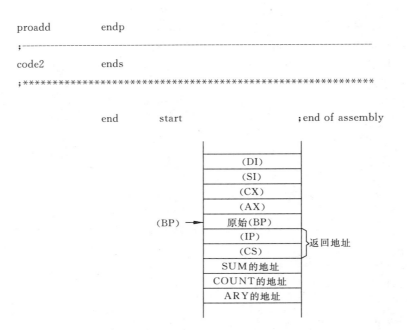

图 6.2　例 6.4 用堆栈传送参数地址时,堆栈最满时的状态

MASM 提供的结构伪操作 STRUC 给用户提供了很大的方便。因篇幅有限,本书不作进一步地说明,如读者需要,请参阅《IBM PC 汇编语言程序设计》一书(清华大学出版社,1991 年 6 月版),或《80x86 汇编语言程序设计》(清华大学出版社 2001 年版)。

以上说明的几种传送参数的方法,读者可以根据具体情况选择使用。

注意：在例子中,后两种方法都用来传送参数地址,实际上它们也可以用来传送参数,在掌握了方法以后,读者可以灵活使用。

5. 多个模块之间的参数传送问题

多个模块的程序相连接时,并不一定要把所有的代码段或数据段分别连接在一起,形成一个大的代码段或数据段。在很多情况下,各程序模块仍有各自的分段,只是通过模块之间的调用来进行工作。有时有些程序模块需要把几个分段的程序连成一个段,在 4.2.2 小节中"1. 完整的段定义伪操作"所介绍的 SEGMENT 伪操作的组合类型提供了这种支持。除段组合外,还必然存在着参数传送问题,在这里要讨论的问题是:如果调用程序和子程序不在同一个程序模块中,那么怎样进行参数传送呢?

(1) 外部符号

从连接的角度看,在源程序中用户定义的符号可以分为局部符号和外部符号两种。我们已经熟悉的在本模块中定义,又在本模块中引用的符号称为局部符号;而另一种在某一个模块中定义,而又在另一个模块中引用的符号称为外部符号。有两个伪操作与外部符号有关。

PUBLIC 伪操作,其格式是:

PUBLIC   symbol [,…]

在一个模块中定义的符号(包括变量、标号、过程名等)在提供给其他模块使用时,必须要用PUBLIC定义该符号为外部符号。

EXTRN伪操作,其格式是:

EXTRN   symbol name : type [,…]

在另一个模块中定义而要在本模块中使用的符号必须使用EXTRN伪操作。如符号为变量,则类型应为byte,word,dword等;如符号为标号或过程名,则类型应为near或far。

有了这两个伪操作就提供了模块间相互访问的可能性。这两个伪操作的使用必须相匹配。连接程序会检查每个模块中的EXTRN语句中的每个符号是否能和与其相连接的其他模块中的PUBLIC语句中的一个符号相匹配。如果不匹配,则应给出出错信息;如果相匹配,则应给予确定值。下面的例子说明各模块中PUBLIC和EXTRN伪操作的匹配情况。

例6.5　三个源模块中的外部符号定义见"例6.5的源模块"程序。连接程序能检查出var4是模块2需要使用的符号,但没有其他模块用PUBLIC来宣布其定义,因而连接将显示出错。在这个例子中,模块3用PUBLIC宣布了lab3的外部定义,但其他模块均未使用该符号,这种不匹配情况由于不影响装入模块的建立,所以并不显示出错。此外,模块1和模块2都定义了局部符号var3,由于局部符号是在汇编时就确定了其二进制值,所以并不影响模块的连接;因而不同模块中的局部符号是允许重名的,但要连接模块的外部符号却不允许重名,如有重名,连接将显示出错。

例6.5的源模块

```
;           source module 1
;
extrn       var2:word,lab2:far
public      var1,lab1
;
data1       segment
            var1    db      ?
            var3    dw      ?
            var4    dw      ?
data1       ends
;
code1       segment
            assume  cs:code1,ds:data1
main        proc    far
start:
            mov     ax,data1
            mov     ds,ax
```

```
                        .
                        .
                        .
        lab1:
                        .
                        .
                        .
                        mov        ax,4c00h
                        int        21h
        main            endp
        code1           ends
        ;
                        end        start


        ;               source module 2
        ;
        extrn           var1:byte,var4:word
        public          var2
        ;
        data2           segment
            var2        dw         0
            var3        db         5 dup(?)
        data2           ends
        ;
        code2           segment
                        assume     cs:code2,ds:data2
                        .
                        .
                        .
        code2           ends
        ;
                        end


        ;               source module 3
        ;
        extrn           lab1:far
        public          lab2,lab3
        ;
        code3           segment
                        assume     cs:code3
                        .
                        .
```

```
lab2:           .
                .
                .
lab3:           .
                .
                .
code3           ends
;
                end
```

外部符号在汇编时是不可能确定其值的,LST 清单中对外部符号记以 E。连接程序可分配段地址、确定外部符号及浮动地址值,连接完成后建立了装入模块,再由装入程序把该模块装入内存等待执行。

(2) 多个模块之间的参数传送方法

可以使用前面已经说过的几种伪操作及其参数来解决参数传送问题。

① 当主程序和子程序不在同一程序模块中时变量的传送方法之一(用例 6.4 说明)

例 6.4 当主程序和子程序不在同一模块时的程序。

```
;PROADD_6_1--EX_6_4_4_1
;************************************************************
;               source module 1

    extrn       proadd:far

;************************************************************
data        segment common                 ;define data segment
  ary           dw          100 dup(?)
  count         dw          100
  sum           dw          ?
data        ends
;************************************************************
code1       segment                         ;define code segment
;------------------------------------------------------------
main        proc        far                 ;main part of program
            assume cs:code1,ds:data

start:      mov         ax,data             ;starting execution address
            mov         ds,ax
```

```
;            .
;            .
;            .
             call       far ptr proadd
;            .
;            .
;            .
             mov        ax,4c00h
             int        21h
main         endp                                      ;end of main part of prog.
;------------------------------------------------------------------------------
code1        ends
;************************************************************
             end        start

;PROADD_6_2--EX6_4_4_2


;************************************************************
;              source module 2

public       proadd


;************************************************************
data         segment common
  ary                   dw         100 dup(?)
  count                 dw         100
  sum                   dw         ?
data         ends
;************************************************************
code2        segment

proadd       proc       far                            ;define subprocedure

             assume cs:code2,ds:data

             mov        ax,data
             mov        ds,ax
;
             push       ax                             ;save the registers
             push       cx
             push       si
             lea        si,ary                         ;put addr of ary in si
```

```
            mov     cx,count              ;put count in cx
            xor     ax,ax                 ;clear ax
next:       add     ax,[si]               ;add the elements of the
            add     si,2                  ;  array to AX
            loop    next
            mov     sum,ax                ;store the result in sum
            pop     si                    ;restore the registers
            pop     cx
            pop     ax
            ret                           ;return
proadd      endp                          ;end subprocedure
;------------------------------------------------------------
code2       ends                          ;end of code segment
;************************************************************
            end                           ;end of assembly
```

在这个例子中,data 段用 common 合并成为一个覆盖段,所以源模块 2 只引用了本模块中的变量,不必作特殊处理。整个程序的外部符号只有 proadd,处理比较简单。

应该注意:由于主程序和子程序已经不在同一程序模块中,所以过程定义及调用都应该是 FAR 类型的,而不应该使用原来的 NEAR 类型。如果以上两个模块的 code 段都使用同一段名并加上 PUBLIC 说明,这样,连接时它们就可以合并为一个段,此时,过程的调用仍可使用 NEAR 属性。

使用公共数据段并不是惟一的办法,我们可以把变量也定义为外部符号,这样就允许其他模块引用在某一模块中定义的变量名。必须注意,我们在引用本模块中的局部变量前,在程序的一开始就用以下两条指令:

```
MOV     AX, DATA_SEG
MOV     DS, AX
```

把数据段地址放入 DS 寄存器中,这样才能保证对局部变量的正确引用。在引用外部符号时也必须把相应的段地址放入段寄存器中。如果程序中要访问的变量处于不同段时,就应动态地改变段寄存器的内容。让我们看下面的例子。

② 当主程序和子程序不在同一程序模块中时变量的传送方法之二(用例 6.6 说明)

例 6.6 有三个源模块见"例 6.6 的程序"。其中模块 1 本身的局部变量都在 DS 段中,而外部变量则在 ES 段中,在程序中动态地改变 ES 寄存器的内容,以达到正确访问各外部变量的目的。如果源模块 1 本身使用 ES 段,或者外部变量较多,为避免动态改变段地址易产生的错误,也可以用例 6.7 所使用的方法。

例 6.6 的程序

```
;                 source module 1
```

```
;*****************************************************
        extrn       var1:word,output:far
        extrn       var2:word
        public      exit
;*****************************************************
local_data          segment
            var     dw      5
                    .
                    .
                    .
local_data          ends
;*****************************************************
code        segment
            assume  cs:code,ds:local_data
;-----------------------------------------------------------------------------------
main        proc    far
start:
            mov     ax,local_data
            mov     ds,ax
                    .
                    .
                    .
            mov     bx,var
            mov     ax,seg var1
            mov     es,ax
            add     bx,es:var1
                    .
                    .
                    .
            mov     ax,seg var2
            mov     es,ax
            sub     es:var2,50
                    .
                    .
                    .
            jmp     output
                    .
                    .
                    .
exit:       mov     ax,4c00h
            int     21h
main        endp
;-----------------------------------------------------------------------------------
```

```
code            ends
;*****************************************************
                end         start

;           source module 2
;*****************************************************
public          var1
;*****************************************************
extdata1        segment
                var1        dw      10
                            .
                            .
                            .
extdata1        ends
;*****************************************************
                .
                .
                .
;*****************************************************
                end

;       source module 3
;*****************************************************
public          var2
extrn           exit:far
;*****************************************************
extdata2        segment
  var2          dw      3
                        .
                        .
                        .
extdata2                ends
;*****************************************************
public          output
;
prognam                 segment
                assume  cs:prognam,ds:extdata2
                        .
                        .
                        .
output:
                jmp     exit
                        .
```

```
                    .
                    .
                    .
prognam       ends
;******************************************************
              end
```

③ 当主程序和子程序不在同一程序模块中时变量的传送方法之三(用例 6.7 来说明)

例 6.7 有两个源模块,见如下程序。

```
;                 source module 1
;******************************************************
global        segment public
  extrn var1:word,var2:word
global        ends
;******************************************************
local_data              segment
                    .
                    .
                    .
local_data              ends
;******************************************************
code          segment
              assume    cs:code,ds:local_data,es:global
;----------------------------------------------------------------------
main          proc      far
start:
              mov       ax,local_data
              mov       ds,ax
              mov       ax,global
              mov       es,ax
                    .
                    .
                    .
              mov       bx,es:var1
              add       es:var2,bx
                    .
                    .
                    .
              mov       ax,4c00h
              int       21h
main          endp
;----------------------------------------------------------------------
code          ends
```

```
;******************************************************
            end     start

;           source module 2
;******************************************************
global      segment     public
  public                var1,var2
  var1      dw          ?
  var2      dw          ?
              .
              .
              .
global      ends
;******************************************************
              .
              .
              .
;******************************************************
            end
```

从以上几个例子可以看出，在掌握了有关外部符号的伪操作及 SEGMENT 伪操作的参数使用方法的情况下，读者可以灵活使用这些工具，并能编制出较好的程序模块来。

6.1.5 增强功能的过程定义伪操作

从 MASM5.1 版开始为用户提供了增强功能的过程定义伪操作，其格式为：

```
procname    PROC    [attributes field][USES register list][, parameter field]
              ⋮
procname    ENDP
```

其中属性字段由以下几项组成：

```
distance    language type    visibility    prologue
```

每一项均为可选，各项之间用一空格或制表符分开。

distance 就用 NEAR 或 FAR，与 6.1.1 小节中所述的类型属性相同。MASM 规定在简化段定义中已指定内存模型情况下，不必再在过程定义中指定类型属性。汇编程序将自动把 TINY，SMALL，COMPACT 和 FLAT 程序中的过程指定为 NEAR；而把 MEDIUM，LARGE 和 HUGE 程序中的过程指定为 FAR。如程序中未指定内存模型，则应由用户自行指定类型属性；如用户未指定，则默认值为 NEAR。

language type 说明当该过程作为某种高级语言程序的子过程时所用的高级语言，如 PASCAL，BASIC，FORTRAN 或 C 等。如在程序的.MODEL 中已说明了所用语言，则此时可以省略。

必须注意：如果过程伪操作中使用了参数字段，但在.model 和 language type 中均未指定一种语言类型的话，则 MASM 将指示出错。为此，即使你的过程并不需要由高级语言调用，也应该指定一种语言类型。

visibility 说明该过程的可见性。可用 Private 或 Public 。如用 Private,则该过程的可见性只能是当前的源文件；如用 Public,则允许其他模块调用该过程。该项的默认是 Public。

prologue 是一个宏的名字，允许用户用宏来控制过程的入口和出口有关的代码。

USES 字段允许用户指定所需保存和恢复的寄存器表，MASM 将在过程入口自动生成 push 指令来保存这些寄存器，并在过程出口的 ret 指令前自动生成 pop 指令来恢复这些寄存器。

参数字段允许用户指定该过程所用参数，其格式为：

identifier:type[, indentifier:type]

其中 identifier 给出参数的符号名，type 给出参数的类型。参数之间用逗号隔开。如果参数太多，一行放不下的话，可以用逗号结束此行，在下一行继续给出参数。

MASM 将自动把这些参数转换为［BP+4］、［BP+6］等形式。

注意：参数排列的顺序与所指定的语言类型有关，PASCAL,BASIC 和 FORTRAN 这三种语言与 C 语言在参数入栈的次序上是相反的。前三种语言是按参数在调用时出现的顺序入栈的，而 C 语言则正好相反，它的第一个参数是最后入栈的，所以其第一个参数正好在返回地址之上，如图 6.3 所示。还需说明的是，图 6.3(3)中的返回地址采用 4 字节，这适用于 FORTRAN 和 PASCAL 的调用情况，它们是采用 FAR 调用的，而 BASIC 是采用 NEAR 调用的，因此返回地址只用 2 个字节，这一点在图中并未显示出来。另外图中所示的局部变量区，在本节后面还会作专门介绍。

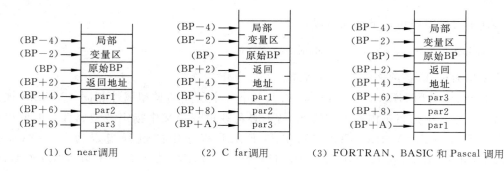

图 6.3 几种高级语言过程调用时的堆栈情况

现在，可以用过程定义伪操作来实现例 6.4 的程序，见"例 6.4 用增强功能的过程定义伪操作的程序实现"程序。汇编程序将自动产生与"例 6.4 采用通过堆栈传送参数地址法的程序实现"完全相同的程序,其中子程序中的指令 push bp 和 mov bp,sp,以及[bp+0ah],[bp+8]和[bp+6]是当建立参数时由汇编程序自动生成的。子程序中的保存寄存器和恢复寄存器的指令也是汇编程序根据 uses 说明而自动生成的。汇编程序并能根据参数情况自动生成 ret 6 的返回指令,因此程序员编程时只要写 ret 就可以了。

例 6.4 用增强功能的过程定义伪操作的程序实现

```
;PROADD_5--EX6_4_5
;************************************************************
        .model    medium
        .data
ary               dw        100 dup(?)
count             dw        100
sum               dw        ?
;************************************************************
        .stack    200h
;************************************************************
        .code     code1
main    proc
start:                                        ;starting execution address
        mov       ax,@data
        mov       ds,ax
;       .
;       .
;       .
        mov       bx,offset ary
        push      bx                          ;push addr of ary onto stack
        mov       bx,offset count
        push      bx                          ;push addr of count onto stack
        mov       bx,offset sum
        push      bx                          ;push addr of sum onto stack
        call      proadd
;       .
;       .
;       .
        mov       ax,4c00h
        int       21h
main    endp
```

```
        ;**********************************************************
                .code       code2
proadd          proc        pascal uses ax cx si di,
                              para:word,parc:word,pars:word
        ;
                mov         si,para                 ;get addr of ary from stack
                mov         di,parc                 ;get addr of count from stack
                mov         cx,[di]                 ;   and put count in CX
                mov         di,pars                 ;get addr of sum from stack
                xor         ax,ax                   ;clear ax
next:           add         ax,[si]                 ;compute sum
                add         si,2
                loop        next
                mov         [di],ax                 ;store result
                ret                                 ;adjust stack and return
proadd          endp
        ;**********************************************************

                end         start                   ;end of assembly
```

增强功能的过程定义伪操作除具有以上功能外,还可在过程中定义局部变量。

局部变量是指在过程内部使用的变量。它是在过程被调用时在堆栈中建立的,在退出过程时被释放。局部变量所建立的区域为图 6.3 堆栈中的局部变量区,它是以[BP－2]、[BP－4]、…的形式被访问的。因此在过程调用期间位于 BP 指针的正偏移量区是返回地址和参数区,而 P 指针的负偏移量区则是局部变量区。

MASM 规定,在过程内可以用 LOCAL 为局部变量申请空间,其格式为:

LOCAL　　vardef [, vardef]

其中变量定义可用的格式为:

label
label：type
label [count]：type

其中未指定类型者,MASM 将使用 WORD。type 可以指定任意合法的类型说明,如 byte,word,dword 等。第三种格式为用户申请数组提供了方便,如可指定 M[20]：sword,sword 表示带符号数的字类型,M 为数组名,其中包括 20 个元素,用户可以用 0～19 的下标来访问该数组。

LOCAL 语句必须紧跟在过程定义伪操作之后,并在任何 80x86 指令或可以产生任何代码的 MASM 语句之前出现。它可以定义多个局部变量,如一行写不下,可用逗号结束前一行,并在下一行继续定义。MASM 将为所定义的变量在堆栈中的 BP 负偏移区生成空间,并对每个局部变量名生成如 [BP－2],[BP－4],…等代码。

下面用例 6.8 来说明局部变量的使用方法。

例 6.8　程序要求把以 ASCII 形式表示的十进制数转换为二进制数。

程序所用的算法很简单:从最低位起,每个数位先将 ASCII 字符转换为数字,然后乘以该位的权(乘法因子),累加后就成为所要求的二进制数。例如,数'12345'的计算步骤是:

	十进制			十六进制	
步骤		乘积	步骤		乘积
5×1	=	5	5×01H	=	5H
4×10	=	40	4×0AH	=	28H
3×100	=	300	3×64H	=	12CH
2×1000	=	2000	2×3E8H	=	7D0H
1×10000	=	10000	1×2710H	=	2710H
求和	=	12345	求和	=	3039H

在程序中的数据区内,定义了二个变量,ASCVAL 存放着以 ASCII 形式表示的十进制数,要求把转换成的二进制数存放在 BINVAL 单元中。在主程序中,把这两个变量的地址作为参数传送给子过程 CONVASCBIN。子过程的过程定义伪操作中除定义参数域外,还定义了两个局部变量:ASCLEN 和 MULFACT,分别存放子过程内部所用的 ASCII 字符串长度和乘法因子。例 6.8 的程序见"例 6.8 的程序实现"。"例 6.8 经汇编后的 CONVASCB IN 子过程程序"中给出了经 MASM 汇编后的 CONVASCBIN 子过程。可以看出汇编程序完成了原先要用户自行编制的保存和恢复寄存器、参数区和局部变量区的建立及释放等工作,为用户提供了方便。图 6.4 给出了在该程序运行过程中,堆栈最满时的状态。如前所述,BP 指针的正偏移区为参数区,而 BP 指针的负偏移区为局部变量区。局部变量区在子程序返回主程序前将自动释放。

例 6.8 的程序实现

```
                .model    small
                .386
                .stack    200h            ;define stack segment
                .data                     ;define data segment
ascval          db        '12345'
binval          dw        ?

                .code                     ;define code segment
main            proc

start:                                    ;starting execution addr
                mov       ax,@data        ;data segment addr
                mov       ds,ax           ;  into DS register
;
```

```
            ;       .
            ;       .
                    lea     bx,ascval           ;put par1
                    push    bx                  ;   onto stack
                    lea     bx,binval           ;put par2
                    push    bx                  ;   onto stack
                    call    convascbin          ;call convert routine
            ;       .
            ;       .
            ;       .
                    mov     ax,4c00h            ;go back to DOS
                    int     21h
            ;
main                endp
            ;
convascbin          proc    pascal uses ax bx cx si di,
                                                par1:word,par2:word
                    local   asclen:word,mulfact:word
                    mov     bx,10               ;mult factor
                    mov     si,par1             ;calculate the length
                    mov     di,par2             ;   of ASCII character
                    sub     di,si               ;      and
                    mov     asclen,di           ;         save it
                    mov     cx,asclen           ;count for loop
                    add     si,asclen           ;address of ascval
                    dec     si
                    mov     mulfact,1           ;initialize mulfact
                    mov     di,par2             ;
                    mov     [di],0              ;initialize binval
next:
                    mov     al,[si]             ;load an ASCII character
                    and     ax,000fh            ;mask off left digit
                    mul     mulfact             ;multiply by factor
                    add     [di],ax             ;add to binary
                    mov     ax,mulfact          ;calculate next
                    mul     bx                  ;   factor
                    mov     mulfact,ax          ;
                    dec     si                  ;last ASCII character?
                    loop    next                ;   no, continue
                    ret                         ;   yes, return
convascbin          endp
            ;
```

        end   main

例 6.8 经汇编后的 CONVASCBIN 子过程程序

```
        push    bp
        mov     bp,sp
        add     sp,0fch
        push    ax
        push    bx
        push    cx
        push    si
        push    di
        mov     bx,0ah
        mov     si,[bp+6]
        mov     di,[bp+4]
        sub     di,si
        mov     [bp-2],di
        mov     cx,[bp-2]
        add     si,[bp-2]
        dec     si
        mov     word ptr[bp-4],0001
        mov     di,[bp+4]
        mov     word ptr[di],0
next:
        mov     al,[si]
        and     ax,000fh
        mul     word ptr[bp-4]
        add     [di],ax
        mov     ax,[bp-4]
        mul     bx
        mov     [bp-4],ax
        dec     si
        loop    next
        pop     di
        pop     si
        pop     cx
        pop     bx
        pop     ax
        mov     sp,bp
        pop     bp
        ret     0004
```

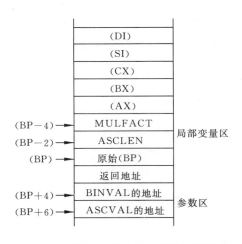

图 6.4　例 6.8 程序运行过程中堆栈最满时的状态

6.2　子程序的嵌套

我们已经知道,一个子程序也可以作为调用程序去调用另一个子程序,这种情况就称为子程序的嵌套。嵌套的层次不限,其层数称为嵌套深度。图 6.5 表示了嵌套深度为 2 时的子程序嵌套情况。

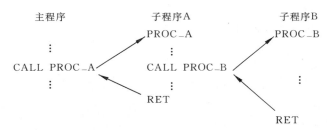

图 6.5　子程序的嵌套

嵌套子程序的设计并没有什么特殊要求,除子程序的调用和返回应正确使用 CALL 和 RET 指令外,要注意寄存器的保存和恢复,以避免各层次子程序之间因寄存器冲突而出错的情况发生。如果程序中使用了堆栈,例如使用堆栈来传送参数等,则对堆栈的操作要格外小心,避免发生因堆栈使用中的问题而造成子程序不能正确返回的错误。

在子程序嵌套的情况下,如果一个子程序调用的子程序就是它自身,这就称为递归调用。这样的子程序称为递归子程序。递归子程序对应于数学上对函数的递归定义,它往往能设计出效率较高的程序,可完成相当复杂的计算,所以它是很有用的。因篇幅所限,本书不加进一步说明,有兴趣的读者可参阅《IBM PC 汇编语言程序设计》一书(清华大学出版社,1991 年 6 月版),或(《80x86 汇编语言程序设计》(清华大学出版社 2001 年版)。

在编制子程序时,特别是在编制嵌套或递归子程序时,堆栈的使用是十分频繁的。在这里顺便说明一下,在堆栈使用过程中,应该注意有关堆栈溢出的问题。由于堆栈区域是在堆栈定义时就确定了的,因而堆栈工作过程中有可能产生溢出。堆栈溢出有两种情况可能发生:如堆栈已满,但还想再存入信息,这种情况称为堆栈上溢;另一种情况是,如堆栈已空,但还想再取出信息,这种情况称为堆栈下溢。不论上溢或下溢,都是不允许的。因此在编制程序时,如果可能发生堆栈溢出,则应在程序中采取保护措施。这可以用给 SP 规定上、下限,在进栈或出栈操作前先做 SP 和边界值的比较,如溢出则作溢出处理,以避免破坏其他存储区或使程序出错的情况发生。

6.3 子程序举例

例 6.9 HEXIDEC 是一个把十六进制数转换成十进制数的程序。要求把从键盘输入的 0～FFFFH 的十六进制正数转换为十进制数并在屏幕上显示出来。

这一例子的功能是和例 6.3 相反的。它由 HEXIBIN 和 BINIDEC 两个主要的子程序组成,由于主程序和子程序在同一个程序模块中,因而省略了对寄存器的保护和恢复工作,子程序之间的参数传送则采用寄存器传送的方式进行。程序可用 Ctrl Break 退出。本程序如下。

十六进制到十进制数转换的程序实现

```
;HEXIDEC--Main program--EX6_9#
;     converts hex on keyboard to dec on screen

;EQU STATEMENTS GO HERE

display         equ     2h              ;video output
key_in          equ     1h              ;keyboard input
doscall         equ     21h             ;DOS interrupt number

;************************************************************
hexidec segment                         ;define code segment
;------------------------------------------------------------
main            proc    far             ;main part of program

    assume cs:hexidec

;Main part of program links subroutines together

start:                                  ;starting execution address

;set up stack for return
                push    ds              ;save old data segment
                sub     ax,ax           ;put zero in AX
```

```
                push    ax                  ;zero on stack

;MAIN PART OF PROGRAM GOES HERE

                call    hexibin             ;keyboard to binary
                call    crlf                ;print cr & lf
;
                call    binidec             ;binary to decimal
                call    crlf                ;print cr & lf
                jmp     main                ;get next input
;
                ret                         ;return to DOS

main            endp                        ;end of main part of program
;------------------------------------------------------------------
hexibin         proc    near                ;define subprocedure
;Subroutine to convert hex on keybd to binary
;   result is left in BX register
                mov     bx,0                ;clear BX for number

;Get digit from keyboard, convert to binary
newchar:
                mov     ah,key_in           ;keyboard input
                int     doscall             ;call DOS
                sub     al,30h              ;ASCII to binary
                jl      exit                ;jump if < 0
                cmp     al,10d              ;is it > 9d ?
                jl      add_to              ;yes, so it's digit
;not digit (0 to 9), may be letter (a to f)
                sub     al,27h              ;convert ASCII to binary
                cmp     al,0ah              ;is it <0a hex?
                jl      exit                ;yes, not letter
                cmp     al,10h              ;is it >0f hex?
                jge     exit                ;yes, not letter
;is hex digit, add to number in BX
add_to:
                mov     cl,4                ;set shift count
                shl     bx,cl               ;rotate BX 4 bits
                mov     ah,0                ;zero out AH
                add     bx,ax               ;add digit to number
                jmp     newchar             ;get next digit
exit:
```

```
                ret                             ;return from hexibin
hexibin         endp                            ;end of subprocedure
;----------------------------------------------------------------
binidec         proc        near
;Subroutine to convert binary number in BX
;   to decimal on console screen

                mov         cx,10000d           ;divide by 10000
                call        dec_div
                mov         cx,1000d            ;divide by 1000
                call        dec_div
                mov         cx,100d             ;divide by 100
                call        dec_div
                mov         cx,10d              ;divide by 10
                call        dec_div
                mov         cx,1d               ;divide by 1
                call        dec_div
                ret                             ;return from binidec
;----------------------------------------------------------------
dec_div proc    near
;Subroutine to divide number in BX by number in CX
;print quotient on screen
;(numberator in DX+AX,denom in CX)

                mov         ax,bx               ;number low half
                mov         dx,0                ;zero out high half
                div         cx                  ;divide by CX
                mov         bx,dx               ;remainder into BX
                mov         dl,al               ;quotient into DL
;print the contents of DL on screen
                add         dl,30h              ;convert to ASCII
                mov         ah,display          ;display function
                int         doscall             ;call DOS
                ret                             ;return from dec_div

dec_div         endp
;----------------------------------------------------------------
binidec         endp
;----------------------------------------------------------------
crlf            proc        near
;print carriage return and linefeed
```

```
              mov      dl,0ah                        ;linefeed
              mov      ah,display                    ;display function
              int      doscall                       ;call DOS
;
              mov      dl,0dh                        ;carriage return
              mov      ah,display                    ;display function
              int      doscall                       ;call DOS
              ret                                    ;return from crlf
crlf          endp
;------------------------------------------------------------
hexidec       ends                                   ;end of code segment
;************************************************************
              end      start                         ;end of assembly
```

例 6.10 一个简单的信息检索系统。在数据区里,有 10 个不同的信息,编号为 0~9,每个信息包括 30 个字符。现在要求编制一个程序:从键盘接收 0~9 之间的一个编号,然后在屏幕上显示出相应编号的信息内容。此程序见下面"例 6.10 的程序实现"。在这个程序里,10 个信息组成一个信息表,对信息表的查找是根据从键盘接收的编号来确定的。

注意:从接收编号后到找到表格中所需区域为止的程序段,这也是查找表格的一种常用方法。此外,程序把显示一个信息编成一个独立功能的子程序 DISPLAY,并把其中常用的显示一个字符的功能也编成一个子程序 DISPCHAR,可以看出这样就使程序的结构更加清晰了。

例 6.10 的程序实现

```
;SEARCH--EX6_10#
;************************************************************
datarea    segment                        ;define data segment
thirty       db         30                ;value for mul instruction
;message table
msg0         db         'I like my IBM-PC-----------------------------'
msg1         db         '8088 programming is fun----------------------'
msg2         db         'Time to buy more diskettes-------------------'
msg3         db         'This program works---------------------------'
msg4         db         'Turn off that printer------------------------'
msg5         db         'I have more memory than you------------------'
msg6         db         'The PSP can be useful------------------------'
msg7         db         'BASIC was easier than this-------------------'
```

```
msg8            db          'DOS is indispensable----------------------------------------------'
msg9            db          'Last massage of the day-------------------------------------------'
;error message
errmsg          db          'error!!!    invalid parameter!!'
datarea ends
;************************************************************
stack           segment
                db          256 dup(0)              ;256 bytes of stack space
    tos         label       word
stack           ends
;************************************************************
prognam         segment                             ;define code segment
;-----------------------------------------------------------------------
main            proc        far                     ;main part of program

        assume cs:prognam,ds:datarea,ss:stack

start:                                               ;starting execution address

;set SS register to current stack segment
                mov         ax,stack
                mov         ss,ax
                mov         sp,offset tos

;set up stack for return
                push        ds                      ;save old data segment
                sub         ax,ax                   ;put zero in AX
                push        ax                      ;save it on stack
;set DS register to current data segment
                mov         ax,datarea              ;datarea segment addr
                mov         ds,ax                   ;   into DS register

;MAIN PART OF PROGRAM GOES HERE

begin:          mov         ah,1                    ;move a parameter from keybd
                int         21h
                sub         al,'0'                  ;convert from ASCII to binary
                jc          error                   ;branch if not numeric(<'0')
                cmp         al,9                    ;check for valid numeric
                ja          error                   ;branch if not valid(>'9')

;select appropriate message from message table
```

```
            mov     bx,offset msg0          ;point to first message
            mul     thirty                  ;(AX)=(AL)*30
            add     bx,ax                   ;point to desired message
            call    display                 ;display the message at [BX]
            jmp     begin
;display error message for invalid parameter
error:      mov     bx,offset errmsg
            call    display                 ;display error message
            ret                             ;return to DOS
;------------------------------------------------------------------------
;Subroutine to display a message on the screen
;Enter with BX-->message to be displayed
;Message is assumed to be 30 characters long

display proc    near
            mov     cx,30                   ;number of char. to display
disp1:      mov     dl,[bx]                 ;get next char. to display
            call    dispchar                ;display it
            inc     bx                      ;point to next char.
            loop    disp1                   ;do it 30 times
            mov     dl,0dh                  ;carriage return
            call    dispchar
            mov     dl,0ah                  ;linefeed
            call    dispchar
            ret                             ;return to caller of
                                            ; 'display'
display endp
;------------------------------------------------------------------------
;Subroutine to display a character on the screen
;Enter with (DL)= character to be displayed
dispchar proc   near
            mov     ah,2                    ;display out a character
            int     21h
            ret                             ;return to caller of
                                            ; 'dispchar'
dispchar endp
;------------------------------------------------------------------------
main        endp                            ;end of main part of program
;------------------------------------------------------------------------
prognam     ends                            ;end of code segment
;************************************************************
```

```
        end       start                        ;end of assembly
```

例 6.11 人名排序程序。先从终端键入最多 30 个人名,当所有人名都进入后,按字母上升的次序将人名排序,并在屏幕上显示已经排好序的人名。程序的总框图如图 6.6 所示,程序则见"例 6.11 人名排序的程序实现"。

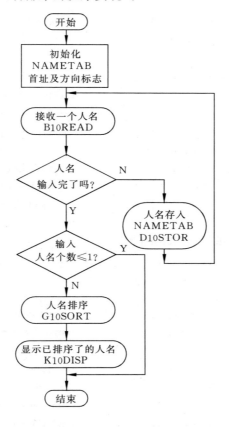

图 6.6 人名排序的程序框图

例 6.11 人名排序的程序实现

```
;NAME_SORT--EX6_11#
;***********************************************
    .model   small
    .stack   40h
;***********************************************
    .data                          ;define data segment
namepar       label    byte         ;name parameter list:
maxnlen       db       21           ;max. length
namelen       db       ?            ;no. chars entered
namefld       db       21 dup(?)    ;name
```

```
        crlf            db      13,10,'$'
        endaddr         dw      ?
        messg1          db      'Name?','$'
        messg2          db      'Sorted names:',13,10,'$'
        namectr         db      0
        nametab         db      30 dup(20 dup(' '))         ;name table
        namesav         db      20 dup(?),13,10,'$'
        swapped         db      0
;***************************************************************
        .code                                               ;define code segment
;---------------------------------------------------------------
begin           proc    far                                 ;main part of program
;set DS and ES register to current data segment
                mov     ax,@data                            ;data segment addr
                mov     ds,ax                               ;   into DS register
                mov     es,ax                               ;   and ES register

;MAIN PART OF PROGRAM GOES HERE
                cld
                lea     di,nametab
a20loop:
                call    b10read                             ;accept name
                cmp     namelen,0                           ;any more names?
                jz      a30                                 ;no, go to sort
                cmp     namectr,30                          ;30 names entered?
                je      a30                                 ;yes, go to sort
                call    d10stor                             ;store entered name in table
                jmp     a20loop                             ;end of input
a30:
                cmp     namectr,1                           ;one or no name entered?
                jbe     a40                                 ;yes, exit
                call    g10sort                             ;sort stored names
                call    k10disp                             ;display sorted names
a40:            mov     ax,4c00h                            ;terminate
                int     21h

begin           endp                                        ;end of main part of program
;---------------------------------------------------------------
;       Accept name as input:

b10read         proc    near
                mov     ah,09
```

```
                lea     dx,messg1           ;display prompt
                int     21h
                mov     ah,0ah
                lea     dx,namepar          ;accept name
                int     21h
                mov     ah,09
                lea     dx,crlf             ;return/linefeed
                int     21h
;
                mov     bh,0                ;clear chars after name
                mov     bl,namelen          ;get count of chars
                mov     cx,21
                sub     cx,bx               ;calc remaining length
b20:            mov     namefld[bx],20h     ;set to blank
                inc     bx
                loop    b20
                ret

b10read         endp
;--------------------------------------------------------------------------
;       Store name in table:

d10stor         proc    near
                inc     namectr             ;add to number of name
                cld
                lea     si,namefld
                mov     cx,10
                rep     movsw               ;move name to table
                ret

d10stor         endp
;--------------------------------------------------------------------------
;       Sort names in table:

g10sort         proc    near
                sub     di,40               ;set up stop address
                mov     endaddr,di
g20:            mov     swapped,0
                lea     si,nametab          ;set up start of table
g30:            mov     cx,20               ;length of compare
                mov     di,si
                add     di,20               ;next name for compare
```

```
                mov     ax,di
                mov     bx,si
                repe    cmpsb                   ;compare name to next
                jbe     g40                     ;no exchange
                call    h10xchg                 ;exchange
        g40:    mov     si,ax
                cmp     si,endaddr              ;end of table?
                jbe     g30                     ;no,continue
                cmp     swapped,0               ;any swaps?
                jnz     g20                     ;yes,continue
                ret                             ;no,end of sort

        g10sort endp
;--------------------------------------------------------------------
;       Exchange table entries:

        h10xchg proc    near
                mov     cx,10
                lea     di,namesav
                mov     si,bx
                rep     movsw                   ;move lower item to save
;
                mov     cx,10
                mov     di,bx
                rep     movsw                   ;move higher item to lower
;
                mov     cx,10
                lea     si,namesav
                rep     movsw                   ;move save to higher item
                mov     swapped,1               ;signal that exchange made
                ret

        h10xchg endp
;--------------------------------------------------------------------
;       Display sorted names:

        k10disp proc    near
                mov     ah,09                   ;display prompt
                lea     dx,messg2
                int     21h
                lea     si,nametab
        k20:    lea     di,namesav              ;initialize start of table
```

```
                mov     cx,10
                rep     movsw
                mov     ah,9
                lea     dx,namesav
                int     21h                 ;display
                dec     namectr             ;is this last one?
                jnz     k20                 ;no,loop
                ret                         ;yes,exit

k10disp         endp
;--------------------------------------------------------------------------
;**************************************************************
                end     begin
```

可以看出，这是一个典型的使用子程序结构的程序。所有具有独立功能的部分都采用了子程序结构，因而主程序设计得使人一目了然。主程序调用 4 个子程序，它们的过程名和功能如下：

① B10READ　接收键入的人名存放在 NAMEPAR 中，并用空格符清除其后的单元。

② D10STOR　把人名从 NAMEPAR 传送到 NAMETAB 中，并用 NAMECTR 计数。

③ G10SORT　用气泡排序算法对人名排序，并用 SWAPPED 作为交换标志控制循环的结束。其中调用子程序 H10XCHG 来交换两个人名串的位置，并设置 SWAPPED 标志。

④ K10DISP　显示已经排序了的人名。

显然，每个子程序的功能都是明确而独立的。可以看出，这样的程序结构是很清晰的。所以，读者在学完本章后，应能善于使用子程序结构来编写程序，这将会取得很好的效果。

本例中的所有主、子程序都在同一个源文件中，因而参数传送问题的处理是简单的，所有子程序都可访问数据段中的变量，而且也省略了寄存器的保存和恢复工作。

例 6.12　位串插入程序。程序要求把一个小于 32 位的位串插入存储器内的一个大位串中的任意位置中去。欲插入的位串存放在 bitsg 中，它是一个右对齐的位串，可称其为子串，其长度用 bitsg_length 为符号名的 = 伪操作来说明。大位串存放在 string 中，并为要插入的子串准备了一个符号名为 sg_end 的双字单元。这里用双字来定义位串的原因在于用双字处理位串可以获得更快的速度。

这一程序主要由移动大位串位位置和插入子串这两个子程序组成，程序框图见图 6.7，程序则见下面"例 6.12 位串插入程序"。主程序中除调用子程序外，主要是作进入子程序条件的判断。一是对子串长度的限制：0 < 子串长度 < 32；二是测试子串插入的位偏移（即在大位串中子串插入的起始位偏移值）与大位串位位置的关系。如果正好在大位串尾，则可直接把子串接在大位串之后；如果已超出大位串的范围，则退出程序不作处理；只有在大位串的范围之内时，才可调用子程序完成本程序所规定的任务。

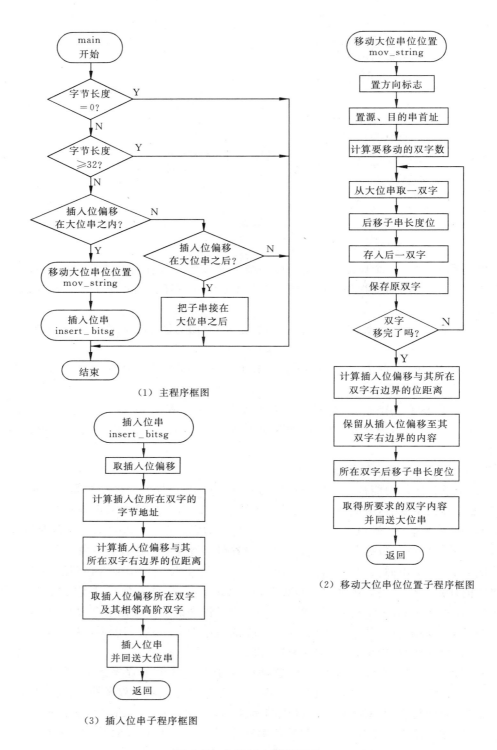

图 6.7 位串插入程序框图

移动大位串位位置子程序要完成的任务是为插入子串而移动大位串的位位置。这一工作分两步完成。

第一步,先移动好插入位偏移所在双字之后的其余双字,这些双字应从尾部开始后移,其后移位数应等于插入串长度(参见图 6.8)。这部分程序是在该子程序的 LOOP 指令之前。其中,要移动的双字数＝大位串双字数 — 插入位偏移所在双字地址。

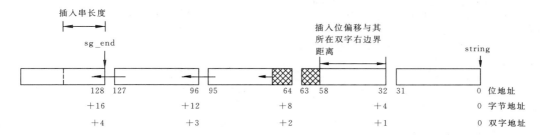

图 6.8 大位串及插入子串举例(位偏移为 58,子串长度为 15 位。)

第二步,移动插入位偏移所在双字。这里要注意的是,插入位偏移之前的各位应保持原位置不变,而其后各位应后移插入串长度位。

在这一子程序中,多处用到双字长移位指令 shld 或 shrd。应该注意该指令执行完后,目的寄存器得到所移入的新内容,而源寄存器则维持原状不变。

插入位串子程序的任务是按指定的插入位偏移值插入子串。也就是,用指定的子串取代原有该位置的值。在程序中,先取出插入位偏移所在的双字和与其相邻的高阶双字,然后用一串移位指令来达到既移入新子串,又不影响子串外原来两个双字的内容。

例 6.12 和例 6.11 一样,主、子程序都在同一个源文件中,因而参数传送和寄存器的保存与恢复工作都简化了。

例 6.12 位串插入程序

```
;Bit string insert--EX6_12#
;****************************************************************
            .model   small
            .386
            .stack   200h              ;define stack segment
;****************************************************************
            .data                      ;define data segment
    bitsg      dd    7fffh             ;substring
    string     dd    12345678h,12345678h,12345678h,12345678h
    sg_end     dd    ?                 ;a larger bit string
    bit_offset dd    58                ;the bit offset of the
                                       ;   start of substring
;****************************************************************
            bitsg_length=15            ;substring length
;****************************************************************
```

```
            .code                              ;define code segment
;------------------------------------------------------------------------
main        proc
start:                                         ;starting execution addr
            mov     ax,@data                   ;data segment addr
            mov     ds,ax                      ;   into DS register
            mov     es,ax                      ;   and ES register
;
            mov     cx,bitsg_length
            cmp     cx,0                       ;bitsg_length=0 ?
            je      exit                       ;exit if zero
            cmp     cx,32                      ;bitsg_length≥32 ?
            jae     exit                       ;exit if≥32
            mov     edi,bit_offset             ;load bit_offset
            mov     ecx,(sg_end-string)/4      ;larger bit string
                                               ;  length(dwords)
            shl     ecx,5                      ;larger bit string
                                               ;  length(bits)
            cmp     edi,ecx                    ;does the larger bit
                                               ;  string contain bit_
                                               ;  offset ?
            ja      exit                       ;exit if not
            jb      move                       ;yes,branch to move string
            mov     esi,bitsg                  ;bit_offset right in the
            mov     sg_end,esi                 ;  end of larger bit string,
            jmp     exit                       ;     insert substring
                                               ;        and exit
move:       call    mov_string                 ;move larger bit string
            call    insert_bitsg               ;insert bit string
;
exit:       mov     ax,4c00h                   ;go back to DOS
            int     21h
;
main        endp
;------------------------------------------------------------------------
mov_string  proc
            sub     eax,eax                    ;put zero in EAX
            std                                ;set direction flag
            mov     si,offset sg_end-4         ;initialize source addr
            mov     di,offset sg_end           ;initialize dest. addr
            mov     ecx,(sg_end-string)/4      ;larger bit string
                                               ;  length(dwords)
```

```
                mov     ebx,bit_offset          ;load bit_offset
                shr     ebx,5                   ;   divide by 32
                sub     ecx,ebx                 ;the number of dwords
                                                ;   to be moved
next:           mov     ebx,[si]                ;move dwords which are
                                                ;   behind the dword contains
                shld    eax,ebx,bitsg_length    ;     the bit_offset to be
                stosd                           ;       inserted
                mov     eax,ebx
                sub     si,4
                loop    next
                sub     ebx,ebx                 ;move dword which contains
                sub     edx,edx                 ;   the bit_offset to be
                mov     ecx,bit_offset          ;     inserted
                and     cl,1fh
                shrd    ebx,eax,cl
                shld    edx,ebx,cl
                shl     eax,bitsg_length
                mov     ebx,-1
                shl     ebx,cl
                and     eax,ebx
                or      eax,edx
                mov     [edi],eax
                ret                             ;return from mov_string
mov_string      endp
;----------------------------------------------------------------------
insert_bitsg    proc
                mov     esi,bitsg               ;load substring
                mov     edi,bit_offset          ;load bit_offset
                mov     ecx,edi
                shr     edi,5                   ;   divide by 32(dwords)
                shl     edi,2                   ;   multiply by 4(byte addr)
                and     cl,1fh                  ;distance between the bit_
                                                ;   offset to be inserted
                                                ;     and the right boundary
                                                ;       of the dword contains
                                                ;         this bit_offset to
                                                ;           be inserted
                mov     eax,string[edi]         ;load low dword into EAX
                mov     edx,string+4[edi]       ;load high dword into EDX
                mov     ebx,eax                 ;insert bit string
                shrd    eax,edx,cl
```

```
            shrd        edx,ebx,cl
            shrd        eax,esi,bitsg_length
            rol         eax,bitsg_length
            mov         ebx,eax
            shld        eax,edx,cl
            shld        edx,ebx,cl
            mov         string[edi],eax
            mov         string+4[edi],edx
            ret                                         ;return from insert_bitsg
insert_bitsg    endp
;--------------------------------------------------------------------------------
;************************************************************************

            end         start
```

习　　题

6.1 下面的程序段有错吗？若有，请指出错误。

```
CRAY    PROC
        PUSH    AX
        ADD     AX,BX
        RET
ENDP    CRAY
```

6.2 已知堆栈寄存器 SS 的内容是 0F0A0H，堆栈指示器 SP 的内容是 00B0H，先执行两条把 8057H 和 0F79BH 分别入栈的 PUSH 指令，然后执行一条 POP 指令。试画出示意图说明堆栈及 SP 内容的变化过程。

6.3 分析下面"6.3题的程序"，画出堆栈最满时各单元的地址及内容。

6.3 题的程序

```
;************************************************************
s_seg       segment     at 1000h                ;define stack segment
            dw          200 dup(?)
   tos      label       word
s_seg       ends
;************************************************************
c_seg       segment                             ;define code segment
            assume      cs:c_seg,ss:s_seg
            mov         ax,s_seg
            mov         ss,ax
            mov         sp,offset tos
```

```
;
                push        ds
                mov         ax,0
                push        ax
                   ⋮
                push        t_addr
                push        ax
                pushf
                   ⋮
                popf
                pop         ax
                pop         t_addr
                ret
c_seg           ends                                ;end of code segment
;****************************************************
                end         c_seg                   ;end of assembly
```

6.4 分析下面"6.4 题的程序"的功能,写出堆栈最满时各单元的地址及内容。

6.4 题的程序

```
;****************************************************
stack           segment at 500h
                dw          128     dup(?)
   tos          label       word
stack           ends
;****************************************************
code            segment                             ;define code segment
;--------------------------------------------------------------------------------
main            proc        far                     ;main part of program
                assume cs:code,ss:stack
start:                                              ;starting execution address
                mov         ax,stack
                mov         ss,ax
                mov         sp,offset tos
;
                push        ds
                sub         ax,ax
                push        ax
;MAIN PART OF PROGRAM GOES HERE
                mov         ax,4321h
                call        htoa
                ret                                 ;return to DOS
main            endp                                ;end of main part of program
;--------------------------------------------------------------------------------
```

			htoa	proc	near	;define subprocedure htoa
				cmp	ax,15	
				jle	bl	
				push	ax	
				push	bp	
				mov	bp,sp	
				mov	bx,[bp+2]	
				and	bx,000fh	
				mov	[bp+2],bx	
				pop	bp	
				mov	cl,4	
				shr	ax,cl	
				call	htoa	
				pop	ax	
			bl:	add	al,30h	
				cmp	al,3ah	
				jl	printit	
				add	al,7h	
			printit:	mov	dl,al	
				mov	ah,2	
				int	21h	
				ret		
			htoa	endp		;end of subprocedure

;--

			code	ends		;end of code segment

;**

				end	start	;end of assembly

6.5 下面是 6.5 题的程序清单，请在清单中填入此程序执行过程中的堆栈变化。

;**

0000				stacksg	segment	
0000	20	[		dw	32 dup(?)	
		????				
		]				
0040				stacksg	ends	

;**

0000				codesg	segment para 'code'	

;--

0000				begin	proc	far
				assume cs:codesg,ss:stacksg		
0000	1E			push	ds	
0001	2B	C0		sub	ax,ax	

0003	50		push	ax
0004	E8 0008 R		call	b10

;--

0007	CB		ret	
0008		begin	endp	

;--

0008		b10	proc	
0008	E8 000C R		call	c10

;--

000B	C3		ret	
000C		b10	endp	

;--

000C		c10	proc	

;--

000C	C3		ret	
000D		c10	endp	

;--

000D		codesg	ends	

;**
         end begin

偏移地址　堆栈

（图：堆栈格子，两组，每组四列六行，下方 SP 指示框）

6.6 写一段子程序 SKIPLINES，完成输出空行的功能。空行的行数在 AX 寄存器中。

6.7 设有 10 个学生的成绩分别是 76,69,84,90,73,88,99,63,100 和 80 分。试编

制一个子程序统计 60～69 分，70～79 分，80～89 分，90～99 分和 100 分的人数并分别存放到 S6,S7,S8,S9 和 S10 单元中。

6.8 编写一个有主程序和子程序结构的程序模块。子程序的参数是一个 N 字节数组的首地址 TABLE,数 N 及字符 CHAR。要求在 N 字节数组中查找字符 CHAR,并记录该字符的出现次数。主程序则要求从键盘接收一串字符以建立字节数组 TABLE,并逐个显示从键盘输入的每个字符 CHAR 以及它在 TABLE 数组中出现的次数。（为简化起见,假设出现次数≤15,可以用十六进制形式把它显示出来）

6.9 编写一个子程序嵌套结构的程序模块,分别从键盘输入姓名及 8 个字符的电话号码,并以一定的格式显示出来。

主程序 TELIST：
  (1) 显示提示符 INPUT NAME：；
  (2) 调用子程序 INPUT_NAME 输入姓名；
  (3) 显示提示符 INPUT A TELEPHONE NUMBER：；
  (4) 调用子程序 INPHONE 输入电话号码；
  (5) 调用子程序 PRINTLINE 显示姓名及电话号码。

子程序 INPUT_NAME：
  (1) 调用键盘输入子程序 GETCHAR,把输入的姓名存放在 INBUF 缓冲区中；
  (2) 把 INBUF 中的姓名移入输出行 OUTNAME。

子程序 INPHONE：
  (1) 调用键盘输入子程序 GETCHAR,把输入的 8 位电话号码存放在 INBUF 缓冲区中；
  (2) 把 INBUF 中的号码移入输出行 OUTPHONE。

子程序 PRINTLINE：
 显示姓名及电话号码,格式为：
   NAME   TEL.
   ×××   ×××

6.10 编写子程序嵌套结构的程序,把整数分别用二进制和八进制形式显示出来。
主程序 BANDO：把整数字变量 VAL1 存入堆栈,并调用子程序 PAIRS；
子程序 PAIRS：从堆栈中取出 VAL1；调用二进制显示程序 OUTBIN 显示出与其等效的二进制数；输出 8 个空格；

 调用八进制显示程序 OUTCT 显示出与其等效的八进制数；调用输出回车及换行符的子程序。

6.11 假定一个名为 MAINPRO 的程序要调用子程序 SUBPRO,试问：
  (1) MAINPRO 中的什么指令告诉汇编程序 SUBPRO 是在外部定义的？
  (2) SUBPRO 怎么知道 MAINPRO 要调用它？

6.12 假定程序 MAINPRO 和 SUBPRO 不在同一模块中,MAINPRO 中定义字节变量 QTY 和字变量 VALUE 和 PRICE。SUBPRO 程序要把 VALUE 除以 QTY,并把商存在 PRICE 中。试问：

(1) MAINPRO 怎么告诉汇编程序外部子程序要调用这三个变量？

(2) SUBPRO 怎么告诉汇编程序这三个变量是在另一个汇编语言程序定义的？

6.13 假设：

(1) 在模块 1 中定义了双字变量 VAR1，首地址为 VAR2 的字节数组和 NEAR 标号 LAB1，它们将由模块 2 和模块 3 所使用；

(2) 在模块 2 中定义了字变量 VAR3 和 FAR 标号 LAB2，而模块 1 中要用到 VAR3，模块 3 中要用到 LAB2；

(3) 在模块 3 中定义了 FAR 标号 LAB3，而模块 2 中要用到它。

试对每个源模块给出必要的 EXTRN 和 PUBLIC 说明。

6.14 主程序 CALLMUL 定义堆栈段、数据段和代码段，并把段寄存器初始化；数据段中定义变量 QTY 和 PRICE；代码段中将 PRICE 装入 AX，QTY 装入 BX，然后调用子程序 SUBMUL。程序 SUBMUL 没有定义任何数据，它只简单地把 AX 中的内容（PRICE）乘以 BX 中的内容（QTY），乘积放在 DX，AX 中。请编制这两个要连接起来的程序。

6.15 试编写一个执行以下计算的子程序 COMPUTE：

$$R \leftarrow X + Y - 3$$

其中 X，Y 及 R 均为字数组。假设 COMPUTE 与其调用程序都在同一代码段中，数据段 D_SEG 中包含 X 和 Y 数组，数据段 E_SEG 中包含 R 数组，同时写出主程序调用 COMPUTE 过程的部分。

如果主程序和 COMPUTE 在同一程序模块中，但不在同一代码段中，程序应如何修改？

如果主程序和 COMPUTE 不在同一程序模块中，程序应如何修改？

第 7 章 高级汇编语言技术

7.1 宏 汇 编

在第 6 章中介绍了子程序结构,并了解到,使用子程序结构具有很多优点:可以节省存储空间及程序设计所花的时间,可提供模块化程序设计的条件,便于程序的调试及修改等。但是,使用子程序也有一些缺点:为转子及返回、保存及恢复寄存器以及参数的传送等都要增加程序的开销,这些操作所消耗的时间以及它们所占用的存储空间,都是为取得子程序结构使程序模块化的优点而增加的额外开销。因此,有时特别在子程序本身较短或者是需要传送的参数较多的情况下,使用宏汇编就更加有利。

7.1.1 宏定义、宏调用和宏展开

宏是源程序中一段有独立功能的程序代码。它只需要在源程序中定义一次,就可以多次调用它,调用时只需要用一个宏指令语句就可以了。

宏定义是用一组伪操作来实现的。其格式是:

```
macro name    MACRO    [dummy parameter list]
              ⋮        (宏定义体)
              ENDM
```

其中 MACRO 和 ENDM 是一对伪操作。这对伪操作之间是宏定义体——是一组有独立功能的程序代码。宏指令名(macro name)给出该宏定义的名称,调用时就使用宏指令名来调用该宏定义。宏指令名的第一个符号必须是字母,其后可以跟字母、数字或下划线字符。其中哑元表(dummy parameter list)给出了宏定义中所用到的形式参数(或称虚参),每个哑元之间用逗号隔开。

经宏定义定义后的宏指令就可以在源程序中调用。这种对宏指令的调用称为宏调用,宏调用的格式是:

```
macro name    [actual parameter list]
```

实元表(actual parameter list)中的每一项为实元,相互之间用逗号隔开。

当源程序被汇编时,汇编程序将对每个宏调用作宏展开。宏展开就是用宏定义体取代源程序中的宏指令名,而且用实元取代宏定义中的哑元。在取代时,实元和哑元是一一对应的,即第一个实元取代第一个哑元,第二个实元取代第二个哑元……依次类推。一般说来,实元的个数应该和哑元的个数相等,但汇编程序并不要求它们必须相等。若实元个数大于哑元个数,则多余的实元不予考虑;若实元个数小于哑元个数,则多余的哑元作"空"处理。

应该注意：宏展开后，即用实元取代哑元后，所得到的语句应该是有效的，否则汇编程序将会指示出错。

下面用一个例子来说明宏定义、宏调用和宏展开的情况。

例 7.1 用宏指令定义两个字操作数相乘，得到一个 16 位的第三个操作数作为结果。宏定义：

MULTIPLY	MACRO	OPR1,OPR2,RESULT
	PUSH	DX
	PUSH	AX
	MOV	AX,OPR1
	IMUL	OPR2
	MOV	RESULT,AX
	POP	AX
	POP	DX
	ENDM	

宏调用：

$\vdots$

MULTIPLY   CX,VAR,XYZ[BX]

$\vdots$

MULTIPLY   240,BX,SAVE

$\vdots$

宏展开：

$\vdots$

1	PUSH	DX
1	PUSH	AX
1	MOV	AX,CX
1	IMUL	VAR
1	MOV	XYZ[BX],AX
1	POP	AX
1	POP	DX

$\vdots$

1	PUSH	DX
1	PUSH	AX
1	MOV	AX,240
1	IMUL	BX
1	MOV	SAVE,AX
1	POP	AX
1	POP	DX

$\vdots$

汇编程序在所展开的指令前加上 1 表示这些指令是由宏展开而得到的(较早的版本用＋符号表示)。从上面的例子可以看出：由于宏指令可以带哑元，调用时可以用实元取代，这就避免了子程序因参数传送带来的麻烦，使宏汇编的使用增加了灵活性。而且，实元可以是常数、寄存器、存储单元名以及用寻址方式能找到的地址或表达式等。从以后的例子中将可看到，实元还可以是指令的操作码或操作码的一部分等，宏汇编的这一特性是子程序所不及的。但是，宏调用的工作方式和子程序调用的工作方式是完全不同的，图 7.1 说明了两者的区别。

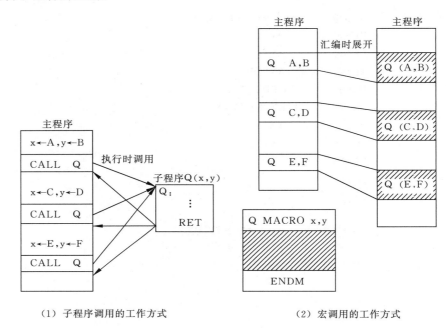

(1) 子程序调用的工作方式　　　　(2) 宏调用的工作方式

图 7.1　子程序调用和宏调用工作方式的区别

可以看出，子程序是在程序执行期间由主程序调用的，它只占有它自身大小的一个空间；而宏调用则是在汇编期间展开的，每调用一次就把宏定义体展开一次，因而它占有的存储空间与调用次数有关，调用次数越多则占有的存储空间也就越大。前面已经提到，用宏汇编可以免去执行时间上的额外开销，但如果宏调用次数较多的话，则其空间上的开销也是应该考虑的因素。因此，读者可以根据具体情况来选择使用方案。一般说来，由于宏汇编可能占用较大的空间，所以代码较长的功能段往往使用子程序而不用宏汇编；而那些较短的且变元较多的功能段，则使用宏汇编就更为合理了。

注意：在程序中，宏定义必须出现在宏调用之前，也就是说必须先定义后调用。因此，往往把宏定义放在程序中所有段的代码之前，即在程序一开始先列出你的程序中所用的所有宏定义，以便在程序中调用。

7.1.2 宏定义中的参数

下面用例子来说明宏定义中的参数。
例 7.2 宏定义可以无变元。
宏定义：

```
SAVEREG     MACRO
            PUSH    AX
            PUSH    BX
            PUSH    CX
            PUSH    DX
            PUSH    SI
            PUSH    DI
            ENDM
```

宏调用：

```
SAVEREG
```

宏展开则将宏定义体的内容全部列出。

例 7.3 变元可以是操作码。
宏定义：

```
FOO         MACRO   P1,P2,P3
            MOV     AX,P1
            P2      P3
            ENDM
```

宏调用：

```
FOO         WORD_VAR,INC,AX
```

宏展开：

```
1           MOV     AX,WORD_VAR
1           INC     AX
```

例 7.4 变元可以是操作码的一部分，但在宏定义体中必须用 & 作为分隔符。
宏定义：

```
LEAP        MACRO   COND,LAB
            J&COND  LAB
            ENDM
```

宏调用：

⋮

```
            LEAP        Z,THERE
                         ⋮
            LEAP        NZ,HERE
                         ⋮
```

宏展开：

```
                         ⋮
   1        JZ          THERE
                         ⋮
   1        JNZ         HERE
                         ⋮
```

&是一个操作符，它在宏定义体中可以作为哑元的前缀，展开时可以把&前后两个符号合并而形成一个符号，这个符号可以是操作码、操作数或是一个字符串。下面用例7.5和例7.6进一步具体说明这个问题。

例7.5

宏定义：

```
   FO           MACRO       P1
                JMP         TA&P1
                ENDM
```

宏调用：

```
   FO           WORD_VAR
```

宏展开：

```
   1            JMP         TAWORD_VAR
```

在这里，如果宏定义写为

```
   FO           MACRO       P1
                JMP         TAP1
                ENDM
```

则在展开时，汇编程序把TAP1看作是一个独立的标号，并不把TAP1中的P1作为哑元看待，这样就不能得到预期的结果。

例7.6 变元是ASCII串。

宏定义：

```
   MSGGEN       MACRO       LAB,NUM,XYZ
                LAB&NUM     DB     'HELLO MR. &XYZ'
                ENDM
```

宏调用：

```
                MSGGEN          MSG,1,TAYLOR
```
宏展开：
```
     1          MSG1            DB      'HELLO MR.TAYLOR'
```

例 7.7 这里再介绍一下宏定义的变元中使用 % 操作符的情况，它的格式是：

```
     %  expression
```

汇编程序把跟在%之后的表达式的值转换成当前基数下的数，在展开期间，用这个数来取代哑元。

宏定义：
```
     MSG             MACRO           COUNT,STRING
                     MSG&COUNT       DB      STRING
                     ENDM
     ERRMSG          MACRO           TEXT
                     CNTR=CNTR+1
                     MSG             %  CNTR,TEXT
                     ENDM
```

宏调用：
```
                ⋮
     CNTR=0
     ERRMSG          'SYNTAX ERROR'
                ⋮
     ERRMSG          'INVALID OPERAND'
                ⋮
```

宏展开：
```
                ⋮
     2          MSG1            DB      'SYNTAX ERROR'
                ⋮
     2          MSG2            DB      'INVALID OPERAND'
                ⋮
```

其中，2 表示它是第二层展开的结果。一般在 LST 清单中，使用隐含的 .XALL 伪操作，即不产生代码的语句在清单中并不列出。为了能看到宏展开后所有的语句，可以在源程序中增加 .LALL 语句，此时就能看到所有的语句了。实际上，本例的展开可分为如下两层：

```
                        ⋮
                    CNTR=0
     1              CNTR=CNTR+1
```

```
1          MSG              %     CNTR, 'SYNTAX ERROR'
2          MSG1             DB    'SYNTAX ERROR'
                  ⋮
1          CNTR=CNTR+1
1          MSG              %     CNTR, 'INVALID OPERAND'
2          MSG2             DB    'INVALID OPERAND'
                  ⋮
```

7.1.3 LOCAL 伪操作

例 7.8 宏定义体内允许使用标号,如

宏定义:

```
ABSOL      MACRO      OPER
           CMP        OPER,0
           JGE        NEXT
           NEG        OPER
NEXT:
           ENDM
```

如果程序中多次调用该宏定义时,则展开后会出现标号的多重定义,这是不能允许的。为此,系统提供了 LOCAL 伪操作,其格式是:

```
LOCAL      list of local labels
```

其中局部标号表内的各标号之间用逗号隔开。汇编程序对 LOCAL 伪操作的局部标号表中的每一个局部标号建立惟一的符号(用??0000～??FFFF)以代替在展开中存在的每一个局部标号。

必须注意:LOCAL 伪操作只能用在宏定义体内,而且它必须是 MACRO 伪操作后的第一个语句,在 MACRO 和 LOCAL 伪操作之间还不允许有注释和分号标志。

本例中的 ABSOL 宏定义在考虑到有多次调用可能性的情况下,应定义为:

```
ABSOL      MACRO      OPER
           LOCAL      NEXT
           CMP        OPER,0
           JGE        NEXT
           NEG        OPER
NEXT:
           ENDM
```

宏调用:

```
                    ⋮
            ABSOL       VAR
                    ⋮
            ABSOL       BX
                    ⋮
```

宏展开：

```
                    ⋮
1           CMP         VAR,0
1           JGE         ??0000
1           NEG         VAR
1??0000:
                    ⋮
1           CMP         BX,0
1           JGE         ??0001
1           NEG         BX
1??0001:
                    ⋮
```

在以上的例子中，宏定义体内只用了一个标号，如宏定义体内的标号数多于一个，则可把它们列在 LOCAL 伪操作之后，如

```
LOCAL    NEXT,OUT,EXIT
```

在宏展开时，汇编程序对第一次宏调用使用??0000 取代 NEXT，用??0001 取代 OUT，用??0002 取代 EXIT。对第二次宏调用将用??0003 取代 NEXT，用??0004 取代 OUT，用??0005 取代 EXIT。

7.1.4 在宏定义内使用宏

例 7.9 宏定义中允许使用宏调用，其限制条件是：必须先定义后调用。

宏定义：

```
DIF         MACRO       X,Y
            MOV         AX,X
            SUB         AX,Y
            ENDM
DIFSQR      MACRO       OPR1,OPR2,RESULT
            PUSH        DX
            PUSH        AX
            DIF         OPR1,OPR2
            IMUL        AX
            MOV         RESULT,AX
```

```
                        POP         AX
                        POP         DX
                        ENDM
```

宏调用：

```
        DIFSQR          VAR1,VAR2,VAR3
```

宏展开：

```
        1               PUSH        DX
        1               PUSH        AX
        2               MOV         AX,VAR1
        2               SUB         AX,VAR2
        1               IMUL        AX
        1               MOV         VAR3,AX
        1               POP         AX
        1               POP         DX
```

在汇编后形成的 LST 清单中，可以看到上述宏展开的结果。

例 7.10 这是又一个宏定义体中使用宏调用的例子。

宏定义：

```
        INT21           MACRO       FUNCTN
                        MOV         AH,FUNCTN
                        INT         21H
                        ENDM
        DISP            MACRO       CHAR
                        MOV         DL,CHAR
                        INT21       02H
                        ENDM
```

宏调用：

```
        DISP            '?'
```

宏展开：

```
        1               MOV         DL,'?'
        2               MOV         AH,02H
        2               INT         21H
```

例 7.11 宏定义体内不仅可以使用宏调用，也可以包含宏定义。

宏定义：

```
        DEFMAC          MACRO       MACNAM,OPERATOR
                        MACNAM      MACRO       X,Y,Z
                        PUSH        AX
```

```
                MOV         AX,X
                OPERATOR    AX,Y
                MOV         Z,AX
                POP         AX
                ENDM
    ENDM
```

其中 MACNAM 是内层的宏定义名,但又是外层宏定义的哑元,所以当调用 DEFMAC 时,就形成一个宏定义。

宏调用:

```
    DEFMAC      ADDITION,ADD
```

形成加法宏定义:

```
    ADDITION    MACRO       X,Y,Z
                PUSH        AX
                MOV         AX,X
                ADD         AX,Y
                MOV         Z,AX
                POP         AX
                ENDM
```

同样,宏调用

```
    DEFMAC      SUBTRACT,SUB
```

可形成减法宏定义。宏调用

```
    DEFMAC      LOGOR,OR
```

可形成逻辑或宏定义等。当然,在形成这些宏定义后,就可以使用宏调用

```
    ADDITION    VAR1,VAR2,VAR3
```

而展开成:

```
    1           PUSH        AX
    1           MOV         AX,VAR1
    1           ADD         AX,VAR2
    1           MOV         VAR3,AX
    1           POP         AX
```

7.1.5 列表伪操作

MASM 提供.XALL,.LALL 和.SALL 来控制汇编清单中宏展开的列出情况。上面已经说明.XALL 为默认情况,它指示清单中只列出产生目标码的宏展开;.LALL 列

出包括注释在内的所有宏展开；.SALL 则不列出任何展开信息。在宏定义中的注释除和一般源程序中采用分号；作标志外，还有一种采用双分号；；的注释，这种；；注释在宏展开时将不予展开，即使在 .LALL 之下在清单中也不会出现。MASM 6.0 分别用 .LISTMACROALL，.LISTMACRO 和 .NOLISTMACRO 来取代 .LALL，.XALL 和 .SALL。

例 7.12 作为宏汇编以及列表伪操作的使用举例见"例 7.12 的源程序"和"例 7.12 代码段的 LST 清单"。

例 7.12 的源程序

```
;defining and using macro--ex7_12_1
;-----------------------------------------------------------------
initz   macro                   ;define macro
        mov     ax,@data        ;initialize segment
        mov     ds,ax           ;  register
        mov     es,ax
        endm                    ;end macro
;
prompt  macro   messge          ;define macro
;       This macro displays any message
;;      Generates code that links to DOS
        mov     ah,09h          ;request display
        lea     dx,messge       ;  prompt
        int     21h
        endm                    ;end macro
;
finish  macro                   ;define macro
        mov     ax,4c00h        ;end processing
        int     21h
        endm                    ;end macro
;-----------------------------------------------------------------
        .model  small
        .386
        .stack  200h            ;define stack segment
        .data                   ;define data segment
messg1  db      'Customer name?',13,10,'$'
messg2  db      'Customer address?',13,10,'$'
;-----------------------------------------------------------------
        .code                   ;define code segment
begin   proc    far
        .sall
```

```
                initz
                .xall
                prompt    messg1
                .lall
                prompt    messg2
                .xall
                finish
begin           endp
;------------------------------------------------------------
                end       begin
```

例7.12 代码段的 LST 清单

```
                        ;defining and using macro--ex7_12_1
                        ;------------------------------------------------------------
                        initz   macro              ;define macro
                                mov    ax,@data    ;initialize segment
                                mov    ds,ax       ;  register
                                mov    es,ax
                                endm               ;end macro
                        ;
                        prompt macro  messge       ;define macro
                        ;      This macro displays any message
                        ;;     Generates code that links to DOS
                                mov    ah,09h      ;request display
                                lea    dx,messge   ;  prompt
                                int    21h
                                endm               ;end macro
                        ;
                        finish  macro              ;define macro
                                mov    ax,4c00h    ;end processing
                                int    21h
                                endm               ;end macro
                        ;------------------------------------------------------------
                                .model  small
                                .386
0200                            .stack  200h       ;define stack segment
0000                            .data              ;define data segment
0000   43 75 73 74 6F 6D 65     messg1 db 'Customer name?',13,10,'$'
       72 20 6E 61 6D 65 3F
       0D 0A 24
0011   43 75 73 74 6F 6D 65     messg2 db 'Customer address?',13,10,'$'
```

```
                    72 20 61 64 64 72 65
                    73 73 3F 0D 0A 24
                                        ;--------------------------------------------------
    0000                                        .code               ;define code segment
    0000                            begin   proc    far
                                            .sall
                                    initz
                                            .xall
                                    prompt messg1
    0007    B4 09                   1       mov     ah,09h      ;request display
    0009    8D 16 0000 R            1       lea     dx,messg1  ;  prompt
    000D    CD 21                   1       int     21h
                                            .lall
                                    prompt messg2
                                    1   ;           This macro displays any message
                                    1   ;
    000F    B4 09                   1       mov     ah,09h      ;request display
    0011    8D 16 0011 R            1       lea     dx,messg2  ;  prompt
    0015    CD 21                   1       int     21h
                                            .xall
                                    finish
    0017    B8 4C00                 1       mov     ax,4c00h    ;end processing
    001A    CD 21                   1       int     21h
    001C                            begin   endp
                                        ;--------------------------------------------------
    001C                                    end     begin
```

7.1.6 宏库的建立与调用

有时在程序里定义了较多宏,或者可以把自己编程中常用的宏定义建立成一个独立的文件,这个只包含若干宏定义的文件可称为宏库,通常用扩展名 MAC 或 INC 来表示。当应用程序中需要用到宏库中的某些宏定义时,只需要在该程序的开始用 INCLUDE 语句说明如下:

INCLUDE   C:\MACRO.MAC

其中 MACRO.MAC 为宏库名,它存在 C 盘中(这里也可以给出所建的宏库名)。汇编程序将把宏库中的所有宏定义都包含在应用程序中。把"例 7.12 的源程序"中的宏定义部分建成宏库 MACRO.MAC,见下面的"为例 7.12 建立的宏库 MACRO.MAC";而把例7.12 的程序编码部分,加上 INCLUDE 语句构成新程序,见"例 7.12 使用宏库情况下程序实现"。其中 INCLUDE 语句在程序的最前面,其实该语句可以不放在程序的最前面,但它必须放在所有的宏调用之前。从"例 7.12 使用宏库情况下的程序的部分 LST 清单"中,可见汇编程序已将宏库中的宏定义都包含在此程序中了,并且完成了宏展开任务。

为例 7.12 建立的宏库 MACRO.MAC

```
;a library of macro sequences--macro.mac
;------------------------------------------------------------
initz    macro                ;define macro
         mov      ax,@data    ;initialize segment
         mov      ds,ax       ;   register
         mov      es,ax
         endm                 ;end macro
;
prompt   macro    messge      ;define macro
;        This macro displays any message
;;       Generates code that links to DOS
         mov      ah,09h      ;request display
         lea      dx,messge   ;   prompt
         int      21h
         endm                 ;end macro
;
finish   macro                ;define macro
         mov      ax,4c00h    ;end processing
         int      21h
         endm                 ;end macro
;------------------------------------------------------------
```

使用宏库情况下的例 7.12 程序实现

```
;using macro--ex7_12_2
;------------------------------------------------------------
         include macro.mac
;------------------------------------------------------------
         .model   small
         .386
         .stack   200h        ;define stack segment
         .data                ;define data segment
messg1   db       'Customer name?',13,10,'$'
messg2   db       'Customer address?',13,10,'$'
;------------------------------------------------------------
         .code                ;define code segment
begin    proc     far
         initz
```

```
                prompt     messg1
                prompt     messg2
                finish
begin   endp
;----------------------------------------------------------------
                end        begin
```

使用宏库情况下的例7.12程序的部分 LST 清单

```
                          ;using macro--ex7_12_2
                          ;----------------------------------------------------------------
                                      include macro.mac
     C                    ;a library of macro sequences--macro.mac

     C                    ;----------------------------------------------------------------
     C    initz     macro                    ;define macro
     C              mov        ax,@data      ;initialize segment
     C              mov        ds,ax         ;    register
     C              mov        es,ax
     C              endm                     ;end macro
     C    ;
     C    prompt    macro      messge        ;define macro
     C    ;                    This macro displays any message
     C    ;;                   Generates code that links to DOS
     C              mov        ah,09h        ;request display
     C              lea        dx,messge     ;    prompt
     C              int        21h
     C              endm                     ;end macro
     C    ;
     C    finish    macro                    ;define macro
     C              mov        ax,4c00h      ;end processing
     C              int        21h
     C              endm                     ;end macro
     C    ;----------------------------------------------------------------
                              .model     small
                              .386
0200                          .stack     200h           ;define stack segment
0000                          .data                     ;define data segment
0000  43 75 73 74 6F 6D 65    messg1 db   'Customer name?',13,10,'$'
      72 20 6E 61 6D 65 3F
      0D 0A 24
```

0011	43 75 73 74 6F 6D 65		messg2 db		'Customer address?',13,10,'$'	
	72 20 61 64 64 72 65					
	73 73 3F 0D 0A 24					

```
                        ;------------------------------------------------
0000                            .code              ;define code segment
0000                    begin   proc    far
                        initz
0000   B8 ---- R    1           mov     ax,@data   ;initialize segment
0003   8E D8        1           mov     ds,ax      ;   register
0005   8E C0        1           mov     es,ax
                        prompt  messg1
0007   B4 09        1           mov     ah,09h     ;request display
0009   8D 16 0000 R 1           lea     dx,messg1  ;   prompt
000D   CD 21        1           int     21h
                        prompt  messg2
000F   B4 09        1           mov     ah,09h     ;request display
0011   8D 16 0011 R 1           lea     dx,messg2  ;   prompt
0015   CD 21        1           int     21h
                        finish
0017   B8 4C00      1           mov     ax,4c00h   ;end processing
001A   CD 21        1           int     21h
001C                    begin   endp
                        ;------------------------------------------------
001C                            end     begin
```

7.1.7 PURGE 伪操作

PURGE 伪操作用来删除不用的宏定义。例如,用 INCLUDE 语句调用宏库时可以用 PURGE 伪操作删除在调用程序中不用的宏定义。当然它只在调用程序中起作用而不会影响宏库的内容。它的格式是:

  PURGE macro_name[,macro_name,...]

例如在"使用宏库情况下的例 7.12 程序实现"中,如果不用 prompt,则可在 INCLUDE 语句后加上 PURGE PROMPT 就可达到目的。此外,在程序中也可以用 PURGE 伪操作来删除一个不再使用的宏定义。删除宏定义的含义是使该宏定义成为空,程序中如果出现了一个已被删除宏定义的宏调用,则汇编程序将不会指示出错,但它将忽略该宏调用,当然也不会予以展开。

7.2 重复汇编

有时汇编语言程序需要连续地重复完成相同的或者几乎完全相同的一组代码,这时可使用重复汇编。

7.2.1 重复伪操作

重复伪操作的格式是：

REPT　　expression
　⋮　　　（重复块）
ENDM

其中表达式的值用来确定重复块的重复次数，表达式中如包含外部或未定义的项则汇编指示出错。

重复伪操作并不一定要用在宏定义体内。下面举例说明重复伪操作的使用方法。

例 7.13

```
X=0
         REPT    10
X=X+1
         DB      X
         ENDM
```

则汇编后产生

```
1        DB      1
1        DB      2
1        DB      3
          ⋮
1        DB      10
```

例 7.14 把字符 A 到 Z 的 ASCII 码填入数组 TABLE。

```
CHAR='A'
TABLE    LABLE   BYTE
         REPT    26
         DB      CHAR
CHAR=CHAR+1
         ENDM
```

经汇编产生

```
1        DB      41H
1        DB      42H
          ⋮
1        DB      5AH
```

例 7.15 用宏定义及重复伪操作把 TAB,TAB+2,TAB+4,⋯,TAB+8 的内容存

入堆栈。

宏定义：

    PUSH_TAB   MACRO   K
                      PUSH   TAB+K
                      ENDM

宏调用：

    I＝0
                      REPT   5
                      PUSH_TAB   ％I
    I＝I+2
                      ENDM

宏展开：

    2   PUSH   TAB+0
    2   PUSH   TAB+2
    2   PUSH   TAB+4
    2   PUSH   TAB+6
    2   PUSH   TAB+8

例 7.16 要求建立一个 8 字的数组，其中每个字的内容是下一个字的地址，而最后一个字的内容是第一个字的地址。

    ARRAY   LABLE WORD
                  REPT   7
                  DW $＋2
                  ENDM
                  DW   ARRAY

经汇编后得

    1   DW $＋2 ⎫
    1   DW $＋2 ⎬ 7个字
                ⋮
    1   DW $＋2 ⎭
               DW   ARRAY

如果数组的首地址为 0034，则建立的 8 字数组如下：

地址	内容
0034	0036
0036	0038
0038	003A
003A	003C

003C	003E
003E	0040
0040	0042
0042	0034

7.2.2 不定重复伪操作

1. IRP 伪操作

格式是：

```
IRP         dummy,⟨argument list⟩
   ⋮              （重复块）
ENDM
```

汇编程序把重复块的代码重复几次,每次重复把重复块中的哑元用自变量表中的一项来取代,下一次取代下一项,重复次数由自变量表中的自变量个数来确定。自变量表必须用尖括号括起来,它可以是常数、符号和字符串等。

IRP 和下面要讲到的 IRPC 都和 REPT 一样,不一定要用在宏定义中。下面用例子说明其使用方法。

例 7.17

```
IRP         X,⟨1,2,3,4,5,6,7,8,9,10⟩
DB          X
ENDM
```

汇编后得：

1	DB	1
1	DB	2
	⋮	
1	DB	10

例 7.18

```
IRP         REG,⟨AX,BX,CX,DX⟩
PUSH        REG
ENDM
```

汇编后得：

1	PUSH	AX
1	PUSH	BX
1	PUSH	CX
1	PUSH	DX

2. IRPC 伪操作

格式为：

```
IRPC        dummy,string（或〈string〉）
  ⋮                （重复块）
ENDM
```

IRPC 和 IRP 类似,但自变量表必须是字符串。重复次数由字符串中的字符个数确定,每次重复用字符串中的下一个字符取代重复块中的哑元。举例说明如下：

例 7.19

```
IRPC        X,01234567
DB          X+1
ENDM
```

汇编后得：

1	DB	1
1	DB	2
	⋮	
1	DB	8

例 7.20

```
        IRPC    K,ABCD
        PUSH    K&X
        ENDM
```

汇编后展开成：

1	PUSH	AX
1	PUSH	BX
1	PUSH	CX
1	PUSH	DX

在 MASM6 中,可用 REPEAT,FOR 和 FORC 取代 REPT,IRP 和 IRPC,以提高程序的可读性。

7.3　条 件 汇 编

汇编程序能根据条件把一段源程序包括在汇编语言程序内或者把它排除在外,这里就用到条件伪操作。条件伪操作的一般格式是：

```
IFxx        argument
  ⋮                        ｝自变量满足给定条件汇编此块
〔ELSE〕
  ⋮                        ｝自变量不满足给定条件汇编此块
ENDIF
```

265

自变量必须在汇编程序第一遍扫视后就成为确定的数值(早期的汇编程序在汇编时,要对源程序作二遍扫视。第一遍扫视建立符号表,第二遍扫视才把汇编语言指令翻译成机器语言指令。从 MASM6 起,已只使用一遍扫视)。条件伪操作中的××表示条件如下:

IF expression	汇编程序求出表达式的值,如此值不为 0,则满足条件。
IFE expression	如求出表达式的值为 0,则满足条件。
IF1	在汇编程序的第一遍扫视期间满足条件。
IF2	在汇编程序的第二遍扫视期间满足条件。
IFDEF symbol	如符号已在程序中定义,或者已用 EXTRN 伪操作说明该符号是在外部定义的,则满足条件。
IFNDEF symbol	如符号未定义或未通过 EXTRN 说明为外部符号则满足条件。
IFB ⟨argument⟩	如自变量为空,则满足条件。
IFNB ⟨argument⟩	如自变量不为空,则满足条件。
IFIDN ⟨arg－1⟩,⟨arg－2⟩	如果字符串 ⟨arg－1⟩ 和字符串 ⟨arg－2⟩ 相同,则满足条件。
IFDIF ⟨arg－1⟩,⟨arg－2⟩	如果字符串 ⟨arg－1⟩ 和字符串 ⟨arg－2⟩ 不相同,则满足条件。

上述 IF 和 IFE 的表达式中可以使用关系操作符 EQ,NE,LT,LE,GT 和 GE。如：IF expression1 EQ expression2 等。

条件伪操作可以用在宏定义体内,也可以用在宏定义体外,并且还允许嵌套任意次。下面举例说明条件伪操作的使用方法。

7.3.1 条件伪操作 IF 的使用举例

例 7.21 宏指令 MAX 把三个变元中的最大值放在 AX 中,而且使变元数不同时产生不同的程序段。

宏定义:

```
    MAX     MACRO   K,A,B,C
            LOCAL   NEXT,OUT
            MOV     AX,A
            IF      K－1
            IF      K－2
            CMP     C,AX
            JLE     NEXT
            MOV     AX,C
            ENDIF
    NEXT:
            CMP     B,AX
            JLE     OUT
```

```
                    MOV     AX,B
                    ENDIF
OUT：
                    ENDM
```

宏调用：

```
                    MAX     1,P
                    MAX     2,P,Q
                    MAX     3,P,Q,R
```

宏展开：

```
                    MAX     1,P
1                   MOV     AX,P
1   ??0001：
                    MAX     2,P,Q
1                   MOV     AX,P
1   ??0002：
1                   CMP     Q,AX
1                   JLE     ??0003
1                   MOV     AX,Q
1   ??0003：
                    MAX     3,P,Q,R
1                   MOV     AX,P
1                   CMP     R,AX
1                   JLE     ??0004
1                   MOV     AX,R
1   ??0004：
1                   CMP     Q,AX
1                   JLE     ??0005
1                   MOV     AX,Q
1   ??0005：
```

例 7.22 宏指令 BRANCH 产生一条转向 X 的转移指令。当它相对于 X 的距离小于 128 字节时产生 JMP SHORT X；否则产生 JMP NEAR PTR X(X 必须位于该转移指令之后，即低地址区)。

宏定义：

```
BRANCH     MACRO    X
           IF       ($-X) LT 128
                    JMP     SHORT X
           ELSE
                    JMP     NEAR PTR X
           ENDIF
```

```
                    ENDM
```
宏调用：
```
                    BRANCH    AA
```
宏展开：
```
    1               JMP       SHORT AA
```
否则产生
```
    1               JMP       NEAR PTR AA
```

读者可以看出在本例的宏定义中使用了关系操作符，这在一些宏定义中经常会用到，希望读者能逐步熟悉它的使用方法。

例 7.23 宏定义可允许递归调用，此时条件伪操作可用来结束宏递归。

宏指令 POWER 可以用来实现 X 和 2^N 相乘。这只需对 X 左移 N 次即可实现，可以设 COUNT 为递归次数的计数值，当该数与 N 相等时就可结束递归调用。

宏定义：
```
    POWER       MACRO    X,N
                SAL      X,1
    COUNT=COUNT+1
                IF       COUNT-N
                POWER    X,N
                ENDIF
                ENDM
```
宏调用：
```
    COUNT=0
                POWER    AX,3
```
宏展开：
```
    1           SAL      AX,1
    2           SAL      AX,1
    3           SAL      AX,1
```

7.3.2 条件伪操作 IF1 的使用举例

例 7.24 仍使用"例 7.12 使用宏库情况下的程序实现"，但在 include 语句前加上 IF1，其含义是使 include 语句只在汇编的第一遍扫视期间出现，这样宏库的拷贝将不在汇编清单中出现，以便节省空间。下面是"使用 IF1 语句的例 7.12 程序实现"和"使用 IF1 语句的例 7.12 程序实现部分 LST 清单"程序。

使用 IF1 语句的例 7.12 程序实现

```
;using if1--ex7_12_3
;------------------------------------------------------------
            if1
                include macro.mac
            endif
;------------------------------------------------------------
            .model   small
            .386
            .stack   200h            ;define stack segment
            .data                    ;define data segment
messg1      db       'Customer name?',13,10,'$'
messg2      db       'Customer address?',13,10,'$'
;------------------------------------------------------------
            .code                    ;define code segment
begin       proc     far
            initz
            prompt   messg1
            prompt   messg2
            finish
begin       endp
;------------------------------------------------------------
            end      begin
```

使用 IF1 语句的例 7.12 程序的部分 LST 清单

```
                        ;using if1--ex7_12_3
                        ;------------------------------------------------------------
                                    endif
                        ;------------------------------------------------------------
                                    .model   small
                                    .386
 0200                               .stack   200h            ;define stack segment
 0000                               .data                    ;define data segment
 0000  43 75 73 74 6F 6D 65    messg1   db     'Customer name?',13,10,'$'
       72 20 6E 61 6D 65 3F
       0D 0A 24
 0011  43 75 73 74 6F 6D 65    messg2   db     'Customer address?',13,10,'$'
       72 20 61 64 64 72 65
       73 73 3F 0D 0A 24
```

```
                              ;------------------------------------------------------------
0000                                  .code              ;define code segment
0000                          begin   proc    far
                                      initz
0000   B8 ---- R          1           mov     ax,@data   ;initialize segment
0003   8E D8              1           mov     ds,ax      ;   register
0005   8E C0              1           mov     es,ax
                                      prompt  messg1
0007   B4 09              1           mov     ah,09h     ;request display
0009   8D 16 0000 R       1           lea     dx,messg1  ;   prompt
000D   CD 21              1           int     21h
                                      prompt  messg2
000F   B4 09              1           mov     ah,09h     ;request display
0011   8D 16 0011 R       1           lea     dx,messg2  ;   prompt
0015   CD 21              1           int     21h
                                      finish
0017   B8 4C00            1           mov     ax,4c00h   ;end processing
001A   CD 21              1           int     21h
001C                          begin   endp
                              ;------------------------------------------------------------
001C                                  end     begin
```

7.3.3 条件伪操作 IFNDEF 的使用举例

例 7.25 这是一个包含宏定义 DIVIDE 在内的程序。下面是"例 7.25 的程序实现"和"7.25 的程序实现的部分 LST 清单"。

DIVIDE 是一个用相继的减法来实现除法的程序,在进入该程序主体之前,使用 IFNDEF 来检查该宏定义的三个参数是否已在程序中得到定义,如其中任一参数未经定义,则将用 CNTR 加以记录,最后用

```
        IF      CNTR
                EXITM
        ENDIF
```

来保证只要有一个参数未被定义则汇编程序将从 EXITM 处退出该宏定义而不作进一步地展开。

例 7.25 的程序实现

```
;Using ifndef--Ex7_25
;------------------------------------------------------------
divide  macro           dividend,divisor,quotient
```

```
        local           comp,out
        cntr=0
; AX=div'nd, BX=div'r, CX=quot't
        ifndef          dividend
; dividend not defined
        cntr=1
        endif
;
        ifndef          divisor
; divisor not defined
        cntr=1
        endif
;
        ifndef          quotient
; quotient not defined
        cntr=1
        endif
;
        if              cntr
; macro expansion terminated
        exitm
        endif
;
        mov             ax,dividend
        mov             bx,divisor
        sub             cx,cx
comp:   cmp             ax,bx
        jb              out
        sub             ax,bx
        inc             cx
        jmp             comp
out:    mov             quotient,cx
        endm
;----------------------------------------------------------------
        if1
        include macro.mac
        endif
        purge           prompt
;----------------------------------------------------------------
        .model          small
        .386
        .stack          200h    ;define stack segment
```

```
        .data                   ;define data segment
divdnd  dw          200         ;dividend
divsor  dw          25          ;divisor
quotnt  dw          ?           ;quotient
;------------------------------------------------------------------
        .code                   ;define code segment
main    proc        far
        .sall
        initz
        .xall
        divide      divdnd,divsor,quotnt
;
        .lall
        divide      divdnd,divsor,quont
;
        .sall
        finish
main    endp
;------------------------------------------------------------------
        end         main
```

例7.25 的部分 LST 清单

```
                        ;Using ifndef--Ex7_25
                        ;------------------------------------------------------------
                            divide   macro   dividend,divisor,quotient
                                local    comp,out
                                cntr=0
                        ;AX=div'nd, BX=div'r, CX=quot't
                                ifndef   dividend
                        ;dividend not defined
                                cntr=1
                                endif
                        ;
                                ifndef   divisor
                        ;divisor not defined
                                cntr=1
                                endif
                        ;
                                ifndef   quotient
                        ;quotient not defined
                                cntr=1
```

```
                        endif
        ;
                        if      cntr
        ; macro expansion terminated
                        exitm
                        endif
        ;
                        mov     ax,dividend
                        mov     bx,divisor
                        sub     cx,cx
                comp: cmp       ax,bx
                        jb      out
                        sub     ax,bx
                        inc     cx
                        jmp     comp
                out:    mov     quotient,cx
                        endm
        ;----------------------------------------------------------------
                        endif
                        purge   prompt
        ;----------------------------------------------------------------
                        .model  small
                        .386
0200                    .stack  200h            ;define stack segment
0000                    .data                   ;define data segment
0000    00C8            divdnd  dw      200     ;dividend
0002    0019            divsor  dw      25      ;divisor
0004    ????            quotnt  dw      ?       ;quotient
        ;----------------------------------------------------------------
0000                            .code           ;define code segment
0000                    main    proc    far
                        .sall
                        initz
                        .xall
                        divide  divdnd,divsor,quotnt
0007    A1 0000 R       1       mov     ax,divdnd
000A    8B 1E 0002 R    1       mov     bx,divsor
000E    2B C9           1       sub     cx,cx
0010    3B C3           1   ??0000: cmp ax,bx
0012    0F 82 001B R    1       jb      ??0001
0016    2B C3           1       sub     ax,bx
0018    41              1       inc     cx
```

0019	EB F5		1		jmp ??0000
001B	89 0E 0004 R		1		??0001: mov quotnt,cx
				;	
					.lall
				divide	divdnd,divsor,quont
=0000			1		cntr=0
			1	; AX=div′nd, BX=div′r, CX=quot′t	
			1		endif
			1	;	
			1		endif
			1	;	
			1		ifndef quont
			1	; quotient not defined	
=0001			1		cntr=1
			1		endif
			1	;	
			1		if cntr
			1	; macro expansion terminated	
			1		exitm
				;	
					.sall
				finish	
0024				main	endp
				;——————————————————————	
0024				end	main

这里给出了一个宏定义中所用的伪操作 EXITM。从上例可见,EXITM 的作用可使汇编程序在宏指令的展开过程中遇到 EXITM 时,将不继续进行宏展开而直接退出该宏定义。从"例 7.25 的部分 LST 清单"中可以看出,第一次 DIVIDE 的宏调用因三个参数均被定义使 CNTR=0,因而不执行 EXITM 而作了宏展开,这样可在程序中实现除法;第二次 DIVIDE 宏调用因商的参数未被定义而使 CNTR=1,因而执行了 EXITM,致使停止其后的宏展开,造成第二次除法操作不能进行。

7.3.4 条件伪操作 IFB 的使用举例

例 7.26 宏指令 GOTO L, X, REL, Y(其中 REL 可以是 Z, NZ, L, NL 等)可以根据不同情况产生无条件转移指令或比较和条件转移指令。
宏定义:

```
    GOTO    MACRO    L,X,REL,Y
            IFB      〈REL〉
            JMP      L
```

```
            ELSE
            MOV         AX,X
            CMP         AX,Y
            J&REL       L
            ENDIF
            ENDM
```

宏调用：

```
                        ⋮
            GOTO        LOOP1, SUM, NZ, 15
                        ⋮
            GOTO        EXIT
                        ⋮
```

宏展开：

```
                        ⋮
1           MOV         AX,SUM
1           CMP         AX,15
1           JNZ         LOOP1
                        ⋮
1           JMP         EXIT
```

例 7.27 宏定义 DISP 可以在给出参数时,在屏幕上显示该字符;不给出参数时,则可显示"空格"。

宏定义：

```
DISP    MACRO       CHAR
        IFB         〈CHAR〉
        MOV         DL,' '
        ELSE
        MOV         DL,CHAR
        ENDIF
        MOV         AH,02H
        INT         21H
        ENDM
```

宏调用：

```
        DISP        'A'
        DISP
```

宏展开：

```
        DISP        'A'
1       MOV         DL,'A'
```

```
        1       MOV             AH,02H
        1       INT             21H
                DISP
        1       MOV             DL,' '
        1       MOV             AH,02H
        1       INT             21H
```

7.3.5 条件伪操作 IFIDN 的使用举例

例 7.28 给出使用 IFIDN 的例子的程序见"例 7.28 的程序实现",其部分 LST 清单则见"例 7.28 的部分 LST 清单"。程序中的 MOVIF 宏定义利用其参数为 B 或 W 而产生 REP MOVSB 或 REP MOVSW 的指令,如不给出参数则产生 REP MOVSB。

例 7.28 的程序实现

```
;using IFIDN--ex7_28
;-----------------------------------------------------------
movif   macro       tag             ;define macro
        ifidn       <&tag>,<b>
        rep         movsb           ;move bytes
        exitm
        endif
;
        ifidn       <&tag>,<w>
        rep         movsw           ;move words
        else
; no B or W tag,default to B
        rep         movsb
        endif
        endm                        ;end macro
;-----------------------------------------------------------
        if1
        include macro.mac
        endif
        purge       prompt
;-----------------------------------------------------------
        .model      small
        .386
        .stack      200h            ;define stack segment
        .data                       ;define data segment
;-----------------------------------------------------------
```

```
                .code                       ;define code segment
begin   proc            far
        .sall
        initz
        .xall
        movif           b
        movif           w
        movif
        .sall
        finish
begin   endp
;--------------------------------------------------------------
        end             begin
```

例7.28 的部分 LST 清单

```
                ;using IFIDN--ex7_28
                ;--------------------------------------------------------------
                movif   macro   tag             ;define macro

                        ifidn   <&tag>,<b>
                        rep     movsb           ;move bytes
                        exitm
                        endif
                ;
                        ifidn   <&tag>,<w>
                        rep     movsw           ;move words
                        else
                ;no B or W tag,default to B
                        rep     movsb
                        endif
                        endm                    ;end macro
                ;--------------------------------------------------------------
                        endif
                        purge   prompt
                ;--------------------------------------------------------------
                        .model  small
                        .386
0200                    .stack  200h            ;define stack segment
0000                    .data                   ;define data segment
                ;--------------------------------------------------------------
0000                    .code                   ;define code segment
```

```
0000                          begin    proc     far
                                       .sall
                                       initz
                                       .xall
                                       movif    b
0007  F3/ A4          1                rep      movsb    ;move bytes
                                       movif    w
0009  F3/ A5          1                rep      movsw    ;move words
                                       movif
000B  F3/ A4          1                rep      movsb
                                       .sall
                                       finish
0012                          begin    endp
                              ;--------------------------------------------------------
0012                                   end      begin
```

以上有关条件汇编的使用举例说明：条件汇编根据条件把一个程序段包括在源程序中或排除在外，也可以根据不同条件选择不同的程序段进入源程序，这就为汇编语言编程提供了很大的便利。

习　　题

7.1 编写一条宏指令 CLRB，完成用空格符将一字符区中的字符取代的工作。字符区首地址及其长度为变元。

7.2 某工厂计算周工资的方法是每小时的工资率 RATE 乘以工作时间 HOUR，另外每工作满十小时加奖金 3 元，工资总数存放在 WAG 中。请将月工资的计算编写成一条宏指令 WAGES，并展开宏调用：

    WAGES R1,42

7.3 给定宏定义如下：

```
        DIF     MACRO   X,Y
                MOV     AX,X
                SUB     AX,Y
                ENDM
        ABSDIF  MACRO   V1,V2,V3
                LOCAL   CONT
                PUSH    AX
                DIF     V1,V2
                CMP     AX,0
                JGE     CONT
                NEG     AX
```

```
CONT:       MOV         V3,AX
            POP         AX
            ENDM
```

试展开以下调用,并判定调用是否有效。

  (1) ABSDIF P1,P2,DISTANCE
  (2) ABSDIF [BX],[SI],X[DI],CX
  (3) ABSDIF [BX][SI],X[BX][SI],240H
  (4) ABSDIF AX,AX,AX

 7.4 试编制宏定义,要求把存储器中的一个用 EOT 字符结尾的字符串传送到另一个存储区中去。

 7.5 宏指令 BIN_SUB 完成多个字节数据连减的功能:

$$RESULT \leftarrow (A-B-C-D-\cdots)$$

要相减的字节数据顺序存放在首地址为 OPERAND 的数据区中,减数的个数存放在 COUNT 单元中,最后结果存入 RESULT 单元。请编写此宏指令。

 7.6 请用宏指令定义一个可显示字符串 GOOD:'GOOD STUDENTS: CLASS X NAME',其中 X 和 NAME 在宏调用时给出。

 7.7 下面的宏指令 CNT 和 INC1 完成相继字存储:

```
CNT         MACRO       A,B
A&B         DW          ?
            ENDM

INC1        MACRO       A,B
            CNT         A,%  B
B=B+1
            ENDM
```

请展开下列宏调用:

```
C=0
            INC1        DATA,C
            INC1        DATA,C
```

 7.8 定义宏指令并展开宏调用。宏指令 JOE 把一串信息'MESSAGE NO. K'存入数据存储区 XK 中。宏调用为:

```
I=0
            JOE         TEXT,I
             ⋮
            JOE         TEXT,I
             ⋮
            JOE         TEXT,I
```

7.9 宏指令 STORE 定义如下：

STORE	MACRO	X,N
	MOV	X+I,I
I=I+1		
	IF	I-N
	STORE	X,N
	ENDIF	
	ENDM	

试展开下列调用：

    I=0

        STORE   TAB,7

7.10 试编写非递归的宏指令，使其完成的工作与题 7.9 的 STORE 相同。

7.11 试编写一段程序完成以下功能，如给定名为 X 的字符串长度大于 5 时，下列指令将汇编 10 次。

    ADD   AX,AX

7.12 定义宏指令 FINSUM：比较两个数 X 和 Y（X、Y 为数，而不是地址），若 X＞Y 则执行 SUM←X+2*Y；否则执行 SUM←2*X+Y。

7.13 试编写一宏定义完成以下功能：如变元 X='VT55'，则汇编 MOV TERMINAL,0；否则汇编 MOV TERMINAL,1。

7.14 对于 DOS 功能调用，所有的功能调用都需要在 AH 寄存器中存放功能码，而其中有一些功能需要在 DX 中放一个值。试定义宏指令 DOS21，要求只有在程序中定义了缓冲区时，汇编为：

    MOV   AH,DOSFUNC
    MOV   DX,OFFSET BUFF
    INT   21H

否则，无 MOV DX,OFFSET BUFF 指令，并展开以下宏调用：

    DOS21   01
    DOS21   0AH,IPFIELD

7.15 编写一段程序，使汇编程序根据 SIGN 的值分别产生不同的指令。

如果 SIGN=0，则用字节变量 DIVD 中的无符号数除以字节变量 SCALE；如果 SIGN=1，则用字节变量 DIVD 中的带符号数除以字节变量 SCALE，结果都存放在字节变量 RESULT 中。

7.16 试编写宏定义 SUMMING，要求求出双字数组中所有元素之和，并把结果保存下来。该宏定义的哑元应为数组首址 ARRAY，数组长度 COUNT 和结果存放单元 RESULT。

7.17 为下列数据段中的数组编制一程序,调用题 7.16 的宏定义 SUMMING,求出该数组中各元素之和。

        DATA   DD   101246,274365,843250,475536
        SUM   DQ   ?

7.18 如把题 7.16 中的宏定义存放在一个宏库中,则题 7.17 的程序应如何修改?

第 8 章 输入输出程序设计

在广泛使用的微型机系统中,外部设备是以实现人机交互和机间通信为目的的一些机电设备。计算机系统通过硬件接口以及 I/O 控制程序对外部设备进行控制,使其能协调地、有效地完成输入输出工作。在对外部设备的控制过程中,主机不可避免地,有时甚至要很频繁地对设备接口进行联络和控制,因此,能直接控制硬件的汇编语言就成了编写高性能 I/O 程序最有效的程序设计语言。本章将以一些常用的 I/O 设备为例,着重讨论 I/O 程序设计,特别是中断程序设计的方法。

8.1 I/O 设备的数据传送方式

8.1.1 CPU 与外设

每种输入输出设备都要通过一个硬件接口或控制器和 CPU 相连。例如,软磁盘通过软盘控制器和 CPU 连接起来;终端显示器通过数据接口和 CPU 连接起来。这些接口和控制器都能支持输入输出指令 IN,OUT 与外部设备交换信息。这些信息包括控制、状态和数据三种不同性质的信息,它们必须按不同的端口地址分别传送。

控制信息输出到 I/O 接口,通知接口和设备要做什么动作。例如,CPU 向 I/O 接口发出启动信号或停止信号以控制外设的启停。

状态信息从 I/O 接口输入到 CPU,表示 I/O 设备当前所处的状态。对于输入设备,通常用准备好(READY)信号来表示外设已准备好输入数据。对于输出设备,通常用忙(BUSY)信号表示设备是否处于空闲状态,如为空闲状态,外设则接收 CPU 送来的信息,如为忙状态,CPU 则要等待。

数据信息是 I/O 设备和 CPU 真正要交换的信息。外设和接口之间的数据信息可以是串行的,也可以是并行的,相应地要使用串行接口或并行接口。不同的 I/O 设备要求传送的数据类型也是不同的,例如向显示器传送的数据必须是 ASCII 码,而不能是二进制形式的数。

8.1.2 直接存储器存取(DMA)方式

80x86 具有一系列简单而又灵活的输入输出方式。在第 3 章已经介绍了一种用 IN 和 OUT 指令直接在端口级上处理输入输出的程序直接控制 I/O 的方式,另外还有中断传送方式和 DMA 方式。程序直接控制输入输出方式和中断传送方式在 8.2 和 8.3 节中专门要讲到,而 DMA 方式因为主要由硬件 DMA 控制器实现其传送功能,所以只在这里做一点简单介绍。

直接存储器存取(direct memory access,DMA)方式,也称为成组数据传送方式。主要是用于一些高速 I/O 设备,如磁带、磁盘、模数转换器(A/D)等设备。这些设备传输字节或字的速率非常快,如磁盘的数据传输率约为每秒 200000 字节,也就是说磁盘与存储器传送一个字节只需 5 微秒。对这类高速 I/O 设备,用执行输入输出指令的方法或完成一次次中断序列的方法来传输字节,将会造成数据的丢失,而 DMA 方式能使 I/O 设备直接和存储器进行成批数据的快速传送。每个字节一到达端口,就直接从接口送到存储器,同样,接口和它的 DMA 控制器也能直接从存储器取出字节并把它送到 I/O 设备去。

DMA 控制器(Intel 8237A)一般包括四个寄存器:控制寄存器、状态寄存器、地址寄存器和字节计数器,这些寄存器在信息传送之前应进行初始化,即系统程序在地址寄存器中设置要传送的数据块的首地址,在字节寄存器中设置要传送的数据长度(字节数),在状态控制寄存器中设置控制字,指出数据是输入还是输出,并启动 DMA 操作。每个字节传送后,地址寄存器增 1,字节计数器减 1。

系统完成 DMA 传送的步骤如下:

(1) DMA 控制器向 CPU 发出 HOLD 信号,请求使用总线。

(2) CPU 发出响应信号 HOLD 给 DMA 控制器,并将总线让出,这时 CPU 放弃了对总线的控制,而 DMA 控制器获得了总线控制权。

(3) 传输数据的存储器地址(在地址寄存器中)通过地址总线发出。

(4) 传输的数据字节通过数据总线进行传送。

(5) 地址寄存器增 1,以指向下一个要传送的字节。

(6) 字节计数器减 1。

(7) 如字节计数器非 0,转向第 3 步。

(8) 否则,DMA 控制器撤销总线请求信号 HOLD,传送结束。

8.2 程序直接控制 I/O 方式

8.2.1 I/O 端口

计算机的外部设备和大容量存储设备都是通过接口连接到系统上,每个接口由一组寄存器组成,这些寄存器都分配有一个称为 I/O 端口的地址编码。计算机的 CPU 和内存就是通过这些端口和外部设备进行通信的。

I/O 接口部件中一般有三种寄存器:一是用作数据缓冲的数据寄存器;二是用作保存设备和接口的状态信息,供 CPU 对外设进行测试的状态寄存器;三是用来保存 CPU 发出的命令以控制接口和设备的操作的命令寄存器。这些寄存器都分配有各自的端口号,CPU 就是通过不同的端口号来选择各种外部设备的。

在 80x86 微机中,I/O 端口编址在一个独立的地址空间中,这个 I/O 空间允许设置 64K(65536)个 8 位端口或 32K(32768)个 16 位端口,这些端口地址实际上只用了其中很小一部分,因为系统中一般只有十几个外部设备和大容量存储设备和主机相连。对不同型号的计算机及其接口,I/O 端口的编号有时不完全相同。表 8.1 列出了部分端口地址(十六进制)。

表 8.1 I/O 端口地址分配

I/O 地址	功 能	I/O 地址	功 能
00～0F	DMA 控制器 8237A	2F8～2FE	2 号串行口(COM2)
20～3F	可编程中断控制器 8259A	320～324	硬盘适配器
40～5F	可编程中断计时器	366～36F	PC 网络
60～63	8255A PPI	372～377	软盘适配器
70～71	CMOS RAM	378～37A	2 号并行口(LPT1 打印机)
81～8F	DMA 页表地址寄存器	380～38F	SDLC 及 BSC 通信
93～9F	DMA 控制器	390～393	Cluster 适配器
A0～A1	可编程中断控制器 2	3A0～3AF	BSC 通信
C0～CE	DMA 通道,内存/传输地址寄存器	3B0～3BF	MDA 视频寄存器
F0～FF	协处理器	3BC～3BE	1 号并行口
170～1F7	硬盘控制器	3C0～3CF	EGA/VGA 视频寄存器
200～20F	游戏控制端口	3D0～3D7	CGA 视频寄存器
278～27A	3 号并行口(LPT2 打印机)	3F0～3F7	软盘控制寄存器
2E0	EGA/VGA 使用	3F8～3FE	1 号串行口(COM1)

8.2.2 I/O 指令

对于一个 I/O 和存储器分离的地址空间系统,80x86 有专门的 I/O 指令与端口进行通信。在第 3 章已介绍了 8086 的 I/O 指令 IN 和 OUT,这两条指令既可以传送字节也可以传送字,并且都有直接端口寻址和间接端口寻址两种方式。

```
    IN      AL,PORT      ;(AL)←(PORT)
    IN      AX,PORT      ;(AX)←(PORT+1:PORT)
    IN      AL,DX        ;(AL)←((DX))
    IN      AX,DX        ;(AX)←((DX)+1:(DX))

    OUT     PORT,AL      ;(PORT)←(AL)
    OUT     PORT,AX      ;(PORT+1:PORT)←(AX)
    OUT     DX,AL        ;((DX))←(AL)
    OUT     DX,AX        ;((DX)+1:(DX))←(AX)
```

以上 IN 和 OUT 指令的前两种方式是直接端口寻址方式,端口地址 PORT 是一个 8 位的立即数,其范围是 0～255。两组指令中的后两种格式是间接寻址方式,端口地址在 DX 中,其范围为 0～65535,这种方式通过对 DX 寄存器的增量可以处理几个连续端口地址的输入输出。

注意：I/O 指令中使用的寄存器必须是 AL 或 AX。

用 IN 指令可以从一个数据寄存器输入数据或从状态寄存器输入接口和外设的状态。例如，下面两条指令能把一个字从端口地址 0028 和 0029 传送到存储器的 DATA-WORD 单元中。

  IN   AX,28H
  MOV   DATA-WORD,AX

又例如，测试某状态寄存器（端口地址为 27H）的第 2 位是否为 1，若为 1，则转移到 ERROR 进行处理。其指令序列为：

  IN   AL,27H
  TEST   AL,00000100B
  JNZ   ERROR

同样，OUT 指令用来输出数据或给一个指定的 I/O 端口传送命令信息。例如，某接口的命令寄存器（端口地址为 126H）的第 7 位控制程组数据传送，那么下面的指令序列将发出成组传送命令：

  MOV   DX,126H
  IN   AL,DX
  OR   AL,80H
  OUT   DX,AL

I/O 指令是 CPU 与外部设备进行通信的最基本途径，即使使用 DOS 功能调用或 BIOS 例行程序，其例行程序本身也是用 IN 和 OUT 指令与外部设备进行数据交换的。例如，当程序请求从键盘输入字符时，程序中安排了一条中断指令 INT 16H，当执行这条指令时，系统将调用 ROM BIOS 的一个键盘例行程序，在这个例行程序中就有一条 IN 指令从端口 60H 输入一个字符到 AL 寄存器。

使用 I/O 指令对端口地址进行直接的输入或输出，比调用 DOS 功能或 BIOS 例行程序更能提高数据的传送速度和吞吐量，但同时也要求程序员对计算机的硬件结构有一定的了解，其程序对硬件的依赖性也大。因此，对于一般的程序设计，还是尽可能使用 DOS 或 BIOS 功能调用。

8.2.3　I/O 程序举例

下面通过几个 I/O 程序的例子，说明使用 I/O 指令直接在端口级上输入输出的方法。

  例 8.1　发声子程序 SOUND。

这是一个最基本的直接控制扬声器发出声音的子程序。程序通过 I/O 指令使设备控制寄存器(I/O 端口地址为 61H)的第 1 位交替为 0 和 1,而端口 61H 的第 1 位和扬声器的脉冲门相连(见图 8.1),当第 1 位由 0 变为 1,延迟一会又由 1 变为 0 时,脉冲门就先打开后关闭,产生了一个脉冲电流。这个脉冲电流被放大后送到扬声器使之发出了声音。61H 端口的第 0 位和一个振荡器(2 号定时器)相连,现在不用振荡器产生声音,所以把第 0 位置零。

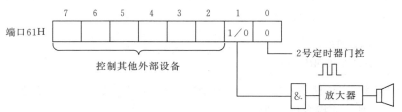

图 8.1 设备控制寄存器

例 8.1 Sound 程序段

```
;-------------------------------------------------------------------
; SOUND — Make a sound according to the frequency and delay
; on entry:  BX：Sound frequency(for example, 6000)
;            CX：Sound delay(for example, 1000)
;-------------------------------------------------------------------
SOUND    PROC    NEAR
         PUSH    AX
         PUSH    DX
         MOV     DX,CX               ;sound duration
         IN      AL,61H              ;get port 61h
         AND     AL,11111100B        ;AND off bits 0,1
TRIG:
         XOR     AL,2                ;toggle bit 1
         OUT     61H,AL              ;output to port 61h
         MOV     CX,BX               ;value of wait
DELAY:
         LOOP    DELAY               ;delayed a while
         DEC     DX                  ;turn on/off  1000 times
         JNE     TRIG
         POP     DX
         POP     AX
         RET
SOUND    ENDP
;-------------------------------------------------------------------
```

Sound 程序中的第 4 条指令 IN AL,61H 取得设备控制寄存器的开关量,然后由第 3

条指令 AND 将第 0 位和第 1 位置零,2～7 位保持不变,XOR 指令将第 1 位置为 1,然后把这个开关量输出到 61H 端口以控制接通扬声器。在第二次循环执行 XOR 指令时,第 1 位又由 1 变为 0,也就是关闭了扬声器,这样在脉冲电流的驱动下,扬声器就发出了声音。

另外两条指令：

```
              MOV    CX, BX
DELAY:        LOOP   DELAY
```

是用来控制脉冲门开关的时间,这个时间值根据 PC 机的主频是可以改变的,主频越快的的机器,这个时间值就应该越大。因为程序里 CX 中的固定值(比如是 6000)控制输出脉冲 1 和 0 的变化,因此扬声器接通和关闭的时间间隔总是相同的,结果发出的声音是一个没有变化的纯音。

通常一个外设的数据端口是 8 位的,而状态与控制信息只需一位或两位,所以不同外设的状态和控制位可共用一个端口。61H 端口的 0、1 位是控制扬声器的,2～7 位分别控制其他外部设备。

例 8.2 打印字符程序 PRT_CHAR。

这是一个采用查询方式的打印字符程序。程序通过反复读取并测试打印机的状态来控制输出。在打印机接口中,数据寄存器的端口地址为 378H,状态寄存器的端口地址为 379H,控制寄存器的端口地址为 37AH。它们各位的含义如图 8.2 所示。

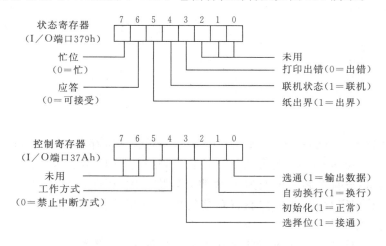

图 8.2 打印机的状态寄存器和控制寄存器

例 8.2 打印字符的程序段

```
TITLE   PRT_CHAR—EXAMPLE 8-2
;Print a message by inquiring printer state
data        segment
            mess    db      'Printer is normal',0dh,0ah
```

```
                count    equ    $ - mess
        data    ends
;------------------------------------------------------------------
        cseg    segment
        main    proc    far
                assume  cs:cseg, ds:data
        start:  mov     si,offset mess          ; message offset
                mov     cx,count                ; count of char.
        next:   mov     dx,379h                 ; state port
        wait:   in      al,dx
                test    al,80h                  ; printer busy?
                Je      wait                    ; yes,test again
                mov     al,[si]                 ; no,read a char.
                mov     dx,378h                 ; data port
                out     dx,al                   ; putout to data port
                mov     dx,37ah                 ; control port
                mov     al,0dh                  ; control code
                out     dx,al                   ; send a strobe=1
                mov     al,0ch                  ; control code
                out     dx,al                   ; strobe=0
                inc     si                      ; addr. Of mess increment
                loop    next                    ; next char.
                mov     ah,4ch                  ; return to DOS
                Int     21h
        main    endp
        cseg    ends
                end     start
```

在例 8.2 打印字符的程序中，使用 TEST 指令对状态寄存器（I/O 端口 379h）的 7 位进行测试，如果 7 位为 0，表示打印机处于忙状态，这时，CPU 不能送出打印数据；所以程序再次循环测试，一直等到 7 位变为 1，表明打印机空闲，程序才从数据区取出一个字符送到打印机的数据寄存器；并由控制寄存器发出一个选通信号（端口 37AH 的 0 位），控制打印机将这个字符打印输出。

这种 CPU 与外部设备交换信息的方式称为查询方式或等待方式。

造成 CPU 必须查询等待的主要原因是许多外设的工作速度比较低。例如，键盘、打印机等外部设备，它们通过按键或打印头的机械动作输入或输出一个数据，其速度是很慢的；而 CPU 执行指令的速度是它的几千倍乃至上万倍，所以 CPU 在接受或发送数据之前必须要了解外设的工作状态，看它是否已经准备好输入或输出。当外设还没有准备好以前，CPU 就要等待，不能做别的操作。为了提高 CPU 的工作效率，可采用中断方式传送数据（关于中断，将在 8.3 节中做详细的介绍）。

有时系统中同时有几个设备要求输入输出数据，那么对每个设备都编写一段执行输入输出数据的程序,然后轮流查询这些设备的准备位,当某一设备准备好允许输入或输出数据时,就调用这个设备的 I/O 程序完成数据传输,否则依次查询下一个设备是否准备好。

例 8.3 CPU 要从 3 个设备轮流输入数据,PROC1,PROC2,PROC3 分别是设备 1,设备 2 和设备 3 的数据输入程序,它们的状态寄存器的端口地址分别用 STAT1,STAT2,STAT3 表示,这三个状态寄存器的 5 位是输入准备位。

例 8.3 轮流查询三个数据输入设备的程序段

```
        ;Round-robin    polling
INPUT:   IN      AL,STAT1        ; check device 1
         TEST    AL,20H          ; if device 1 is ready
         JZ      DEV2            ; no,goto device 2
         CALL    FAR PTR PROC1   ; yes,device 1 input data
DEV2:    IN      AL,STAT2        ; check device 2
         TEST    AL,20H          ; if device 2 is ready
         JZ      DEV3            ; no,goto device 3
         CALL    FAR PTR PROC2   ; yes,device 2 input data
DEV3:    IN      AL,STAT3        ; check device 3
         TEST    AL,20H          ; if device 3 is ready
         JZ      NO_INPUT        ; no,goto no-input
         CALL    FAR PTR PROC3   ; yes,device 3 input data
NO_INPUT: ……
```

查询方式的优点是:可以用程序安排几个输入输出设备的先后优先次序,最先查询的设备,其工作的优先级也最高。修改程序中的查询次序,实际上也就修改了设备的优先级。查询方式的缺点:前面提到的在查询过程中,浪费了 CPU 原本可执行大量指令的时间,而且由询问转向相应的处理程序的时间较长,尤其在设备比较多的情况下。

8.3　中断传送方式

中断是 CPU 和外部设备进行 I/O 的有效方法。这种 I/O 方式一直被大多数计算机所采用,它可以避免因反复查询外部设备的状态而浪费时间,从而提高了 CPU 的效率。

中断是一种使 CPU 中止正在执行的程序而转去处理特殊事件的操作。这些引起中断的事件称为中断源,它们可能是来自外部设备的 I/O 请求,也可能是计算机的一些异常事故或其他内部原因。由外设控制器或协处理器(8087/80287)引起的中断一般称为硬件中断或外中断,由程序中安排的中断指令 INT 产生的中断,或由 CPU 的某些错误结果产生的中断称为软件中断或内中断。80x86 的中断源如图 8.3 所示。图中引线端标示的数字为系统分配的中断类型号。

连到 CPU 的非屏蔽中断(NMI)是为电源错、内存或 I/O 总线的奇偶等异常事件的中断保留的,它不受中断允许标志 IF 的屏蔽,而且在整个系统中只能有一个非屏蔽中断,

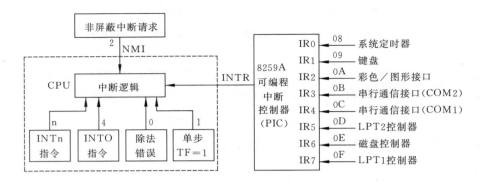

图 8.3　80x86 中断源

其中断类型号为 2。

外部设备的中断是通过 Intel 8259A 可编程中断控制器(PIC)连到主机上的。CPU 通过一组 I/O 端口控制 8259A，而 8259A 则通过 INTR 管脚给 CPU 传送中断信号。多个 8259A PIC 可以树形结构连在一起，从而使大量的外部设备顺序连接到 CPU 的中断系统上。外部设备和 8259A PIC 的连法是由设计人员规定好的，这种外中断类型的分配由硬件连线实现，因而软件不能对其修改。图 8.3 中外设与 8259A PIC 的连法是 80x86 的标准连法。内中断不是由连线接到硬件上的，中断 20H 到 3FH 用于调用 DOS 功能例行程序，其他中断号小于 20H 或大于 3FH 的中断，用于调用 IBM PC ROM BIOS 或一些应用软件，这些内容在以后的章节中将要陆续讲到。

8.3.1　8086 的中断分类

1. 软件中断

软件中断又称为内中断，它通常由三种情况引起：

(1) 由中断指令 INT 引起。
(2) 由于 CPU 的某些错误而引起。
(3) 为调试程序(DEBUG)设置的中断。

下面分别介绍这几种软件中断。

(1) 中断指令 INT 引起的内中断

CPU 执行完一条 INT n 指令后，会立即产生中断，并且调用系统中相应的中断处理程序来完成中断功能，中断指令的操作数 n 指出中断类型号。

例如，若要对存储器的容量进行测试，可在程序中安排一条中断指令：

　　INT　12H

当 CPU 执行这条指令时，立即产生了一个中断，并从中断向量表的 0：48H 开始的四个字节单元中取出段地址和偏移地址；然后转去执行相应的中断处理程序以完成对存储器的测试，返回调用程序后，AX 寄存器中的数据即为存储器的大小。

INT 指令可以指定 0～0FFH 中的任何类型号。除系统占用的类型号之外，用户还可利用为用户保留的类型号扩充新的中断处理功能。

(2) 处理 CPU 某些错误的中断

CPU 在执行程序时,还会发现一些运算中出现的错误。为了能及时处理这些错误,CPU 就以中断的方式中止正在运行的程序,待程序员改正错误后,重新运行程序。

① 除法错中断　除法错的中断类型号为 0。

在执行除法指令时,若发现除数为 0 或商超过了寄存器所能表达的范围,则立即产生一个类型为 0 的中断。

② 溢出中断　如果溢出标志 OF 置 1,有一条专门的指令 INTO 来中断发生溢出的算术操作,如 OF 为 0,则 INTO 指令不产生中断,CPU 继续运行原程序。溢出中断处理程序的主要功能是打印出一个出错信息,在处理程序结束时,不返回原程序继续运行,而是把控制交给操作系统。

溢出中断的中断类型号为 4,下面的指令用来测试加法的溢出:

ADD　　　AX,VALUE
INTO

(3) 为调试程序(DEBUG)设置的中断

一个新的程序编制好以后,必须要上机调试才能正确可靠地工作。在调试程序时,为了检查中间结果或寻找程序中的问题,往往要在程序中设置断点或进行单步工作(一次只执行一条指令),这些功能都是由中断系统来实现的。

① 单步中断　单步是一种很有用的调试方法。当标志位 TF 置为 1 时,每条指令执行后,CPU 自动产生类型号为 1 的中断——单步中断。产生单步中断时,CPU 同样自动地将 FLAGS,CS 和 IP 的内容保存入栈,然后清除 TF 和 IF,于是,当进入单步中断处理程序后,就不是处于单步方式了,它将按正常方式运行中断处理程序。在单步处理程序结束时,原来的 FLAGS 从堆栈中取回,又把 CPU 重新置成单步方式。

使用单步中断可以一条指令一条指令地跟踪程序的流程,观察 CPU 每执行一条指令后,各个寄存器及有关存储单元的变化,从而指出和确定产生错误的原因。

② 断点中断　断点中断也是供 DEBUG 调试程序使用的,它的中断类型号为 3。通常调试程序时,把程序按功能分成几段,然后每段设一个断点。当 CPU 执行到断点时便产生中断,这时程序员可以检查各寄存器及有关存储单元的内容。

断点可以设置在程序的任何地方,设置断点实际上是把一条断点指令 INT 3 插入程序中,CPU 每执行到断点处的 INT 3 指令便产生一个中断。

在上述内中断中,INT 指令和 INTO 指令产生的中断,以及除法错中断都不能被禁止,并且比任何外部中断的优先权都高。

2. 硬件中断

硬件中断来自处理机的外部条件,如 I/O 设备或其他处理机等,以完全随机的方式中断现行程序而转向另一处理程序。硬件中断又称为外中断。

从图 8.3 可以看出,硬件中断主要有两种来源:一种是非屏蔽中断(NMI),另一种是来自各种外部设备的中断。由外部设备的请求引起的中断也称为可屏蔽中断。微型计算机配置的外部设备一般有硬磁盘(disk),软磁盘(floppy disk),显示器(CRT)和各种打印机(line printer)等。这些外部设备通过 8259A 可编程中断控制器和 CPU 相连。8259A 可编程中断控制器可接收来自外设的中断请求信号,并把中断源的中断类型号送 CPU,如果 CPU 响应该外设的中断请求,就自动转入相应的中断处理程序。但是从外设发出中断请求到 CPU 响应中断,有两个控制条件是起决定性作用的,一是该外设的中断请求是否屏蔽,另一个是 CPU 是否允许响应中断。这两个条件分别由 8259A 的中断屏蔽寄存器(IMR)和标志寄存器(FLAGS)中的中断允许位 IF 控制。

中断屏蔽寄存器的 I/O 端口地址是 21H,它的 8 位对应控制 8 个外部设备(见图 8.4(a)),通过设置这个寄存器的某位为 0 或为 1 来允许或禁止某外部设备的中断。某位为 0 表示允许某种外设中断请求,某位为 1 表示某种外设的中断请求被屏蔽(禁止)。

例如,只允许键盘中断,可设置如下中断屏蔽字:

```
MOV      AL, 11111101B
OUT      21H, AL
```

如果系统重新增设键盘中断,则可用下列指令实现:

```
IN       AL, 21H
AND      AL, 11111101B
OUT      21H, AL
```

在编写中断程序时,应在主程序的初始化部分设置好中断屏蔽寄存器,以确定允许用中断方式工作的外部设备。

外部设备向 CPU 发出中断请求,CPU 是否响应还与标志寄存器中的中断标志位 IF 有关。如果 IF=0,CPU 就禁止响应任何外设的中断,也就是说,CPU 将不会产生中断来处理外设的请求。如果 IF=1,则允许 CPU 响应外设的中断请求。下面两条指令能设置和清除 IF 位。

```
STI      设置中断允许位(IF=1)
CLI      清除中断允许位(IF=0)
```

允许 CPU 响应外设的中断请求(IF=1)也叫做开中断,反之叫做关中断(IF=0)。

已经知道,当任何类型的中断发生时,当前的 FLAGS 要保存入栈,然后清除 IF 位,进入中断处理程序。如果允许在一个中断处理程序的执行过程中发生硬中断,则必须用一条 STI 指令开中断。当执行到中断返回指令 IRET,又取出 FLAGS 先前的值,其中 IF 为 1,CPU 将允许硬中断再次发生。

有一种特殊的硬件中断和 IF 标志位无关,这就是非屏蔽中断。非屏蔽中断的类型号为 2,CPU 不能禁止非屏蔽中断;如果系统使用了这种类型的中断,那么 CPU 总会响应的,所以非屏蔽中断主要用于一些紧急的意外处理,如电源掉电等。另外计算机内部的实时钟希望能不停地计时,所以也可以把非屏蔽中断提供给实时钟。

在一次中断处理结束之前,还应给 8259A 可编程中断控制器的中断命令寄存器发出中断结束命令(end of interrupt, EOI)。中断命令寄存器的 I/O 端口地址为 20H(见图

8.4(b)),它的各控制位可动态地控制中断处理过程,其中 L2~L0 三位指定 IR0~IR7 中具有最低优先级的中断请求。6 位(Set Level)和 7 位(Rotate)控制 IR0~IR7 的中断优先级的顺序。5 位(EOI)是中断结束位,当 EOI 位为 1 时,当前正在处理的中断请求就被清除。所以在中断处理完成后,必须把中断结束位置为 1,否则以后将屏蔽掉对同级中断或低级中断的处理。当然在必要的时候,在中断处理程序中也可利用 EOI 命令清除当前的中断请求,使得在中断处理的过程中又能响应同级和低级中断。

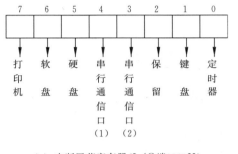
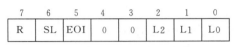

(a) 中断屏蔽寄存器(I/O端口21H)　　(b) 中断命令寄存器(I/O端口20H)

图 8.4　中断屏蔽寄存器和中断命令寄存器

结束硬件中断用下面的指令:

MOV　　　AL,20H
OUT　　　20H,AL

8.3.2　中断向量表

每种中断都给安排一个中断类型号。80x86 中断系统能处理 256 种类型的中断,类型号为 0~0FFH。如图 8.3 所示的中断源,系统时钟的中断类型为 08,键盘为 09,软中断中的除法错误的中断类型为 0,等等。每种类型的中断都由相应的中断处理程序来处理,中断向量表就是各类型中断处理程序的入口地址表。

我们知道存储器的低 1.5KB,地址从 0 段 0000 ~ 5FFH 为系统占用,其中最低的 1KB,地址从 0000 ~ 3FFH 存放中断向量。中断向量表中的 256 项中断向量对应 256 种中断类型,每项占用四个字节。其中两个字节存放中断处理程序的段地址(16 位),另两个字节存放偏移地址(16 位)。因为各处理程序的段地址和偏移地址在中断向量表中按中断类型号顺序存放(如图 8.5 所示),所以每类中断向量的地址可由中断类型号乘以 4 计算出来。例如,报警中断的中断类型为 4AH,它的中断向量地址为 4AH×4=128H,即 128H,129H 两字节存放的是报警中断处理程序的偏移地址;12AH,12BH 两字节存放的是报警中断处理程序的段地址,取出段地址和偏移地址放入 CS 和 IP,CPU 就可以转入相应的中断处理程序。

图 8.5 中断向量表

图 8.6 以 BIOS 中断 INT 4AH 为例,表示出中断操作的 5 个步骤：

① 取中断类型号；
② 计算中断向量地址；
③ 取中断向量,偏移地址送 IP,段地址送 CS；
④ 转入中断处理程序；
⑤ 中断返回到 INT 指令的下一条指令。

采用向量中断的方法,大大加快了中断处理的速度。因为计算机可直接通过中断向量表转向相应的处理程序,而不需要 CPU 去逐个检测和确定中断原因。

表 8.2 列出了 80x86 各类型中断在中断向量表中的地址。

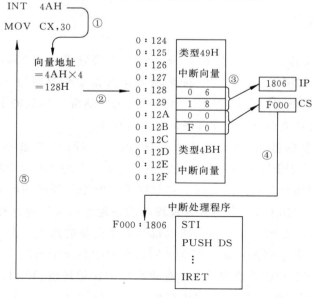

图 8.6 中断操作步骤

表 8.2　中断向量表地址分配

地　址	中断类型号		地　址	中断类型号	
0～7F	0～1F	BIOS 中断向量	1C0～1DF	70～77	I/O 设备中断向量
80～FF	20～3F	DOS 中断向量	1E0～1FF	78～7F	保留
100～17F	40～5F	扩充 BIOS 中断向量	200～3C3	80～FD	BASIC
180～19F	60～67	用户中断向量	3C4～3FF	F1～FF	保留
1A0～1BF	68～6F	保留			

用户可以利用保留的中断类型号扩充自己需要的中断功能,对新增加的中断功能要在中断向量表中建立相应的中断向量。前面已经讨论了中断类型和中断向量地址的对应关系,下面编写指令来为中断类型 N 设置中断向量。

```
        MOV     AX, 0
        MOV     ES, AX                      ;set to base interrupt vectors
        MOV     BX, N * 4                   ;offset of type N interrupt
        MOV     AX, OFFSET INTHAND
        MOV     ES:WORD PTR [BX], AX        ;set addr of INTHAND
        MOV     AX, SEG INTHAND
        MOV     ES：WORD PTR[BX+2], AX
        ⋮
INHAND:                                     ;interrupt processing routine
        ⋮
        IRET
```

如果新的中断功能只供自己使用,或用自己编写的中断处理程序代替系统中的某个中断处理功能时,要注意保存原中断向量。在设置自己的中断向量时,应先保存原中断向量再设置新的中断向量,在程序结束之前恢复原中断向量。

实际上,在检查或设置任何中断向量时,总是避免直接使用中断向量的绝对地址,而是使用 DOS 功能调用(21H)存取中断向量。

设置中断向量
　　把由 AL 指定的中断类型的中断向量 DS：DX 放在中断向量表中
预置：AH = 25H
　　AL = 中断类型号
　　DS：DX = 中断向量
执行：INT 21H

取中断向量
　　把由 AL 指定的中断类型的中断向量从中断向量表中取到 ES：BX 中
预置：AH = 35H
　　AL = 中断类型号

执行：INT 21H
返回时送：ES：BX = 中断向量

例 8.4 使用 DOS 功能调用存取中断向量。

```
            ⋮
    MOV     AL, N                       ;type N interrupt
    MOV     AH, 35H                     ;get interrupt vector
    INT     21H
    PUSH    ES                          ;save the old base and
    PUSH    BX                          ;   offset of interrupt N
    PUSH    DS
    MOV     AX, SEG INTHAND
    MOV     DS, AX                      ;base of INTHAND in DS
    MOV     DX, OFFSET INTHAND          ;offset in DX
    MOV     AL, N                       ;type N
    MOV     AH, 25H                     ;set interrupt vector
    INT     21H
    POP     DS
            ⋮
    POP     DX                          ;restore the old offset
    POP     DS                          ;   and base of interrupt
    MOV     AL, N                       ;type N
    MOV     AH, 25H                     ;set interrupt vector
    INT     21H
    RET                                 ;return
;
INTHAND: ⋮                              ;interrupt processing routine
            ⋮
    IRET
```

8.3.3 中断过程

当中断发生时,由中断机构自动完成下列动作:
(1) 取中断类型号 N;
(2) 标志寄存器(FLAGS)内容入栈;
(3) 当前代码段寄存器(CS)内容入栈;
(4) 当前指令计数器(IP)内容入栈;
(5) 禁止硬件中断和单步中断(IF=0,TF=0);
(6) 从中断向量表中取 4×N 的字节内容送 IP,取 4×N+2 中的字节内容送 CS;
(7) 转中断处理程序。

中断发生的过程很像我们所熟悉的子程序调用,不同的是在保护中断现场时,除了保存返回地址 CS∶IP 之外,还保存了标志寄存器(FLAGS)的内容。因为标志寄存器记录了中断发生时,程序指令运行的结果特征,当 CPU 处理完中断请求返回原程序时,要保证原程序工作的连续性和正确性,所以中断发生时的 FLAGS 内容也要保存起来。另一个不同点是,在中断发生时,CPU 还自动清除了 IF 位和 TF 位,这样设计的目的是使 CPU 转入中断处理程序后,不允许再产生新的中断,如果在执行中断处理程序的过程中,还允许外部的中断,可通过 STI 指令再把 IF 置为 1。

编写中断处理程序和编写子程序一样,所使用的汇编语言指令没有特殊限制,只是中断程序返回时使用 IRET 指令。这条指令的工作步骤和中断发生时的工作步骤正好相反。它首先把 IP,CS 和 FLAGS 的内容出栈,然后返回到中断发生时紧接着的下一条指令。图 8.7 为中断过程的示意图。

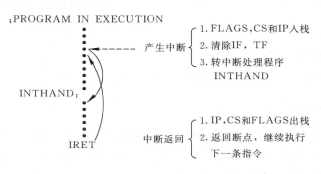

图 8.7 中断过程

8.3.4 中断优先级和中断嵌套

在 80x86 系统中,有软件中断、硬件中断等多个中断源。当多个中断源同时向 CPU 请求中断时,CPU 应如何处理呢?办法是给各种中断源事先安排一个中断优先级次序,当多个中断源同时申请中断时,CPU 先比较它们的优先级(priority),然后从优先级高到优先级低的次序来依次处理各个中断源的中断请求。

8086 规定中断的优先级次序为:

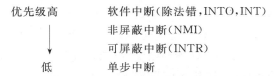

可屏蔽中断的优先权又分为八级,在正常的优先级方式下,优先级次序是:

IR0,IR1,IR2,IR3,IR4,IR5,IR6,IR7

也就是说,按图 8.3 的连接,定时器的优先级最高,键盘其次,打印机的优先权最低。

在 8.3.1 小节中已经提到 8259A 的中断命令寄存器(图 8.4)的 6 位和 7 位能控制各中断请求端的优先次序。在发出一个 EOI 命令时,7 位(R)和 6 位(SL)有四种组合,其含义如下:

R	SL	
0	0	正常优先级方式
0	1	清除由 L2～L0 指定的中断请求
1	0	各中断优先级依次左循环一个位置
1	1	各中断优先级依次循环到由 L2～L0 指定的中断请求到达最低优先级位置上

硬件中断的优先级次序一般在正常优先级方式下(R＝0,SL＝0),但在必要的情况下,设置中断命令寄存器能改变 IR0～IR7 的优先级次序。例如,IR0～IR7 原为正常的优先级次序,现在要使 IR4 成为最低级的中断请求,则给端口 20H 送命令码:11100100,即 R＝1,SL＝1,EOI＝1,L2L1L0＝100,这样,各中断优先级就依次循环到 IR4 为最低优先级的位置上:

IR5,IR6,IR7,IR0,IR1,IR2,IR3,IR4

如果再送一个命令码:10100000,则优先级次序再向左循环一个位置,成为:

IR6,IR7,IR0,IR1,IR2,IR3,IR4,IR5

在下面的章节中,如无特别说明,则中断优先级是指正常方式下的中断优先级。

正在运行的中断处理程序,又被其他中断源中断,这种情况叫做中断嵌套。80x86 没有规定中断嵌套的深度(中断程序又被中断的层次),但在实际使用时,多重的中断嵌套要受到堆栈容量的限制,所以在编写中断程序时,一定要考虑有足够的堆栈单元来保存多次中断的断点及各寄存器的内容。

一个正在执行的中断处理程序,在开中断(IF＝1)的情况下,能被优先级高于它的中断源中断,但如果要被同级或低级的中断源中断,则必须发出 EOI 命令,清除正在执行的中断请求,才能响应同级或低级的中断。

下面举一个例子来说明在正常优先级方式下,优先级中断和中断嵌套发生时的处理过程,如图 8.8 所示。

假定在主程序的执行过程中,IR2 和 IR4 的中断请求同时发生,而后 IR1 的中断请求又到达,最后 IR3 的中断请求也到达。

首先,CPU 响应优先级高的 IR2,转去处理 IR2 的中断处理程序。进入 IR2 处理程序后,IF 被置为 1。当 IR1 的中断请求到达后,因 IR1 的优先级高于 IR2,CPU 就立即中断 IR2 的程序,转去执行 IR1 的处理程序。在 IR1 处理程序中,由指令发出了 EOI 命令,结束了 IR1 的中断请求。返回 IR2 处理程序后,同样由于发出 EOI 命令清除了 IR2 的中断请求,所以在较低级的中断请求 IR4 到达后,即转向处理 IR4 的中断请求。在 IR4 处理程序的执行过程中,IR3 的中断请求到达,当判断 IF 已被置为 1,则又中断了 IR4 的程序,转去执行 IR3 的程序。在 IR3 程序中,也发出了开中断指令(STI)和中断结束命令(EOI),最后 IRET 指令使其返回到 IR4 程序,IR4 在返回 IR2 之前也发出了 EOI 命令,结束了 IR4 的中断请求。IR2 中断请求在前面已被清除,所以 IR4 执行完后,IR2 继续执行直到返回主程序。

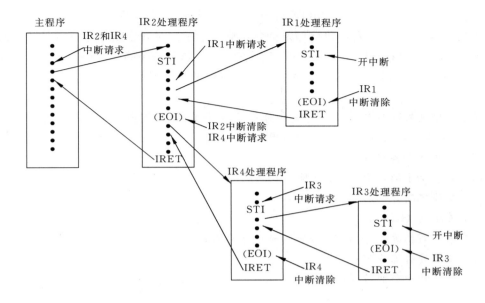

图 8.8 正常优先级方式下的典型中断序列

8.3.5 中断处理程序

通过前面几节对中断的介绍,对如何编写中断程序已经有了一些了解,现在把它们归纳在一起就更清楚了。

下面是主程序为响应中断所做的准备工作以及硬件(包括 CPU 和外设接口)自动完成的动作:

主程序 { (1)设置中断向量
(2)设置设备的中断屏蔽位
(3)设置 CPU 的中断允许位 IF(开中断)

硬件 { (4)外设接口送中断请求给 CPU
(5)当前指令执行完后,CPU 送响应信号给外设接口
(6)CPU 接收中断类型号
(7)当前的 FLAGS,CS 和 IP 保存入栈
(8)清除 IF 和 TF
(9)中断向量送 IP 和 CS

至此,CS 和 IP 寄存器取得了中断处理程序的段地址和偏移地址,CPU 就把控制转给中断处理程序。这里要注意的是设备发到 CPU 的中断请求信号在时间上是随机的,只要未被屏蔽的设备本身的状态是准备好或空闲的,它就会向 CPU 请求中断,如果此时 CPU 正在执行一条指令,那么就要等这条指令执行完后,才响应中断。对加封锁的指令(如 LOCK MOV AX,BX)应看作为一条指令处理;对加重复前缀的串指令(如 REP MOVSB),也要作为一个整体来处理,但不是把串操作全部重复执行完,而是执行一次重

复和串指令即可响应中断。对 MOV 指令和 POP 指令，如果处理对象是段寄存器时，那么本条指令执行完后，接着再执行一条指令才响应中断。对开中断指令 STI 和中断返回指令 IRET，也是要在 STI 或 IRET 指令执行完后，再执行一条指令才响应中断。以上是几种特殊情况，对一般指令，只要一条指令的执行周期结束即可响应中断。

注意：中断处理程序的编写方法和标准子程序很类似，下面是编写中断处理子程序的步骤，但它与子程序编写也有一些不同之处。

（1）保存寄存器内容；
（2）如允许中断嵌套，则开中断（STI）；
（3）处理中断；
（4）关中断（CLI）；
（5）送中断结束命令（EOI）给中断命令寄存器；
（6）恢复寄存器内容；
（7）返回被中断的程序（IRET）。

进入中断处理程序时，IF 和 TF 已经被清除，这样在执行中断处理程序的过程中，将不再响应其他外设的中断请求，如果这个中断处理程序允许其他设备中断，则需用 STI 指令把 IF 位置 1。中断结束命令（EOI）在程序的什么地方发出，这要看程序员是否要求在其处理过程中允许同级或低级中断。一般设备希望一次中断的处理过程最好是完整的，所以只在中断处理结束之前发出 EOI 命令。

处理中断部分是中断处理程序的主体部分，它要完成的任务是各种各样的，这与实际应用有关。如果它的任务是处理某种错误的，一般要求显示输出一系列出错信息。如果它是对一个 I/O 设备进行服务的，就按其端口地址接收或发送一个单位（字节或字）的数据。

注意：CPU 产生一次中断，I/O 设备只完成一个字节（或字）的输入输出，所以中断处理程序所用的指针变量或数据变量一般应设置存储单元来保存。

下面通过几个实例来说明中断程序设计的方法。

例 8.5 编写一个中断处理程序，要求在主程序运行过程中，每隔 10 秒响铃一次，同时在屏幕上显示出信息"The bell is ring!"

在系统定时器（中断类型为 8）的中断处理程序中，有一条中断指令 INT 1CH，时钟中断每发生一次（约每秒中断 18.2 次）都要嵌套调用一次中断类型 1CH 的处理程序。在 ROM BIOS 例程中，1CH 的处理程序只有一条 IRET 指令，实际上它并没有做任何工作，只是为用户提供了一个中断类型号。如果用户有某种定时周期性的工作需要完成，就可以利用系统定时器的中断间隔，用自己设计的处理程序来代替原有的 1CH 中断程序。

1CH 作为用户使用的中断类型，可能已被其他功能的程序所引用，所以在编写新的中断程序时，应做下述工作：

(1) 在主程序的初始化部分,先保存当前1CH的中断向量,再设置新的中断向量。
(2) 在主程序的结束部分恢复保存的1CH中断向量。

```
;**************************************************
;eg8-5.asm
;Purpose: ring and display a message every 10 seconds.
;**************************************************
            .model    small
;--------------------------------------------------------------------
            .stack
;--------------------------------------------------------------------
            .data
count       dw        1
msg         db        'The bell is ringing!', 0dh, 0ah, '$'
;--------------------------------------------------------------------
            .code
; Main program
main        proc      far
start:
            mov       ax, @data             ;allot data segment
            mov       ds, ax

; save old interrupt vector
            mov       al, 1ch               ;al<=vector number
            mov       ah, 35h               ;to get interrupt vector
            int       21h                   ;call DOS
            push      es                    ;save registers for restore
            push      bx
            push      ds

;set new interrupt vector
            mov       dx, offset ring       ;dx<=offset of procedure ring
            mov       ax, seg ring          ;ax<=segment of procedure ring
            mov       ds, ax                ;ds<=ax
            mov       al, 1ch               ;al<=vector#
            mov       ah, 25h               ;to set interrupt vector
            int       21h                   ;call DOS

            pop       ds                    ;restore ds
            in        al, 21h               ;set interrupt mask bits
            and       al, 11111110b
            out       21h, al
```

```
                sti

                mov     di, 20000
delay:          mov     si, 30000
delay1:         dec     si
                jnz     delay1
                dec     di
                jnz     delay
;restore old interrupt vector
                pop     dx                          ;restore registers
                pop     ds
                mov     al, 1ch                     ;al<=vector#
                mov     ah, 25h                     ;to restore interrupt
                int     21h                         ;call DOS

                mov     ax, 4c00h                   ;exit
                int     21h
main            endp
;--------------------------------------------------------------------------------
; Procedure ring
; Purpose: ring every 10 seconds when substituted for interrupt 1ch
ring            proc    near
                push    ds                          ;save the working registers
                push    ax
                push    cx
                push    dx

                mov     ax, @data                   ;allot data segment
                mov     ds, ax
                sti

;siren if it is time for ring
                dec     count                       ;count for ring interval
                jnz     exit                        ;exit if not for ring time

                mov     dx, offset msg              ;dx<=offset of msg
                mov     ah, 09h                     ;to display msg
                int     21h                         ;call DOS

                mov     dx, 100                     ;dx<=turn on/off times(100)
                in      al, 61h                     ;get port 61h
                and     al, 0fch                    ;mask bits 0,1
```

```
sound:
          xor      al, 02                      ;toggle bit 1
          out      61h, al                     ;output to port 61h

          mov      cx, 1400h                   ;value of wait
wait1:    loop     wait1
          dec      dx                          ;control turn on/off 10 times
          jne      sound
          mov      count, 182                  ;control ring interval delay(10s)
exit:     cli
          mov      al,20h                      ;set EOI
          mov      20h,al
          pop      dx                          ;restore the reg.
          pop      cx
          pop      ax
          pop      ds
          iret                                 ;interrupt return
ring      endp
;------------------------------------------------------------------
          end      start                       ;end assemble
```

例 8.6 在配置了键盘输入(中断类型 09)和打印机输出(中断类型 0FH)两种外部设备的 80x86 中断系统中,要求从键盘上接收字符,同时对 32 字节的输入缓冲区进行测试,如果缓冲区已满,则键盘挂起(禁止键盘中断输入),由打印机输出一个提示信息。

键盘和打印机分别由中断屏蔽寄存器(21H)的 1 位和 7 位控制。键盘的输入寄存器端口地址为 60H,控制寄存器的端口地址为 61H。打印机输出寄存器的端口地址为 378H,打印机控制寄存器的端口地址为 37AH。

在这种特定情况下,只要求打印机在键盘输入缓冲区满了后,打印出提示信息,因此它可以在屏蔽键盘中断的同时,设置打印机的中断屏蔽位。另外,在中断处理程序中用到的一些指针及计数值要保存在指定的存储单元中,每次进入中断,取出指针及计数值,退出中断时,再把修改后的指针及计数值保存起来。

这个中断程序包括以下几部分:

MAIN　　初始化部分。保存 09 和 0FH 的原中断向量,设置新的中断向量。主程序用有限循环来模拟。主程序结束时,恢复原中断向量。

KBDINT　　键盘中断处理程序。接收按键的扫描码并保存在缓冲区中,如果输入的字符数超过 32,则屏蔽键盘中断,允许打印机中断,并调用 INIT_PRT 子程序初始化打印机。

INIT_PRT　　初始化打印机。启动适配器,发出选通信号。

PRTINT　　打印机中断处理程序。按照指针取出打印机字符送到输出寄存器,发出选通信号。

DISPLAY_HEX　　用十六进制显示 AL 中的代码。

```asm
;**************************************************************
;eg8-6.asm
;Purpose: accept keyboard input and print messages on the printer
;**************************************************************
                .model      small
;--------------------------------------------------------------------
                .stack
;--------------------------------------------------------------------
                .data
old_ip09        dw          ?
old_cs09        dw          ?
old_ip0f        dw          ?
old_cs0f        dw          ?
count           dw          ?
buffer          db          20h dup(' ')
buf_p           dw          ?
start_msg       db          0ah, 0dh, 'RUN', 0ah, 0dh, '$'
end_msg         db          0ah, 0dh, 'END', 0ah, 0dh, '$'
full_msg        db          'Buffer full!', 0ah, 0dh,
;--------------------------------------------------------------------
                .code
; Main program
main            proc        far
start:
                mov         ax, @data               ;ds<=data segment
                mov         ds, ax

; initialize
                lea         ax, buffer              ;buf_p<=buffer address
                mov         buf_p, ax
                mov         count, 0                ;count<=0

; save old interrupt 09h
                mov         al, 09h                 ;al<=vector#
                mov         ah, 35h                 ;to get interrupt vector
                int         21h                     ;call DOS
                mov         old_cs09, es            ;save registers for restore
                mov         old_ip09, bx
                push        ds                      ;save ds
```

```
;set new interrupt 09h
        lea     dx, kbdint              ;dx<=offset of procedure kbdint
        mov     ax, seg kbdint          ;ax<=segment of procedure kbdint
        mov     ds, ax                  ;ds<=ax
        mov     al, 09h                 ;al<=intrrupt number
        mov     ah, 25h                 ;to set interrupt vector
        int     21h                     ;call DOS
        pop     ds                      ;restore ds

;set keyboard interrupt mask bits
        in      al, 21h
        and     al, 0fdh
        out     21h, al

;save old interrupt 0fh
        mov     al, 0fh                 ;al<=vector #
        mov     ah, 35h                 ;to get interrupt vector
        int     21h                     ;call DOS
        mov     old_cs0f, es            ;save registers for restore
        mov     old_ip0f, bx
        push    ds                      ;save ds

;set new interrupt 0fh
        lea     dx, prtint              ;dx<=offset of procedure prtint
        mov     ax, seg prtint          ;ax<=segment of procedure prtint
        mov     ds, ax                  ;ds<=ax
        mov     al, 0fh                 ;al<=vector #
        mov     ah, 25h                 ;to set interrupt vector
        int     21h                     ;call DOS
        pop     ds                      ;restore ds

        mov     ah, 09h                 ;print start message
        lea     dx, start_msg
        int     21h

        sti
        mov     di, 20000               ;main process
mainp:  mov     si, 30000
mainp1: dec     si
        jnz     mainp1
        dec     di
```

```
                jnz         mainp

                mov         ah, 09h                     ;print end msg of main process
                lea         dx, end_msg
                int         21h

                cli
;restore old interrupt 09h
                push        ds                          ;save ds
                mov         dx, old_ip09                ;ds:dx<=old handler address
                mov         ax, old_cs09
                mov         ds, ax
                mov         al, 09h                     ;al<=vector #
                mov         ah, 25h                     ;to restore interrupt vector
                int         21h                         ;call DOS
                pop         ds                          ;restore ds

; restore old interrupt 0fh
                push        ds                          ;save ds
                mov         dx, old_ip0f                ;ds:dx<=old address
                mov         ax, old_cs0f
                mov         ds, ax
                mov         al, 0fh                     ;al<=vector #
                mov         ah, 25h                     ;to restore interrupt
                int         21h                         ;call DOS
                pop         ds                          ;restore ds

; enable keyboard interrupt
                in          al, 21h
                and         al, 0fdh
                out         21h, al

                sti
                mov         ax, 4c00h
                int         21h
main            endp
;--------------------------------------------------------------------------------
; Interrupt Handler kbd
;               Purpose: fill buffer until full when substituted for interrupt 09h
kbdint          proc        near
                push        ax                          ; save registers
```

```
                push        bx

                cld                             ;direction: forward
                in          al, 60h             ;read a character
                push        ax                  ;save it
                in          al, 61h             ;get the control port
                mov         ah, al              ;save the value in ah
                or          al, 80h             ;reset bits for kbd
                out         61h, al             ;send out
                xchg        ah, al              ;restore control value
                out         61h, al             ;kdb has been reset

                pop         ax                  ;restore scan code
                test        al, 80h             ;press or release?
                jnz         return1             ;ignore when release

                mov         bx, buf_p           ;bx<=buffer pointer
                mov         [bx], al            ;store in buffer
                call        display_hex         ;display in hex
                inc         bx                  ;move pointer
                inc         count               ;count characters
                mov         buf_p, bx           ;save the pointer
check:
        cmp     count, 20h                      ;judge whether full
                jb          return1
                in          al, 21h
                or          al, 02              ;mask kdb bits
                and         al, 7fh             ;enable prt bits
                out         21h, al
                call        init_prt            ;initiate printer
return1:
cli
                mov         al, 20h             ;end of interrupt
                out         20h, al
;restore registers
                pop         bx
                pop         ax
                iret                            ;interrupt return
kbdint          endp
;------------------------------------------------------------------------------
; Interrupt Handler prtint
; Purpose: print characters when substituted for interrupt 0fh
```

```
prtint      proc    near
            push    ax                      ;save registers
            push    bx
            push    dx

            mov     bx, buf_p               ;bx<=buffer pointer
            mov     al, [bx]                ;get from buffer
            mov     dx, 378h                ;printer data port
            out     dx, al                  ;output a character

            push    ax                      ;save ax
            mov     dx, 37ah                ;printer control port
            mov     al, 1dh                 ;al<=control code
            out     dx, al                  ;send out
            jmp     $+2                     ;slight delay
            mov     al, 1ch                 ;al<=control code
            out     dx, al                  ;send out
            pop     ax                      ;restore ax

            inc     bx                      ;move pointer
            mov     buf_p, bx               ;save the pointer
            cmp     al, 0ah                 ;end of message?
            jnz     return2

            in      al, 21h                 ;disable printer
            or      al, 80h                 ;   interrupt
            out     21h, al

return2:
            mov     al, 20h                 ;end of interrupt
            out     20h, al

            pop     dx                      ;restore registers
            pop     bx
            pop     ax
            iret                            ;interrupt return
prtint      endp
;------------------------------------------------------------------

; Procedure init_prt
init_prt    proc    near
            push    ax                      ;save registers
```

```
            push        bx
            push        dx

            cli
            lea         bx, full_msg            ;bx<=offset of full_msg
            mov         buf_p, bx               ;save full_msg address

            mov         dx, 378h                ;printer data port
            mov         al, 0dh                 ;CR
            out         dx, al                  ;output a character

            mov         dx, 37ah                ;printer control port
            mov         al, 1dh                 ;al<=control code
            out         dx, al                  ;send out
            jmp         $+2                     ;slight delay
            mov         al, 1ch                 ;al<=control code
            out         dx, al                  ;send out

            pop         dx                      ;restore registers
            pop         bx
            pop         ax

            ret
init_prt    endp
;-----------------------------------------------------------------
display_hex proc        near                    ;display char with hex
            push        ax
            push        cx
            push        dx
            mov         ch, 2                   ;number of digits
            mov         cl, 4
nextb:
            rol         al, cl                  ;highest of bits to lowest
            push        ax
            mov         dl, al
            and         dl, 0fh                 ;mask off left digit
            or          dl, 30h                 ;convert to ASCII
            cmp         dl, 3ah
            jl          dispit
            add         dl, 7h                  ;digit is A to F
dispit:
            mov         ah, 2                   ;display character
```

```
            int     21h
            pop     ax
            dec     ch                      ;done 2 digits?
            jnz     nextb                   ;not yet
            mov     ah, 2
            mov     dl,','                  ;display ','
            int     21h
            pop     dx
            pop     cx
            pop     ax
            ret                             ;return from display_hex
display_hex endp
;--------------------------------------------------------------
            end     start
```

例8.7 除数为0时的软件中断(类型0)处理程序

此程序分成两个主要部分：初始化部分和中断处理部分。

初始化部分(Init)设置新的0型中断向量，显示一条信息，然后完成终止和驻留后退出程序。这种特殊的退出是用INT 21H的功能31H，它将保留程序所占用的内存，从而使这些内存单元不被以后的应用程序破坏。

中断处理程序(Zdiv)在发生一个被零除中断时接收控制。中断处理程序先保存有关寄存器的值，然后打印出信息询问用户是退出程序(quit)还是继续(continue)。若键入"C"要求继续执行程序，则处理程序恢复所有寄存器并执行IRET返回主程序(显示一个标记符♯)，当然此时除法的操作结果应是无效的。若键入"Q"要求退出，则从处理程序直接返回DOS(无标记符显示)。这里返回DOS，是用INT 21H的功能4CH，该功能是惟一不依赖于任何段寄存器内容的终止功能。例如，CS寄存器不必指向PSP所在的段。该功能的另一个优点是能在AL中返回一个表明程序是否正常终止的出口代码。系统出口代码的含义为：00——正常终止；01——用Ctrl_C终止；02——严重设备错误引起终止；03——用功能调用31H终止；0FFH——CPU错误引起终止。下面是处理除数为0错误的中断处理程序清单。

```
;************************************************************
;eg8-7.asm
;Purpose: zero_division handler
;************************************************************
            .model   small
;--------------------------------------------------------------
            .stack
;--------------------------------------------------------------
            .code
; Main program
```

```
main            proc        near
;reset interrupt vector 0
                lea         dx, zdiv                    ;set interrupt vector
                mov         ax, seg zdiv
                mov         ds, ax
                mov         al, 0                       ;interrupt number
                mov         ah, 25h                     ;to reset interrupt vector
                int         21h                         ;call DOS

;print introduction message
                mov         ax, @code                   ;ds<=code segment
                mov         ds, ax
                mov         dx, ok_msg                  ;print introduction
                mov         ah, 9
                int         21h
;simulate zero_division condition
                mov         ax, 1
                mov         dl, 0
                div         dl

; display '#' after return from interrupt handler zdiv
                mov         ah, 2                       ;to display a character
                mov         dl, '#'                     ;dl<='#'
                int         21h                         ;call DOS

;exit and reside in memory
                mov         ah, 31h                     ;to exit and reside
                mov         al, 0                       ;return code: 0
                mov         dx, ((prog_len+15)/16)+16   ;dx<=memory paragraph
                                                        ;    for residence
                int         21h                         ;call DOS
main            endp
;--------------------------------------------------------------------------------
; Interrupt Handler zdiv
zdiv            proc        far
                push        ax                          ;save registers
                push        bx
                push        cx
                push        dx
                push        si
                push        di
                push        bp
```

```
                push    ds
                push    es
                sti
prt_warn:
                mov     ax, @code
                mov     ds, ax
                mov     dx, offset warn_msg         ;print warning
                mov     ah, 9
                int     21h
input:
                mov     ah, 1                       ;to accept keyboard input
                int     21h                         ;call DOS
                cmp     al, 'c'                     ;judge whether 'c'
                je      continue
                cmp     al, 'q'                     ;judege whether 'q'
                je      exit

                mov     dx, offset beep             ;beep when illegal input
                mov     ah, 09
                int     21h
                jmp     prt_warn
exit:
                mov     ax, 4cffh
                int     21h
continue:
                mov     dx, offset crlf             ;print CR & LF
                mov     ah, 09
                int     21h
                cli

                pop     es                          ;restore registers
                pop     ds
                pop     bp
                pop     di
                pop     si
                pop     dx
                pop     cx
                pop     bx
                pop     ax
                iret                                ;interrupt return
zdiv            endp
;--------------------------------------------------------------------------
```

```
; Data area
ok_msg          db          0dh, 0ah, 'Zero-division Handler installed!'
                db          0dh, 0ah, '$'
warn_msg        db          'Zero-division detected，', 07h
                db          'Continue or Quit(c/q)? $'
beep            db          07h, '$'
crlf            db          0dh, 0ah, '$'

prog_len        equ         $-main
;------------------------------------------------------------
                end         main
```

习　　题

8.1 写出分配给下列中断类型号在中断向量表中的物理地址。
　　　(1)　INT　12H　　　(2)　INT　8
8.2 用 CALL 指令来模拟实现 INT 21H 显示字符 T 的功能。
8.3 写出指令将一个字节数据输出到端口 25H。
8.4 写出指令将一个字数据从端口 1000H 输入。
8.5 假定串行通信口的输入数据寄存器的端口地址为 50H,状态寄存器的端口地址为 51H,状态寄存器各位为 1 时含义如下：

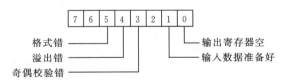

请编写一程序:输入一串字符并存入缓冲区 BUFF,同时检验输入的正确性,如有错,则转出错处理程序 ERR_ROUT。

8.6 试编写程序,它轮流测试两个设备的状态寄存器,只要一个状态寄存器的第 0 位为 1,则与其相应的设备就输入一个字符;如果其中任一状态寄存器的第 3 位为 1,则整个输入过程结束。两个状态寄存器的端口地址分别是 0024 和 0036,与其相应的数据输入寄存器的端口则为 0026 和 0038,输入字符分别存入首地址为 BUFF1 和 BUFF2 的存储区中。

8.7 假设外部设备有一台硬币兑换器,其状态寄存器的端口地址为 0006,数据输入寄存器的端口地址为 0005,数据输出寄存器的端口地址为 0007。试用查询方式编制一程序,该程序作空闲循环等待纸币输入,当状态寄存器的第 2 位为 1 时,表示有纸币输入。此时可以数据输入寄存器输入的代码中测出纸币的品种,壹角纸币的代码为 01,贰角纸币为 02,伍角纸币则为 03,然后程序在等待状态寄存器的第 3 位变为 1 后,把应兑换的伍分硬币数(用十六进制)从数据输出寄存器输出。

8.8 给定(SP)=0100,(SS)=0300,(FLAGS)=0240,以下存储单元的内容为(00020)=0040,(00022)=0100,在段地址为0900及偏移地址为00A0的单元中有一条中断指令INT 8,试问执行INT 8指令后,SP,SS,IP,FLAGS的内容是什么？栈顶的三个字是什么？

8.9 类型14的中断向量在存储器的哪些单元里？

8.10 假设中断类型9的中断处理程序的首地址为INT_ROUT,试写出主程序中为建立这一中断向量而编制的程序段。

8.11 编写指令序列,使类型1CH的中断向量指向中断处理程序SHOW_CLOCK。

8.12 如设备D1,D2,D3,D4,D5是按优先级次序排列的,设备D1的优先级最高。而中断请求的次序如下所示,试给出各设备的中断处理程序的运行次序。假设所有的中断处理程序开始后就有STI指令。

(1) 设备3和4同时发出中断请求；

(2) 在设备3的中断处理程序完成之前,设备2发出中断请求；

(3) 在设备4的中断处理程序未发出中断结束命令(EOI)之前,设备5发出中断请求；

(4) 以上所有中断处理程序完成并返回主程序,设备1,3,5同时发出中断请求。

8.13 在8.12题中假设所有的中断处理程序中都没有STI指令,而它们的IRET指令都可以由于FLAGS出栈而使IF置1,则各设备的中断处理程序的运行次序应是怎样的？

8.14 试编制一程序,要求测出任意程序的运行时间,并把结果打印出来。

第 9 章 BIOS 和 DOS 中断

在存储器系统中,从地址 0FE000H 开始的 8KB ROM(只读存储器)中装有基本输入输出系统(basic input /output system,BIOS)例行程序。驻留在 ROM 中的 BIOS 给 PC 系列的不同微处理器提供了兼容的系统加电自检、引导装入、主要 I/O 设备的处理程序以及接口控制等功能模块来处理所有的系统中断。使用 BIOS 功能调用,给程序员编程带来很大方便。程序员不必了解硬件操作的具体细节,直接使用指令设置参数,并中断调用 BIOS 例行程序,所以利用 BIOS 功能编写的程序简洁,可读性好,而且易于移植。

磁盘操作系统(disk operating system,DOS)是 PC 机上最重要的操作系统,它是由软盘或硬盘提供的。它的两个 DOS 模块 IBMBIO.COM 和 IBMDOS.COM 使 BIOS 用起来更方便。因为 DOS 模块提供了更多更必要的测试,使 DOS 操作比使用相应功能的 BIOS 操作更简易,而且 DOS 对硬件的依赖性更少些。

IBMBIO.COM 是一个输入输出设备处理程序,它提供了 DOS 到 ROM BIOS 的低级接口,它完成将数据从外设读入内存,或把数据从内存写到外设去的工作。

IBMDOS.COM 包括一个文件管理程序和一些处理程序,在 DOS 下运行的程序可以调用这些处理程序。为了完成 DOS 功能调用,IBMDOS.COM 把信息传送给 IBMBIO.COM,形成一个或多个 BIOS 调用。它们之间的关系如图 9.1 所示。

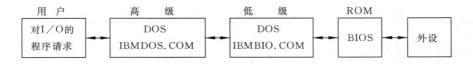

图 9.1 DOS 模块和 ROM BIOS 的关系

在一些情况下,既能选择 DOS 中断也能选择 BIOS 中断来执行同样的功能。例如,打印机输出一个字符的功能,可用 DOS 中断 21H 的功能 5,也可用 BIOS 中断 17H 的功能 0。因为 BIOS 比 DOS 更靠近硬件,因此建议尽可能地使用 DOS 功能,但在少数情况下必须使用 BIOS 功能。例如,BIOS 中断 17H 的功能 2 为读打印机状态,它就没有等效的 DOS 功能。

DOS 中断能处理大多数的 I/O,但有一些功能还没有提供,如声音控制等,这就要考虑用 I/O 指令在端口级上编程,或使用高级语言编程。

表 9.1 和表 9.2 列出了 IBM PC 系统主要的 BIOS 中断类型和 DOS 中断类型。

DOS 功能与 BIOS 功能都通过软件中断调用。在中断调用前需要把功能号装入 AH 寄存器,把子功能号装入 AL 寄存器,除此而外,通常还需在 CPU 寄存器中提供专门的调用参数。一般地说,调用 DOS 或 BIOS 功能时,有以下几个基本步骤:

表 9.1 BIOS 中断类型

CPU 中断类型	8259 中断类型	BIOS 中断类型	用户应用程序和数据表指针
			用户应用程序
0 除法错	8 系统定时器(IRQ0)	10 显示器 I/O	1B 键盘终止地址(Ctrl_Break)
1 单步	9 键盘(IRQ1)	11 取设备信息	4A 报警(用户闹钟)
2 非屏蔽中断	A 彩色/图形接口(IRQ2)	12 取内存容量	
3 断点	B COM2 控制器(IRQ3)	13 磁盘 I/O	1C 定时器
4 溢出	C COM1 控制器(IRQ4)	14 RS-232 串行口 I/O	
5 打印屏幕	D LPT2 控制器(IRQ5)	15 磁带 I/O	数据表指针
6 保留	E 磁盘控制器(IRQ6)	16 键盘 I/O	1D 显示器参数表
7 保留	F LPTI 控制器(IRQ7)	17 打印机 I/O	1E 软盘参数表
		18 ROM BASIC	1F 图形字符扩展码
		19 引导装入程序	41 0# 硬盘参数表
		1A 时钟	46 1# 硬盘参数表
		40 软盘 BIOS	49 指向键盘增强服务变换表

表 9.2 DOS 中断类型

20	程序终止	27	结束并驻留内存
21	功能调用	28	键盘忙循环
22	终止地址	29	快速写字符
23	Ctrl_C 中断向量	2A	网络接口
24	严重错误向量	2E	执行命令
25	绝对磁盘读	2F	多路转接接口
26	绝对磁盘写	30～3F	保留给 DOS

(1) 将调用参数装入指定的寄存器中；
(2) 如需功能号,把它装入 AH；
(3) 如需子功能号,把它装入 AL；
(4) 按中断号调用 DOS 或 BIOS 中断；
(5) 检查返回参数是否正确。

9.1　键盘 I/O

键盘是计算机最基本的一种输入设备,用来输入信息,以达到人机对话的目的。键盘主要由三种基本类型的键组成：

(1) 字符数字键。如字母 A(a)～Z(z),数字 0～9 以及％,$,♯等常用字符。
(2) 扩展功能键。如 Home,End,Backspace,Arrows,Return,Delete,Insert,PgUp,pgD 以及程序功能键 F1～F10 等。
(3) 和其他键组合使用的控制键。如 Alt,Ctrl 和 Shift 等。

字符数字键给计算机传送一个 ASCII 码字符,而扩展功能键产生一个动作,如按下

Home 键能把光标移到屏幕的左上角，End 键使光标移到屏幕上文本的末尾，使用组合控制键能改变其他键所产生的字符码。

键盘和主机通过五芯电缆相连，这五根线分别是电源线、地线、复位线以及键盘数据线和键盘时钟线。PC 机系列的键盘触点电路按 16 行×8 列的矩阵来排列，用单片机 Intel 8048 来控制对键盘的扫描。按键的识别采用行列扫描法，即根据对行线和列线的扫描结果来确定闭合键的位置，并通过键盘数据线将闭合键所对应的扫描码(8 位)送往主机。

9.1.1 字符码与扫描码

当在键盘上"按下"或"放开"一个键时，如果键盘中断是允许的(21H 端口第 1 位 = 0)，就会产生一个类型 9 的中断，并转入到 BIOS 的键盘中断处理程序。该处理程序从 8255 可编程序外围接口芯片的输入端口 60H 读取一个字节，这个字节的低 7 位是键的扫描码。最高位为 0 或为 1，分别表示键是"按下"状态还是"放开"状态。按下时，取得的字节称为通码，放开时取得的字节称为断码。如按下 Esc 键时产生一个通码为 01H (0000001B)，放开 Esc 键时产生一个断码为 81H(10000001B)。

键盘上的每个键都对应一个扫描码，从 01(Esc)到 83(Del)，或从 01H 到 53H，所以根据扫描码就能惟一地确定哪一个键改变了状态。表 9.3 是键盘上每个键对应的扫描码(十六进制)。

表 9.3 IBM 键盘的扫描码表

键	扫描码	键	扫描码	键	扫描码	键	扫描码
Esc	01	U and u	16	| and \	2B	F6	40
! and 1	02	I and i	17	Z and z	2C	F7	41
@ and 2	03	O and o	18	X and x	2D	F8	42
# and 3	04	P and p	19	C and c	2E	F9	43
$ and 4	05	{ and [	1A	V and v	2F	F10	44
% and 5	06	} and]	1B	B and b	30	NumLock	45
^ and 6	07	Enter	1C	N and n	31	ScrollLock	46
& and 7	08	Ctrl	1D	M and m	32	7 and Home	47
* and 8	09	A and a	1E	< and ,	33	8 and ↑	48
(and 9	0A	S and s	1F	> and .	34	9 and PgUp	49
) and 0	0B	D and d	20	? and /	35	－(灰色)	4A
_ and -	0C	F and f	21	Shift(右)	36	4 and ←	4B
+ and =	0D	G and g	22	PrtSc	37	5(小键盘)	4C
Backspace	0E	H and h	23	Alt	38	6 and →	4D
Tab	0F	J and j	24	Space	39	＋(灰色)	4E
Q and q	10	K and k	25	CapsLock	3A	1 and End	4F
W and w	11	L and l	26	F1	3B	2 and ↓	50
E and e	12	: and ;	27	F2	3C	3 and PgDn	51
R and r	13	" and '	28	F3	3D	0 and Ins	52
T and t	14	~ and `	29	F4	3E	. and Del	53
Y and y	15	Shift(左)	2A	F5	3F		

BIOS 键盘处理程序将取得的扫描码转换成相应的字符码,大部分键的字符码是一个标准的 ASCII 码,没有相应 ASCII 码的键。如 Alt 和功能键(F1 ~ F10),字符码为 0,还有一些非 ASCII 码键产生一个指定的操作,如打印屏幕内容等。转换成的字符码以及扫描码存储在 ROM BIOS 数据区的键盘缓冲区 KB_BUFFER 中。

```
0040:001A    BUFF_HEAD       DW    ?              ;键盘缓冲区的首地址
0040:001C    BUFF_TAIL       DW    ?              ;键盘缓冲区的末地址
0040:001E    KB_BUFFER       DW    16 DUP(?)      ;16 个输入量的空间
0040:003E    KB_BUFFER_END   LABEL WORD
```

这个缓冲区是一个先进先出的循环队列,BUFF_HEAD 和 BUFF_TAIL 是缓冲区的两个地址指针。当 HEAD 指针和 TAIL 指针相等时,说明缓冲区空。当 CPU 想要得到键盘输入时,就调用 BIOS 键盘例行程序,它按其接收时的次序从缓冲区取出字符和扫描码,回送给 CPU。缓冲区的大小可适应最快的打字员,但如果缓冲区已满又按下一个键,BIOS 不处理这个键,只发出"嘀"的响声。

我们可以用 BIOS 中断,也可以用 DOS 中断和键盘通信,下面我们分别讨论这两种键盘中断。

9.1.2 BIOS 键盘中断

类型 16 的中断提供了基本的键盘操作,它的中断处理程序包括 3 个不同的功能,分别根据 AH 寄存器的内容来选择(见表 9.4)。

表 9.4 BIOS 键盘中断(INT 16H)

AH	功　　能	返回参数
0	从键盘读一字符	AL=字符码
		AH=扫描码
1	读键盘缓冲区的字符	如 ZF=0
		AL=字符码
		AH=扫描码
		如 ZF=1,缓冲区空
2	取键盘状态字节	AL=键盘状态字节

利用 INT 16H 调用键盘 I/O ROM 例行程序时,先在 AH 中存放一个功能编号 0,1 或 2。例如,要查看按键的扫描码和 ASCII 码,可以调用中断类型 16H 的 0 功能,该功能把扫描码回送到 AH 中,把 ASCII 码回送到 AL 中;然后调用二进制转换十六进制的子程序 BINIHEX,把 AH 和 AL 中的内容打印出来。其指令序列为:

```
MOV     AH,0        ; read character function
INT     16H         ; keyboard I/O ROM call
MOV     BX,AX       ; move AX to BX
```

```
        CALL    BINIHEX         ; print scan code & char
```

前面已经提到 Shift，Ctrl，Alt，Num Lock，Scroll，Ins 和 Caps Lock 这些键不具有 ASCII 码，但按动了它们能改变其他键所产生的代码。那么如何能判断这些键按动与否呢？INT 16H 的 AH＝2 的功能可以把表示这些键状态的字节——键盘状态字节（KB_FLAG）回送到 AL 寄存器。图 9.2 标出了 KB_FLAG 各位表示的状态信息，其中高 4 位指出各种键盘方式（Ins，Caps Lock，Num Lock，Scroll）是 ON(1) 还是 OFF(0)；低 4 位表示 Alt，Ctrl 和 Shift 键是否按动。这 8 个键有时又被称为变换键。

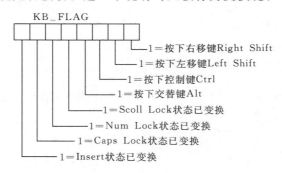

图 9.2 键盘状态字节

例 9.1 的程序可以读取 KB_FLAG 的内容，如果要显示出各位的状态，可调用 BINIHEX 子程序来显示 KB_FLAG 的十六进制内容。

例 9.1 读取键盘状态字节，并以十六进制打印出来。

```
AGAIN：
        MOV     AH,02H          ; shift status function
        INT     16H             ; call kbd ROM routine
        MOV     BX,AX           ; put result in BX
        CALL    BINIHEX         ; print out result
        MOV     DL,0DH          ; print CR
        MOV     AH,02H
        INT     21H
        JMP     AGAIN           ; repeat
```

9.1.3 DOS 键盘功能调用

在 9.1.2 小节介绍了 BIOS 键盘中断（16H），它能同时回送字符码和扫描码，这在使用功能键和变换键的程序中是很重要的。但对一般的键盘操作，最好使用适应能力更强的由 INT 21H 中断提供的键盘功能调用。

表 9.5 列出了与键盘输入有关的 DOS 21H 功能调用，它包括把单字符读入 AL 和把一个字符串读入存储器等功能。在编写程序时，定会感到使用 DOS 21H 键盘功能调用非常方便。

表 9.5　DOS 键盘操作(INT 21H)

AH	功　能	调用参数	返回参数
1	从键盘输入一个字符并回显在屏幕上		AL=字符
6	读键盘字符	DL=0FFH	若有字符可取,AL=字符
			ZF=0
			若无字符可取,AL=0
			ZF=1
7	从键盘输入一个字符,不回显		AL=字符
8	从键盘输入一个字符,不回显,检测 Ctrl-Break		AL=字符
A	输入字符到缓冲区	DS:DX=缓冲区首址	
B	读键盘状态		AL=0FFH 有键入
			AL=00 无键入
C	清除键盘缓冲区,并调用一种键盘功能	AL=键盘功能号 (1,6,7,8 或 A)	

1. 单字符输入

　　DOS 调用 INT 21H 的功能 1 能等待从键盘输入一个字符,并在视频显示器上回显。当得到字符并已显示时,本功能就返回其 ASCII 码;如果该字符是扩展 ASCII 字符,需要调用本功能两次,第一次返回 0,第二次返回所按键的扫描码。使用 01H 功能时,如果按下 Ctrl_C 或 Ctrl_Break,DOS 在返回前调用 INT 23H 并结束程序。

　　INT 21H 功能 07 和 08 的输入操作与功能 01H 相似,不同的是输入字符不回显,使用功能 7,不进行 Ctrl_C 或 Ctrl_Break 的检查处理。08 功能与 01H 一样,支持 Ctrl_C 或 Ctrl_Break 的中断处理。

　　INT 21H 的 06 功能直接读写控制台,当(DL)= 00H ～ 0FEH 时,请求输出字符。(DL)=0FFH 时,请求读键盘字符。该功能有时被称作原始 I/O 操作,它不带回显地读键盘字符,不对 Ctrl_C 或 Ctrl_Break 进行特殊处理,而是将其直接传递给调用程序,不转到中断处理程序。该功能是仅有的能正确读出 Alt 组合键输入的 DOS 功能,而功能 1,7 和 8 对 Alt 组合键的输入可能会产生误解。

　　在交互程序中常常需要用户对一个提示做出应答,或通过输入一个字母或数字对菜单的各项进行选择,这时就要用到 21H 的单字符输入功能。例如,程序显示出一串信息,要求回答 Y 或 N。回答 Y,程序将转入标号为 YES 的程序段,而回答 N 使程序转入标号为 NO 的程序段,按下其他键程序就等待。这样的工作由例 9.2 的程序段来完成。

　　例 9.2　接收键盘输入并对其进行测试。

```
GET_KEY:  MOV    AH,1            ;Read a key with echo
          INT    21H
```

```
            CMP     AL,'Y'              ; Is it Y ?
            JE      YES                 ; If so,jump to YES
            CMP     AL,'N'              ; Is it N ?
            JE      NO                  ; If so,jump to NO
            JNE     GET_KEY             ; otherwise,wait for Y or N
```

测试 Y,N 或其他字母、数字和符号可直接把它们写在 CMP 指令中,用单引号括起来。但是如果想检测 Enter(Return)键,就要在指令中写出它的 ASCII 码 0D(十六进制)或 13(十进制)。例如,要求程序在按下 Return 键后才继续运行。

例 9.3 检测键盘输入的字符是否是回车键。

```
WAIT_HERE: MOV     AH,7                ; Wait for enter
           INT     21H
           CMP     AL,0DH              ; Is it enter ?
           JNE     WAIT_HERE           ; no,wait for next
```

例 9.3 中用 AH=7 代替 AH=1,差别只是不把按下的键显示出来,或不执行键的特定功能。

如果要求程序能接收功能键或数字组合键必须进行两次 DOS 调用,第一次回送 00,第二次回送扫描码。例如,程序显示出一个菜单,要求用户通过键入 F1,F2 或 F3 来选择 1,2 或 3 项,按其他键则产生错误信息。程序的应答检测部分如例 9.4。

例 9.4 检测键盘输入的功能键。

```
           MOV     AH,7                ; Wait for key
           INT     21H
           CMP     AL,0                ; Is it a function key ?
           JE      GET_EC              ; yes,read the scan code
           JMP     ERROR               ; no,display error message
GET-EC:    MOV     AH,7
           INT     21H
           CMP     AL,3BH              ; F1?
           JE      OPTION1
           CMP     AL,3CH              ; F2?
           JE      OPTION2
           CMP     AL,3DH              ; F3?
           JE      OPTION3
           JMP     ERROR               ; Invalid key,display error message
```

2. 输入字符串

在许多应用程序中,要求用户输入姓名、地址或其他字符串,中断 21H 的功能 A 能从键盘读入一串字符并把它存入用户定义的缓冲区中。缓冲区的第一个字节保存最大字符数,这个最大字符数由用户程序给出。如果键入的字符数比此数大,PC 机就会发出

"嘟嘟"声,而且光标不再向右移动。由于缓冲区的最大字符数仅用一个字节来表示,所以缓冲区的逻辑上限为 255 字节。

第二个字节是实际输入字符的个数,这个数据是由功能 A 填入的,而不是由用户填入。在这两个字节之后,字符串就按字节存入缓冲区,最后结束字符串的回车 0DH 还要占用一个字节,因此整个缓冲区的字节空间应为最大字符数(包括 Return 在内)加 2。例如,在数据区定义的字符缓冲区如下:

```
        MAXLEN      DB    32
        ACTLEN      DB    ?
        STRING      DB    32   DUP(?)
```

输入字符串的指令如下:

```
        LEA    DX,   MAXLEN        ; Make DX point to buffer
        MOV    AH,   0AH           ; Input the string
        INT    21H
```

如果键入如下字符串:

By brooks too broad for leaping ↙

此时缓冲区 MAXLEN 的各字节的存储情况如图 9.3 所示。

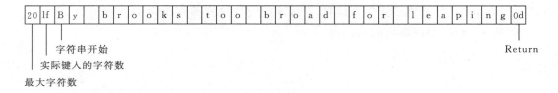

图 9.3 字符缓冲区

INT 21H 的功能 A 把实际字符数(不包括 Return)填入缓冲区的第二个字节,并保持 DS:DX 指向缓冲区的第一个字节。如果想把实际字符数放入 CX 寄存器,并把指针 (DS:DX)指向字符串的第一个字符,例 9.5 的程序可以完成这个工作。

例 9.5 输入字符串程序

```
;****************************************************************
;eg9-5.asm
;Purpose: read a string from keyboard   ——
;         this procedure read up to 50 keys
;****************************************************************
        .model    small
;----------------------------------------------------------------
        .stack
;----------------------------------------------------------------
        .data
```

```
user_string      db      50, 0, 50 dup(?)
```
;--
```
                 .code
; Main program
read_keys        proc    far
                 mov     ax, @data           ;ds<=data segment
                 mov     ds, ax

                 lea     dx, user_string     ;read string
                 mov     ah, 0ah
                 int     21h

                 sub     ch, ch              ;cx<=character number
                 mov     cl, user_string+1
                 add     dx, 2               ;make DX point to string
exit:
                 mov     ax, 4c00h
                 int     21h
read_keys        endp
```
;--
```
                 end     read_keys
```

3. 清除键盘缓冲区

从键盘输入的字符实际上先放在一个16字节的键盘缓冲区内,功能1,7,8和0AH都是从键盘缓冲区取得输入字符的DOS功能。

INT 21H的功能0CH能清除键盘缓冲区,然后执行在AL中指定的功能。AL指定的功能可以是1,6,7,8或0AH,使用0CH功能可以使程序在输入一个字符之前,将以前键入的字符清除掉。

功能0CH的用法如下:

```
MOV    AH, 0CH
MOV    AL, 08H
INT    21H
```

这几条指令实际提供的输入功能是8,它不回显,但要检测Ctrl_C或Ctrl_Break,如果不想用Ctrl_C或Ctrl_Break来结束程序,可以用功能7代替功能8。

使用0CH功能的好处是可以避免由于偶然超前键入的字符而出现的错误。例如,在格式化磁盘时,程序员在格式化磁盘程序开始运行时,又超前键入了一个字符,当程序询问使用者是否确实要清除磁盘数据时,若利用0CH功能读取键盘,则先清除缓冲区,再接收用户的回答,这样就可以防止由于刚才超前键入的字符而引起的错误动作。

4. 检验键盘状态

DOS 21H的功能0BH能检验一个键是否被按动,如果按下一个键,则在AL寄存器中放入 0FFH；如没有按下键，则在 AL 中放 00，无论哪种情况都将继续执行程序中的下一条指令。

注意：该功能并不返回实际字符码，仅提供一种是否按键的提示。有时这是一种不可少的功能，例如希望程序保持运行状态，同时又检验键盘，看用户是否按下任意一个键来终止程序或退出循环。例 9.6 指令序列的特点是，在未按键之前，程序总是不断循环执行，只要按下任何一个键，程序就退出循环并返回。

例 9.6　某程序在执行过程中检测是否有键盘输入。

```
SOUNDER:    ⋮
            ⋮                   ; Sound the tone
            MOV    AH,0BH       ; get kbd status
            INT    21H          ; call DOS
            INC    AL           ; if AL not 0ffh, then
            JNZ    SOUNDER      ; no key pressed
            RET                 ; key pressed return
```

9.2　显示器 I/O

显示器通过显示适配器与 PC 机相连。显示器可以简单地分为单色显示器和彩色显示器。随着显示技术的发展，显示器的种类也更加丰富，常见的显示器有阴极射线管(CRT)、存储管式显示器、光栅扫描显示器、液晶显示器、等离子显示器、场效发光显示器等。目前，微机系统广泛使用的是光栅扫描显示器，它的显示原理与电视机相似，是以光栅扫描的方式控制像素点阵的亮度来显现字符和图形的，它也分为单色和彩色显示器。

显示适配器也称为显示卡，是计算机和显示器的接口。早期的 PC 机通常使用两种显示适配器，一种是单色显示适配器(monochrome display adaptor,MDA)，一种是彩色图形适配器(color graphics adaptor,CGA)。MDA 连接单色显示器，它只能显示 ASCII 码字符，字符由标准字母，数字和各种符号组成；还有一些简单的图形，如菱形、矩形及笑脸符等。CGA 可用在彩色显示器上，能以红、绿、蓝彩色显示以点绘制的图形以及 ASCII 码字符。1984 年 IBM 公司基于 PC 和 PS/2 系列计算机开发了增强型图形适配器(enhanced graphics adaptor,EGA)图形标准，1987 年又开发了视频图形阵列适配器(video graphics array,VGA)，这两种显示适配器的分辨率和彩色功能比 MDA 和 CGA 有很大提高，可以设置为多种字符方式和图形方式，可以驱动单色显示器和彩色显示器。EDA、VGA 在字符显示方式下是与 MDA 和 CGA 兼容的。有关 EGA/VGA 的彩色图形功能，将在第 10 章中专门介绍，本节只介绍与字符显示相关的 BIOS 和 DOS 功能调用。

9.2.1　字符属性

显示器的屏幕通常划分为行和列的一个二维系统，适配器就在行列组成的网格位置上显示字符,如图 9.4 所示。例如,屏幕以 25 行 80 列的格式来显示字符,一副屏幕上就有 2000 个字符(25×80),0 行 0 列相对于屏幕左上角的位置,24 行 79 列相对于右下角的位置。

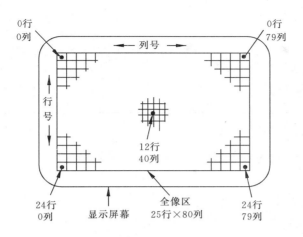

图 9.4　显示屏幕上的字符位置

对应屏幕上的每个字符位置,主存空间都有相应的存储单元与之对应,因此可以说显示屏幕是"存储器映像"的。这种存储器的映像,使显示器电路很容易知道那个单元的内容对应屏幕上的那个位置,也能使程序员从行列值算出主存地址空间中的显示存储区的地址。

对应显示屏幕上的每个字符,在存储器中由连续的两个字节表示,一个字节保存 ASCII 码,另一个字节保存字符的属性。在屏幕上处理字母、数字以及一些字符图形称为文本方式。在文本方式下,属性字节对单色显示和彩色显示都是有效的。

1. 单色字符显示

对单色显示,字符的属性定义了字符的显示特性,如字符是否闪烁,是否加强亮度,是否反相显示。单色显示属性字节的各位功能如图 9.5 所示。

属性可以有不同的组合,例如可以在屏幕上显示白底黑字(反相显示)代替通常的黑底白字。正常的属性是 07(二进制 00000111),即背景为黑色(000),前景为白色(111),而闪烁位为正常(0),加强亮度位也是正常(0)。为改变成反相显示,必须使背景为白色(111),前景为黑色(000),所以属性字节的值应为 70,即二进制 01110000。如果想要黑底白字及闪烁显示,属性值应为 87(10000111)。背景为 000,前景为 001,这种组合可产生下划线。

属性值可以任意组合,表 9.6 是一些单色显示的属性。

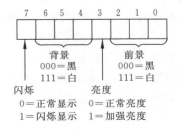

图 9.5　单色显示的属性字节

屏幕上的字符可以按相同的属性显示,也可以按不同的属性显示,如果你设置的属性为 00H,字符就显示不出来。

表 9.6　单色显示的属性

属性值（二进制）	属性值（十六进制）	显 示 效 果
00000000	00	无显示
00000001	01	黑底白字，下划线
00000111	07	黑底白字，正常显示
00001111	0F	黑底白字，高亮度
01110000	70	白底黑字，反相显示
10000111	87	黑底白字，闪烁
11110000	F0	白底黑字，反相闪烁

2. 彩色字符显示

在显示彩色文本时，属性字节能够选择前景（显示的字符）和背景的颜色。每个字符可以选择 16 种颜色中的一种，背景有 8 种颜色可以选择。图 9.6 是 16 色文本方式显示的属性字节。前景的 16 种颜色由位 0～3 组合，RGB 分别表示红、绿、蓝，BL 表示闪烁，I 为亮度，闪烁和亮度只应用于前景。表 9.7 列出了 16 色字符方式颜色的组合。

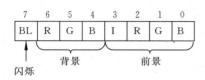

图 9.6　16 色方式下的属性字节

表 9.7　16 种颜色的组合

颜色	I R G B	颜色	I R G B	颜色	I R G B	颜色	I R G B
黑	0 0 0 0	灰	1 0 0 0	红	0 1 0 0	浅红	1 1 0 0
蓝	0 0 0 1	浅蓝	1 0 0 1	品红	0 1 0 1	浅品红	1 1 0 1
绿	0 0 1 0	浅绿	1 0 1 0	棕	0 1 1 0	黄	1 1 1 0
青	0 0 1 1	浅青	1 0 1 1	灰白	0 1 1 1	白	1 1 1 1

显示屏幕的背景颜色只能是表 9.7 中 I 为 0 的 8 种颜色。如果前景和背景是同一种颜色，显示出的字符是看不见的，但属性字节中的位 7 可以使字符闪烁（BL=1）。

3. 显示存储器

对于所有的显示适配器，文本方式下显示字符的原理都是一样的，所不同的是各种适配器的视频显示存储器（又称为显存）的起始地址不同：对 MDA，显存的起始地址为 B000∶0000；对 CGA、EGA、VGA 是 B800∶0000。每个字符的 ASCII 码和属性码字节存放于连续的两个字节中，图 9.7 表明了显示存储器单元与屏幕上字符的对应关系。

在 25×80 的文本显示方式下，屏幕可有 2000 个字符位置，因每个字符需要用两个字节来表示，所以显存容量需要 4KB（只使用 4000B）。如果显存有 16KB，则可保存 4 屏幕的字符数据，通常称为 4 页数据。对 CGA，EGA 和 VGA 的 80 列显示方式，0 页在显存中的起始地址是 B800∶0000，1 页是 B800∶1000，2 页是 B800∶2000，3 页是 B800∶3000。

屏幕上某一字符位置在显存中的偏移地址可由下列公式算出：

$$\text{Char\_offset} = \text{Page\_offset} + ((\text{row} \times \text{width}) + \text{column}) \times \text{byte}$$

在这个公式中,Page_offset 是页偏移地址,width 是每行可显示的字符数,在 25×80 的字符显示方式下,width=80,Byte 是表示一个字符所用的字节数,在字符显示方式中,byte=2,row 和 column 是相对于屏幕左上角位置(0,0)的行列坐标。例 9.7 定义了一个宏 Video_addr 计算屏幕位置在显存中的地址。

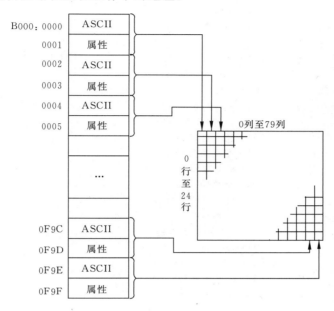

图 9.7　显示存储器单元与显示屏幕上字符对应关系

例 9.7　计算屏幕上某一字符位置所对应的显存地址。

```
Video_addr    MACRO width,page_num
;input:DH= row,DL= column
;output:DI = Char_offset
        push      ax
        mov       al,width
        mul       dh
        xor       dh,dh
        add       ax,dx
        shl       ax,1
        add       ax,page_num * 1000h
        mov       di,ax
        pop       ax
        ENDM
```

9.2.2　BIOS 显示中断

表 9.8 列出了中断类型 10H 的部分显示操作及所用的寄存器。

表 9.8 类型 10H 的显示操作

AH	功 能	调 用 参 数	返回参数／注释
1	置光标类型	$(CH)_{0\sim3}$＝光标开始行 $(CL)_{0\sim3}$＝光标结束行	
2	置光标位置	BH＝页号 DH＝行 DL＝列	
3	读光标位置	BH＝页号	CH/CL＝光标开始/结束行 DH/DL＝行/列
5	置当前显示页	AL＝页号	
6	屏幕初始化或上卷	AL＝上卷行数 AL＝0 全屏幕为空白 BH＝卷入行属性 CH＝左上角行号 CL＝左上角列号 DH＝右下角行号 DL＝右下角列号	
7	屏幕初始化或下卷	AL＝下卷行数 AL＝0 全屏幕为空白 BH＝卷入行属性 CH＝左上角行号 CL＝左上角列号 DH＝右下角行号 DL＝右下角列号	
8	读光标位置的属性和字符	BH＝显示页	AH＝属性 AL＝字符
9	在光标位置显示字符及其属性	BH＝显示页 AL＝字符 BL＝属性 CX＝字符重复次数	
A	在光标位置只显示字符	BH＝显示页 AL＝字符 CX＝字符重复次数	
E	显示字符 （光标前移）	AL＝字符 BL＝前景色	光标跟随字符移动
13	显示字符串	ES:BP＝串地址 CX＝串长度 DH,DL＝起始行列 BH＝页号 AL＝0,BL＝属性 串:Char,char,…,char AL＝1,BL＝属性 串:Char,char,…,char AL＝2 串:Char,attr,…,char,attr AL＝3 串:Char,attr,…,char,attr	 光标返回起始位置 光标跟随移动 光标返回起始位置 光标跟随串移动

1. 控制光标

光标在屏幕上指示字符的显示位置,它不是 ASCII 字符表中的字符。计算机有专门的硬件来控制光标,我们熟悉的光标符一般是一个下划线或方块符。

利用 INT 10H 的功能 1 使光标显现或关闭。这个功能也控制光标行的开始和结束,也就是说控制光标的大小。表示光标行开始和结束的数据分别放在 CH 和 CL 的低 4 位(0~3),当 CH 的第 4 位为 1 时,光标不显现出来(关闭);当第 4 位为 0 时,光标在屏幕上显现出来。单色显示器的光标大小的范围从 0 到 13。

INT 10H 的功能 2 设置光标位置。光标位置的行号设在 DH 寄存器中,列号设在 DL 中。在 24×80 的显示方式中,坐标设在(0,0)是屏幕的左上角,(24,79)是屏幕的右下角。BH 中必须包含被输出的页号,对单色显示器来说,页号总是 0。

例 9.8 置光标开始行为 5,结束行为 7,并把它设置到第 5 行第 6 列。

```
    MOV     CH,5        ;Beginning of cursor and turn on
    MOV     CL,7        ;End of cursor
    MOV     AH,1        ;define cursor
    INT     10H         ;call BIOS

    MOV     DH,4        ;row 5
    MOV     DL,5        ;column 6
    MOV     BH,0        ;page 0
    MOV     AH,2        ;place cursor
    INT     10H         ;call BIOS routine
```

2. 读光标位置

INT 10H 功能 3 是读光标位置,页号必须在 BH 中指定。此功能把光标位置的行号回送给 DH,列号回送给 DL。光标大小的参数填入 CH 和 CL,也就是说,在 CH 和 CL 中回送的是用功能 1 设置的光标参数。

例 9.9 读 0 页的当前光标位置。

```
    MOV     AH,3
    MOV     BH,0
    INT     10H
```

3. 选择显示页

INT 10H 的功能 5 可由程序确定显存中的显示区域。ROM BIOS 将 CGA 的显存分为 4 页,每页 25×80 个字符,或分为 8 页,每页 25×40 个字符。每一页的起始地址在 1KB 的边界。这 4 页的起始地址分别为 B800:0000,B800:1000,B800:2000,B800:3000。

例 9.10 选择显示页。

```
    MOV     AL,vpage    ;AL = video page number
    MOV     AH,5        ;function number
    INT     10H         ;call BIOS
```

4. 清屏和卷屏

INT 10H 功能 6 能使屏幕内容上卷指定的行,这个功能需要设置 7 个参数。如果屏幕的起始行列不为(0,0),结束的行列不为(24,79),则屏幕只有指定的一部分具有上卷的功能,这个屏幕上的部分区域叫做窗口(Window),像这样的窗口可以在屏幕上设置多个,这些窗口都可独立使用。如果上卷超过指定窗口的顶部,这些行的内容就消失,出现在窗口底部的新行被填为空格,其属性由 BH 寄存器决定。

如果 AL=0,则实际完成的工作是清除屏幕的功能,它将按 AL 中的 Blank 字符(0)使指定的窗口为空白。

例 9.11 编写清除全屏幕的子程序。

```
;****************************************************************
;eg9-11.asm
;Purpose: clear the screen
;Usage:   call clear_screen
;****************************************************************
clear_screen    proc    near
;save registers
    push    ax
    push    bx
    push    cx
    push    dx
;clear screen
    mov     ah, 6           ;to scroll up screen
    mov     al, 0           ;blank screen
    mov     bh, 7           ;blank line
    mov     ch, 0           ;upper left row
    mov     cl, 0           ;upper left column
    mov     dh, 24          ;lower right row
    mov     dl, 79          ;lower right column
    int     10h             ;call video BIOS

;locate cursor
    mov     dx, 0
    mov     ah, 2           ;to locate cursor
    int     10h             ;call video BIOS

;restore registers
    pop     dx
    pop     cx
    pop     bx
    pop     ax
    ret
clear_screen    endp
;----------------------------------------------------------------
```

10H 的功能 7 和功能 6 类似,也能使屏幕(或窗口)初始化或使屏幕(或窗口)的内容下卷指定的行,其他参数的设置与功能 6 一样。请看下面的例 9.12。

例 9.12   清除左上角为(0,0),右下角为(24,39)的窗口,初始化为反相显示,该窗口相当于全屏幕的左半部分。

```
    MOV    AH,7        ;scroll downward function
    MOV    AL,0        ;code to blank screen
    MOV    BH,70H      ;reverse video attribute
    MOV    CH,0        ;upper left row
    MOV    CL,0        ;upper left column
    MOV    DH,24       ;lower right row
    MOV    DL,39       ;lower right column
    INT    10H         ;video ROM call
```

下面编写一个完整的程序(例 9.13)在 PC 机上运行。此程序在屏幕的中间建立一个 20 列宽和 9 行高的窗口,然后把键盘输入的内容在这个窗口显示出来。键入的字符将被显示在窗口的最下面一行,每当输入 20 个字符,该行就向上卷动,9 行字符输入完后,顶端行的内容消失。

例 9.13   在屏幕中心的小窗口显示字符。

```
;******************************************************************
;eg9-13.asm
;Purpose:display characters in a window until Esc pressed
;******************************************************************
        .model   small
;-------------------------------------------------------------------
        .stack
;-------------------------------------------------------------------
        .code
Esc_key      equ    1bh              ;ASCII of Esc key
win_ulc      equ    30               ;window upper left column
win_ulr      equ    8                ;window upper left row
win_lrc      equ    50               ;window lower right column
win_lrr      equ    16               ;window lower right row
win_width    equ    20               ;width of window

        include cls.inc               ;clear the screen

; Main program
main         proc    far
             call    clear_screen     ;clear the screen
locate:
             mov     ah, 2            ;to locate cursor
```

```
            mov     dh, win_lrr         ;dx <= start position
            mov     dl, win_ulc
            mov     bh, 0               ;page# <= 0
            int     10h                 ;call video BIOS
    ; get characters from kbd
            mov     cx, win_width       ;cx <= width of window
    get_char:
            mov     ah, 1               ;to accept input
            int     21h                 ;call DOS
            cmp     al, Esc_key         ;judge whether Esc pressed
            jz      exit
            loop    get_char            ;get another

    ;scroll up
            mov     ah, 6               ;scroll up function
            mov     al, 1               ;number of scroll lines
            mov     ch, win_ulr         ;upper left row
            mov     cl, win_ulc         ;upper left column
            mov     dh, win_lrr         ;lower right row
            mov     dl, win_lrc         ;lower right column
            mov     bh, 7               ;attribute: normal
            int     10h                 ;call video BIOS
            jmp     locate
    exit:
            mov     ax, 4c00h
            int     21h
    main    endp
    ;--------------------------------------------------------------------------------
            end     main
```

例9.13 的程序使用了几种 ROM 显示例行程序：清除屏幕、光标定位和上卷。如果在屏幕上同时有几个窗口工作，就要分别清除它们，这可通过设置不同的左上角坐标和右下角坐标来完成。

5. 字符显示

10H 的功能 9 和功能 0A 都能把一个字符送到显示屏幕，然后光标返回到它的初始位置，所以在当前光标位置上写一个字符之后，必须用 INT 10H 的功能 02 移动光标到下一个字符位置上。

这两种功能的区别是，AH＝9 的功能把字符及其属性输出到当前光标位置上，而 AH＝0AH 的功能只输出字符，它的属性值就是这一位置上先前已具有的属性。0AH 功能在使用单色显示器时特别方便，因为此时我们很少改变显示字符的属性。10H 的功能 8 可读取当前光标位置的字符及属性。

例 9.14 置光标到 0 显示页的(20,25)位置,并以正常属性显示一个星号'＊'。

```
MOV     AH,2        ;set cursor position
MOV     BH,0        ;page 0
MOV     DH,20       ;row 20
MOV     DL,25       ;column 25
INT     10H         ;video ROM call
MOV     AH,9        ;write character
MOV     AL,'＊'     ;character '＊'
MOV     BH,0        ;page 0
MOV     BL,7        ;normal attribute
MOV     CX,1        ;number of repeat char
INT     10H
```

例 9.15 在 0 显示页的(11,0)位置读取字符和属性。

```
MOV     AH,2        ;set cursor position
MOV     BH,0        ;page 0
MOV     DH,11       ;row 11
MOV     DL,0        ;column 0
INT     10H         ;video ROM call
MOV     AH,8        ;read char and attr
MOV     BH,0        ;page 0
INT     10H         ;video ROM call
```

6. 彩色和字符串显示

在编写字符显示程序时,彩色显示和单色显示类似。例如,利用 BIOS 10H 的 09 功能显示彩色字符时,BL 中设置的数据应为前景和背景的属性值,属性值的典型组合如表 9.9。

表 9.9　属性字节的典型组合

显示颜色	位								十六进制
	7	6	5	4	3	2	1	0	
	BL	R	G	B	I	R	G	B	
黑底黑字	0	0	0	0	0	0	0	0	00h
黑底蓝字	0	0	0	0	0	0	0	1	01h
蓝底红字	0	0	0	1	0	1	0	0	14h
绿底青字	0	0	1	0	0	0	1	1	23h
灰白底浅品红字	0	1	1	1	1	1	0	1	7Dh
绿底灰字闪烁	1	0	1	0	1	0	0	0	A8h

例 9.16 在品红背景下,显示 5 个浅绿色闪烁的星号。

```
MOV     AH,09       ;display a char and attr
MOV     AL,'＊'     ;asterisk
MOV     BH,0        ;page 0
```

```
        MOV     BL,0DAH              ;color attribute
        MOV     CX,05                ;five times
        INT     10H                  ;call BIOS
```

使用 INT 10H 的 13H 功能显示字符串有 4 种方式,前两种方式(AL=0,1)要指定整个显示字符串的属性,后两种方式(AL=2,3)必须指定每个字符的属性,通过例 9.17 和例 9.18 来了解它的用法。

例 9.17 在屏幕上以红底蓝字显示字符串:"WORLD SCENERY"。

```
STRING      DB      'WORLD SCENERY'
LEN_STR     EQU     $-STRING
            ⋮
            MOV     AL,3                 ;select 80×25 color text
            MOV     AH,0                 ;change mode
            INT     10H
            MOV     BP,SEG STRING        ;base of string
            MOV     ES,BP
            MOV     BP,OFFSET STRING     ;offset of string
            MOV     CX,LEN_STR           ;character count
            MOV     DX,0                 ;start at top left corner
            MOV     BL,41H               ;use blue-on-red lettering
            MOV     AL,0                 ;make curson return
            MOV     AH,13H               ;display the string
            INT     10H                  ;call BIOS
```

例 9.18 在屏幕上以红底蓝字显示"WORLD",然后分别以红底绿字和红底蓝字相间地显示"SCENERY"。

```
STRING1     BD      'WORLD'
STRING2     DB      'S',42H,'C',41H,'E',42H,'N',41H
            DB      'E',42H,'R',41H,'Y',42H
LEN_STR2    EQU     $-STRING2
            ⋮
            MOV     AL,3                 ;select 80×25 color text
            MOV     AH,0                 ;change mode
            INT     10H
            MOV     BP,SEG STRING 1      ;base of first string
            MOV     ES,BP
            MOV     BP,OFFSET STRING1    ;offset of string1
            MOV     CX,STRING2-STRING1   ;character count
            MOV     DX,0                 ;start at top left corner
            MOV     BL,41H               ;use blue-on-red lettering
            MOV     AL,1                 ;make cursor return
            MOV     AH,13H               ;display the string
```

```
        INT     10H                         ;call BIOS
        MOV     AH,3                        ;read the cursor positions
        INT     10H
        MOV     BP,OFFSET STRING2           ;offset of string 2
        MOV     CX,LEN_STR2                 ;character count
        MOV     AL,3                        ;display char and attr
        MOV     AH,13H                      ;display the string 2
        INT     10H                         ;call BIOS
```

9.2.3 DOS 显示功能调用

表 9.10 为 INT 21H 的显示操作,其中有两个是显示单字符的功能,另一个是显示字符串功能,这些功能都自动向前移动光标。

表 9.10 INT 21H 显示操作

AH	功　能	调　用　参　数	
2	显示一个字符(检验 Ctrl_Break)	DL＝字符	;光标跟随字符移动
6	显示一个字符(不检验 Ctrl_Break)	DL＝字符	;光标跟随字符移动
9	显示字符串	DS:DX＝串地址	串必须以 $ 结束,光标跟随串移动

AH＝9 的功能是显示字符串,它要求被显示输出的字符必须以 $ 字符(24H)作为定界符,此功能是用 $ 作为标记来计算串的长度的。有些 ASCII 码,如控制码,不能出现在该字符串中。显示字符串时,如果希望光标能自动换行,那么可在字符串结束之前加上回车和换行的 ASCII 码,如下例定义的字符串:

```
        MESSAGE    DB    'The sort operation is finished.',13,10,'$'
```

要显示输出的信息一般定义在数据段,输出上面定义的字符串 MASSAGE 的指令为:

```
        MOV     AH,9
        MOV     DX,SEG MESSAGE
        MOV     DS,DX
        MOV     DX,OFFSET MESSAGE
        INT     21H
```

使用赋值伪操作可以使程序的可读性更好,另外也可以根据显示格式的要求使用 TAB 符,TAB 符的 ASCII 码为 09。

```
        CR      EQU     13      (或 CR    EQU    0DH)
        LF      EQU     10      (或 LF    EQU    0AH)
        TAB     EQU     09
        MESSAGE    DB    TAB,'The sort operation is finished.'
```

```
            DB    CR,LF,'$'
```

使用 INT 21H 显示字符串,一定要在显示串之后加上定界符 $,丢失定界符可能会在屏幕上引起意想不到的后果。例 9.19 的程序定义了一个显示字符串的宏指令 PRINT,在后面的程序中多次要用到它。

例 9.19 编写一个显示字符串的宏定义文件。

```
;********************************************************************
;eg9-19.asm
;Purpose: print a string in given address
;Usage:   PRINT str_addr
;Entry:   macro parameter str_addr = string address
;********************************************************************
print   macro   str_addr
        push    dx                      ;save registers
        push    ax
;print the string
        mov     dx, offset str_addr     ;dx<=address of the string
        mov     ah, 09                  ;to display string
        int     21h                     ;call DOS
;restore registers
        pop     ax
        pop     dx
        endm
;--------------------------------------------------------------------
```

9.3 打印机 I/O

打印机是计算机的主要硬拷贝设备。按照印字原理分为字模式、针式、喷墨式、热转印式、激光式、LED 式、LCS 式、荧光式、电灼式、磁式和离子式等多种。字模式和针式属于击打式打印机,其余的为非击打式印字机。字模式打印机是以全字符的方式输出打印,打印质量好,但它不能同时提供高质量文本和高质量图像的打印要求,现在基本上已被淘汰。针式打印机和喷墨、激光打印机等都以点阵式构成文字和图像的方式进行打印,所以又同属于点阵式打印机(dot matrix printers)。它们的质量主要用点密度(又称分辨率)来衡量。单位为 DPI(印点/英寸)。其密度越高,分辨率越高,字符/图形就越逼真,越清晰。

随着图形和图像处理技术的高速发展。与彩色显示系统配套的各种彩色打印机也得到了发展。另外,国内用户还要求能打印出一种或几种符合国际汉字字形点阵要求的打印机。目前,使用最广泛的针式打印机、激光打印机和喷墨打印机都有彩色和汉字打印功能的产品问世。

打印机通常以串行或并行口与计算机接口,通过并行接口,打印机一次从处理器接收

8位代码;通过串行接口,打印机每次从处理器接收一位代码。PC 机一般适用并行接口,所有打印机使用 ASCII 标准。打印机一般都具有能存储几千个字符的缓存器。

打印字符/图形要求软件将字符/图形的输出转化为打印机的控制码,这些软件通常称为打印机驱动程序。打印机驱动程序一般由开发商提供,特定的驱动程序使打印机能识别并处理从处理器来的信号,如换页、换行或列表符(tab)等,也能使处理器理解打印机发出的信号,如忙或纸出界等。

不同类型的打印机可以响应从处理器来的不同的信号,这给打印机与接口的程序设计造成一些困难。所以在编写打印机程序之前,必须先了解连接在计算机上的打印机的型号,认真查阅打印机的技术手册。但是就打印机的处理过程而言,它比屏幕处理和磁盘处理都要简单,它只涉及到很少的一些操作,既能用 DOS INT 21H 来实现,也能用 BIOS INT 17H 来实现。表 9.11 是有关打印机 I/O 的中断操作。

表 9.11 打印机 I/O 中断

INT	AH	功　能	调用参数	返回参数
21H	5	打印一个字符	DL = 字符	
17H	0	打印一个字符并回送状态字节	AL = 字符 DX = 打印机号	AH = 状态字节
17H	1	初始化打印机回送状态字节	DX = 打印机号	AH = 状态字节
17H	2	回送状态字节	DX = 打印机号	AH = 状态字节

9.3.1 DOS 打印功能

INT 21H 的功能 5 把一个字符送到打印机,字符必须放在 DL 寄存器中,这是惟一的 DOS 打印功能。如果需要回车、换行等打印功能,必须由汇编语言程序送出回车、换行等字符码。例 9.20 的功能段是送一个字符给打印机,为了连续打印,还指定了打印的字符数。当然也可以用指定的结束符代替计数控制的方法。

例 9.20 调用 DOS 功能打印字符。

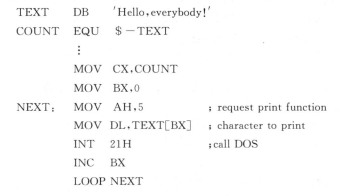

例 9.20 的指令也适用于发送打印控制字符。例如 TEXT 字符串定义如下：

TEXT   DB   0CH,'Hello,everybody!',0DH,0AH,0AH

字符串中的第一个字符是换页码(0CH)，最后两个字符是换行码(0AH)。用上面的指令把 TEXT 字符串在打印机上输出，则字符串打印在新一页的顶部，并与下文有两个空行的距离。下面介绍打印机的标准控制字符和特殊控制字符。

9.3.2 打印机的控制字符

1. 标准控制字符

打印机的标准控制字符如表 9.12。

表 9.12　打印机标准控制字符

十进制	十六进制	功　能
08	08	空格
09	09	水平 Tab（横表）
10	0A	换行
11	0B	垂直 Tab（纵表）
12	0C	换页
13	0D	回车

水平 Tab(09H)仅当打印机有此功能，并被置成打印机 Tab 状态时才能实现，否则打印机不执行此命令，或者用多个空格来代替 Tab。

换行命令使打印机向前空走一行，若连续用两次换行命令，则会空出两行。

当打印机加电启动后，打印头在一页纸的顶部位置。打印机打印时，记下所打印的行数，并检查是否到了一页的最大行数(如 55 行/页)，如打印了最后一行，就执行一个换页命令(0CH)，然后再把行计数器置成 0。

一般显示器遇到显示文件中的 Tab 字符(09)，就把当前的光标位置移到 8,16,24,…字符位置上。但许多打印机并不认识 Tab 字符，若要打印一个 ASCII 码文件(如汇编源程序)，就必须检验送到打印机的每个字符，若该字符是 Tab，就要插入空格，使下一个字符的位置在 8,16,…。形成 Tab 终止位置的方法用下面三个例子来说明。

当前打印的位置：	1	9	21
二进制数：	00000001	00001001	0C010101
清最右边三位：	00000000	00001000	0C010000
加 8：	00001000	00010000	0C011000
新的 Tab 终止位置：	8	16	24

例 9.21　编写一个能打印 ASCII 文件的子程序(PRTASC)，它的基本功能是把已读

到输入缓冲区(file_buffer)中的字符送到一个打印区(prt_buffer),并检查行尾和文件尾,处理换行符和 Tab 符。

```
;***************************************************************
;eg9-21.asm
;Purpose: print an ASCII file which is supposed
;         to be already loaded into file_buffer
;***************************************************************
        .model    small
;---------------------------------------------------------------
        .stack
;---------------------------------------------------------------
        .data

count           dw    0                    ;count char num in a line
file_buffer     db    'Hello', 09h         ;file buffer
                db    'Hello', 0dh, 0ah
                db    'Hello world', 1ah
                db    487 dup(' ')

prt_buffer      db    82 dup(' ')          ;print buffer
;---------------------------------------------------------------
        .code
; Main program
main            proc  far
start:
                mov   ax, @data            ;ds<=data segment
                mov   ds, ax
                mov   ax, @data            ;es<=data segment
                mov   es, ax

                call  prt_file             ;print the ASCII file
                mov   ax, 4c00h
                int   21h
main            endp
;---------------------------------------------------------------
; Procedure prt_file
; Purpose: print an ASCII file in file_buffer
;---------------------------------------------------------------
prt_file  proc  near
;initialize
```

```
            cld                             ;set left to right
            lea   si, file_buffer           ;si<=address of file_buffer
load_di:
            lea   di, prt_buffer            ;di<=address of prt_buffer
            mov   count, 0                  ;clear count
load_dx:
            lea   dx, file_buffer+512       ;dx<=end of file_buffer
            cmp   si, dx
            je    prt_lastline              ;print the last line
            mov   bx, count                 ;bx<=count
            cmp   bx, 80                    ;end of a print line?
            jb    not_endl

;dispose when reach end of line
            mov   word ptr[di+bx], 0d0ah    ;insert CR/LF
            add   count, 2                  ;reset count
            call  prt_line                  ;print the line
            mov   count, 0                  ;clear count
            mov   bx, 0

not_endl:
            lodsb                           ;al<=[si], increase si
            mov   [di+bx], al               ;copy char to prt_buffer
            inc   bx                        ;reset bx(count)
            cmp   al, 1ah                   ;end of file?
            je    prt_lastline
            cmp   al, 0ah                   ;end of line?
            jne   handle_tab                ;handle Tab character
            call  prt_line                  ;print the line
            jmp   load_di

;dispose if meeting Tab character
handle_tab:
            cmp   al, 09h                   ;Tab char?
            jne   not_tab
            dec   bx
            mov   byte ptr[di+bx], 20h      ;Tab->blank

;get the correct position of the character right after Tab
            and   bx, 0fff8h                ;clear rightmost 3 bits
            add   bx, 8                     ; & add 8-->Tab stop
```

```
not_tab:
          mov   count, bx              ;refresh count
          jmp   load_dx

;print the last line
prt_lastline:
          mov   bx, count              ;refresh bx
          mov   byte ptr[di+bx], 0ch   ;insert page break
          call  prt_line               ;print the line
          ret
prt_file  endp
;------------------------------------------------------------
; Procedure prt_line
; Purpose: print a line in prt_buffer
;------------------------------------------------------------
prt_line  proc  near
;initialize
          mov   cx, count              ;cx<=length of line
          inc   cx
          mov   bx, 0                  ;bx: pointer in prt_buffer

prt_char:
          mov   ah, 5                  ;to send print request
          mov   dl, prt_buffer[bx]     ;dl<=character to be printed
          int   21h                    ;call DOS
          inc   bx                     ;move pointer
          loop  prt_char               ;print another character

;clear prt_buffer
          mov   ax, 2020h              ;to fill with blank
          mov   cx, 41                 ;cx<=buffer size
          lea   di, prt_buffer         ;di<=address of prt_buffer
          rep   stosw                  ;fill
          ret
prt_line  endp
;------------------------------------------------------------
          end   start
```

2. 特殊的打印命令

前面已经讨论了打印机基本控制命令的使用。还有一些特殊命令如表 9.13 所示。

表 9.13 特殊的打印命令

十进制	十六进制	功　能
15	0F	设置紧缩方式
14	0E	设置扩展方式
18	12	取消紧缩方式
20	14	取消扩展方式

一些命令需要和 Esc(1BH)字符一起使用,这些命令是:

```
1BH    30H     设置每英寸为 8 行
1BH    32H     设置每英寸为 6 行
1BH    45H     设置加重打印方式
1BH    46H     取消加重打印方式
```

可以用下面两种不同的方式把命令码发送给打印机:

(1) 在数据区中定义命令码。下述数据区中的命令是设置紧缩方式,每寸 8 行,打印一个标题,并发送回车和换行字符。

```
HEAD    DB    0FH,1BH,30H,'Title …',0DH,0AH
```

(2) 直接用指令方式

```
MOV    AH,5         ;请求打印
MOV    DL,0FH       ;设置紧缩方式
INT    21H
```

上面的指令使以后打印的字符都是以紧缩方式打印,只有当程序发送取消此方式的命令后,才变成正常的方式进行打印。

上述这些特殊命令并不适用于所有型号的打印机,这就需要查阅打印机的手册,看其是否具有执行这些特殊命令的功能。

9.3.3　BIOS 打印功能

BIOS 17H 中断指令提供了由 AH 寄存器指定的三种不同的操作。

BIOS 中断 17H 的功能 0 是打印一个字符的功能。要打印输出的字符放在 AL 中,打印机号放在 DX 中,BIOS 最多允许连接三台打印机,机号分别为 0,1 和 2。如果只有一台打印机,那么就是 0 号打印机,打印机的状态信息被回送到 AH 寄存器。

```
MOV    AH,0         ; request print
MOV    AL,char      ; character to be printed
MOV    DX,0         ; select printer 0#
INT    17H          ; call BIOS
```

17H 的功能 1 初始化打印机,并回送打印机状态到 AH 寄存器。如果把打印机开关关上然后又打开,打印机各部分就复位到初始值。此功能和打开打印机时的作用一样。

在每个程序的初始化部分可以用 17H 的功能 1 来初始化打印机。

```
    MOV   AH, 01        ; request initialize printer
    MOV   DX, 0         ; select printer 0#
    INT   17H           ; call BIOS
```

这个操作要发送一个换页符,因此这个操作能把打印机头设置在一页的顶部。对于大多数打印机,只要一接通电源,就会自动地初始化打印机。

BIOS 17H 的功能 2 把状态字节读入 AH 寄存器。打印机的状态字节如图 9.8 所示。

打印机忙(printer busy)表示打印机正在接收数据,或正在打印,或处于脱机状态。应答位(acknowledge)表示打印机已发出一个表明它已经接收到数据的信号。选择位(select)表示打印机是联机的。超时位(time out)表示打印机发出忙信号很长一段时间了,系统将不再给它传送数据。表示打印出错的是第 5 位(纸出界)或第 3 位(I/O 错)为 1。如果打印机

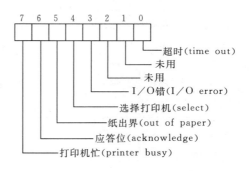

图 9.8 打印机的状态字节

没有接上电源,没有装上纸或没有联机,而打印程序已开始运行,这时显示器的指示光标会不停地闪烁,当接通打印机的电源后,某些输出数据就会丢失。如果在打印程序中先安排指令测试打印机的状态,则 BIOS 操作就会返送回状态码,DOS 打印操作是自动进行测试的,但对各种情况都显示一个"纸出界"的信息。当打印机接通电源后,即开始正常打印,而且不丢失任何数据。

例 9.22 应用 BIOS 和 DOS 功能调用,编写一个简单的打字程序(TYPER)。它要求把从键盘上接收的字符显示在屏幕上,并由打印机输出,在键盘上按下 Esc 键,即退出程序。

```
;*************************************************************
;eg9-22.asm
;Purpose: display and print characters input on keyboard
;*************************************************************
        .model   small
;-------------------------------------------------------------
        .stack
;-------------------------------------------------------------
        .data
;introduction message
intr_msg    db   'You are using a typer simulator.'
            db   'To quit this program, press Esc', 13, 10, '$'
prompt_msg  db   9eh, 10h, '$'

key_esc     equ  1bh                   ;key Esc
```

```
key_cr      equ    0dh                 ;key CR
key_lf      equ    0ah                 ;key LF
;------------------------------------------------------------------------
        .code
        print   macro   str_addr
            push    dx                  ;save regsters
            push    ax
            mov     dx,str_addr
            mov     ah,09               ;display string func.
            int     21h
            pop     ax                  ;restore regsters
            pop     dx
        endm
;------------------------------------------------------------------------
        include    cls.inc              ;clear the screen
        main    proc    far
            sti
            cld
            mov     ah,0                ;initialize printer
            mov     dx,0
            int     17h

            call    clear_screen        ;clear the screen

;print introduction message and prompt
            mov     ax,@data            ;ds<=code segment
            mov     ds,ax

            mov     dx,0
            mov     ah,2                ;locate the cursor
            int     10h

            print   intr_msg            ;print introduction
            print   prompt_msg          ;print prompt

;accept and check keyboard input
get_char:
            mov     ah,1                ;to accept keyboard input
            int     21h                 ;call DOS
            cmp     al,0
            jz      get_char            ;judge whether non-ASC key
            cmp     al,key_esc          ;judge whether Esc key
```

```
            jz      exit

;print the normal input character
            mov     dl, al                  ;print a character
            mov     ah, 5
            int     21h

;handle CR/LF condition
            cmp     al, key_cr              ;judge whether CR key
            jnz     get_char
;if CR pressed, LF ensues
            mov     dl, key_lf
            mov     ah, 2
            int     21h                     ;display LF
            mov     ah, 5
            int     21h                     ;print LF
            print   prompt_msg              ;print prompt
            jmp     get_char
exit:
            mov     ax, 4c00h               ;return to DOS
            int     21h

main        endp
;--------------------------------------------------------------------------------
            end     main
```

9.4 串行通信口 I/O

计算机传输数据有并行和串行两种方式。在并行数据传输方式中,使用 8 条或更多的导线来传送数据,例如计算机通过电缆将字节并行地传送给打印机或硬盘。虽然并行传输方式的速度很快,但由于信号衰减或失真等原因,并行传输的距离不能太长,一般在几米的范围内。为了在相距几百米甚至几公里的两个系统之间传输数据,就要采用串行通信的方式。

在串行通信方式中,通信接口每次由 CPU 得到 8 位的数据,然后串行地通过一条线路,每次发送一位将该数据发送出去。

串行通信采用两种方式:同步方式和异步方式。同步传输数据时,一次传送一个字节,而异步传输数据是一次传送一个数据(字符)块。可以使用这两种方式直接编写通信软件,但许多厂商也为串行通信生产了专用的 IC 芯片,以完成串行通信中对数据的打包、拆包、接收、发送等工作。这种芯片通称为通用异步收发器(universal asynchronous receiver-transmitter,UART)和通用同-异步收发器(universal synchronous-asynchronous receiver-transmitter,USART)。它位于串行适配器板,负责将每个字节转换为 1 和 0 的数据流,并指出每个字节的开始和结束。

9.4.1 串行通信接口

在串行通信方式中,单条数据线代替了并行通信的 8 位数据线,这不仅使串行通信更为廉价,而且也使远距离的两台计算机可以通过电话线通信。

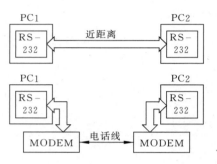

图 9.9 两台 PC 机串行通信的连接方式

串行通信时,数据字节从微机的 8 位数据总线上获取,并通过一个并行入串行出的移位寄存器转换成串行位,然后一位位地串行传送。同样,在接收端也必须有一个串行入并行出的移位寄存器来接收串行数据,并组合(打包)成一个字节,并行地送至系统的接收端。如果通过电话线远距离传输信息,必须把表示 0 和 1 的数字信号转换为能在电话线上传输的音频信号,或者从音频信号转换为数字信号。这种转换的工作由一个叫调制解调器(modem)的装置来完成。对近距离的传输,就不需要经过调制解调器,可直接在数据线上传送信号。例如,键盘和母板之间就是直接传送信号的。图 9.9 是两种连接方式的示意图,图中 RS-232 为标准串行通信接口。

1. 串行通信基础

以串行方式进入数据线的是由 0 和 1 组成的数据,那么发送者和接收者如何识别这些连续不断的数字信号呢?一个数据从哪里开始,又在哪里结束?一个串行传送的字符由多少位构成?这些都需要在计算机通信之前对数据的格式进行约定。

在同步通信方式中,每个字符都需要加上起始位和终止位,这个过程称为"组帧(framing)"。像 ASCII 字符这样的数据一般由 1 位起始位和 1 位终止位之间的一组位序列来表示。起始位总是 1 位;而终止位可以是 1 位或 2 位。起始位由 0(低电平)表示;终止位由 1(高电平)表示。图 9.10 是组帧后的 ASCII 字符 a(01100001B),它有 1 位起始位和 2 位终止位。

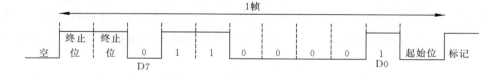

图 9.10 组帧后的 ASCII"a"(61H)

注意:传输由起始位开始,接着的字符从低位(D0)到高位(D7)的顺序送出,最后 2 位终止位表示字符 a 结束。在没有传输信号时,1(高电平)称为标记(Mark),0(低电平)称为间隔(space)。

在某些系统中,为了保证传输数据的正确性,字符字节的校验位也包括在数据帧内。这就是说对每个字符(7位或8位)除起始位和终止位外,还有1位校验位。在奇校验的情况下,数据位(包括校验位)应有奇数个1,同样偶校验下的数据位(包括校验位)应有偶数个1。例如,ASCII"A"(01000001B)的偶校验位应是0。UART芯片允许在编程时设定校验方式为奇校验、偶校验或无校验。校验位在字符最高位(D7)之后,紧接着是终止位。

串行通信的数据传输率用 b/s(bits per second)来表示。另外还有一种表示信号传输速度的单位是波特率(baud rate)。波特率和 b/s 不一定相等,这是因为波特率是一种信号调制单位,它定义为每秒钟传输的离散信号的数目。所谓离散信号,就是指不均匀的、不连续的也不相关的信号。在调制解调器中,如果采用4种相位,而每种相位代表2个数位,这时按 b/s 计算的传输率就是波特率的2倍。在计算机里,因为只允许1(高电平)和0(低电平)两种信号,所以 b/s 和波特率是完全相同的。

计算机系统的数据传输率取决于系统配置的通信端口。例如,早期的 IBM PC/XT 采用 100~9 600 b/s 的速度来传输,近几年的 PC,PS,80x86 及兼容机,其传输率可高达 19 200 b/s。

必须注意:在同步串行数据通信中,波特率通常限制在 100000 b/s。

例 9.23 (1) 计算串行传输 5 页,每页 80×25 个字符总共需要多少位?假设每个字符 8 位,1 位起始位和 1 位终止位。

(2) 计算传输上述 5 页数据所花费的时间。

数据传输率分别为:① 2400 b/s ② 9600 b/s

解答:(1) 10b/字符,80×25 字符/页×10 位/字符×5 页 = 100000 位

(2) ① 100000b/2400 b/s= 41.67s

② 100000b/9600 b/s=10.4s

2. RS 232 串行通信接口

为了兼容各厂家生产的数据通信设备,1960 年电子工业协会(Electronics Industries Association,EIA)制定了 RS-232 接口标准,以后又陆续发布了修订版本 RS-232A,RS-232B 和 RS-232C。目前广泛应用于个人计算机上的是 RS-232 串行接口,它安装于 PC 机内的通信适配器板上。这个标准串行接口既可用于近程或远程的数据通信,又可用来连接附加的一些外部设备,如具有串行接口的绘图仪、数字化仪、汉字或西文终端、各式打印机等。

然而由于制定串行接口标准早在制定 TTL 逻辑系列之前,因此,输入输出电平不与 TTL 兼容。因为这个原因,连接 RS-232 到微机系统必须经过电平转换,如图 9.11 所示。使用了 MC1488 从 TTL 转换到 RS-232 电平,用 MC1489 从 RS-232 转换到 TTL 电平。MC1488 和 MC1489 集成电路芯片通常称为线路驱动器和线路接收器。

3. IBM PC 通信端口

在微机系统中,可以有两个或多个串行端口连接到不同外设上进行通信,如 IBM PC

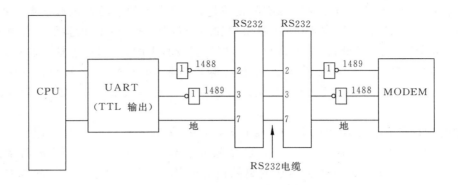

图 9.11 UART 与 RS232 的连接

和 80x86 兼容机可以连接 4 个通信端口,它们的编号为 COM1～4(相应的 BIOS 编号为 COM0～3),但程序每次只能对其中一个端口进行存取。当微机加电时,由加电自检程序(power-on self-test,POST)来测试通用异步收发器(UART)的 4 个 COM 端口是否存在。如果微机系统设置了 COM 端口,则设置的每个 COM 端口的 I/O 地址就写到 BIOS 数据区的 0040:000～0040:0007 字节,每个 COM 地址占用 2 个字节,比如 0040:0000 和 0040:0001 保存的是 COM0 的 I/O 端口基地址,0040:0002 和 C040:0003 保存 COM1 的 I/O 端口基地址等。如果系统没有连接串行端口,BIOS 数据区的这几个单元内容就成为 0。可以用 DEBUG 查看 UART 的端口地址。

```
C>DEBUG
 - d 0040:0000 L08
0040:0000   F8 03 F8 02 00 00 00 00
```

上例查看的结果表明系统有两个 COM 端口,COM0 的端口基地址为 03F8H,COM1 的端口基地址为 02F8H。

保存在 BIOS 数据区的 COM 端口地址之所以称为基地址,这是因为每个 COM 端口都包括一组 8 位的寄存器。程序员可以通过这些寄存器来进行编程,以控制数据按设定的格式发送或接收。BIOS 数据区中保存的只是每个 COM 端口的第一个寄存器的 I/O 地址,其他寄存器的地址按递增的顺序来排列。比如对于 UART 的 COM1 来说,第一个寄存器的 I/O 地址为 3F8H,第二个寄存器的 I/O 地址为 3F9H,如此等等。

9.4.2 串行口功能调用

DOS 和 BIOS 都提供了存取 PC 机串行 COM 端口的功能,BIOS INT 14H 为串行数据通信功能,在 MS-DOS 或 PC-DOS 下,可以用 MODE 命令对 COM 端口的数据大小,波特率等参数进行初始化。

1. DOS 串行通信口功能

使用 DOS 命令 MODE 可以设置串行通信参数,如数据的字长、波特率、校验位和终止位数。对 PC/XT、PC/AT,BIOS 支持的波特率为 110,150,300,600,1200,2400,4800

和9600。对 IBM PS 及兼容机，BIOS 还支持19 200波特。

设置串行通信参数命令的一般格式为：

MODE   COMm：b,p,d,s

这里 m 表示 COM 的端口号(1～4)，b 是波特率，用波特率数高两位数字来表示，如11 表示 110 波特，96 表示 9 600波特，19 表示 19 200波特（对 IBM PC 及兼容机）。p 是校验位（N 为无校验，O 为奇校验，E 为偶校验），d 表示数据的字长（5,6,7,8 位、默认值为 7 位），s 是终止位的位数（1,1.5 或 2 位）。如果传输率为 110 波特，则终止位必须是 2 位，其他情况终止位可以是 1 位。如果直接通过 BIOS 和 8250 UART 编程，传输率则可以超过19 200波特。

例如下面的命令：

MODE   COM1：24, O,8, 1

这个 MODE 命令为 COM1 设置的参数是：2 400波特率，奇校验，8 位字长及 1 位终止位。

必须注意：COM 端口由 BIOS 分配为 0～3 号，但是 DOS 使用的编号为 1～4 号。

DOS 手册中称串行通信接口为辅助设备。INT 21H 功能 03H 是从辅助设备（COM1）读一个字符到寄存器 AL。功能 04H 将 DL 寄存器中的字符传送给串行设备，如果输出设备正忙，该功能调用等待，直到设备准备好接收字符。

表 9.14   DOS 串行通信口功能

AH	功　能	调用参数	返回参数
3	从串行通信口读一个字符		AL＝输入的 8 位数据
4	向串行通信口写一个字符	DL＝输出的 8 位数据	

注意：在多数 DOS 系统中，串行设备没有缓冲和中断，如果串行通信口或其他辅助设备送的数据比程序处理数据快，字符可能丢失。

下面是两个使用 DOS 串行口 I/O 的例子。

例 9.24   从串行通信口输入一字符并存入 INPUT_CHAR 单元中。

```
        MOV     AH,3
        INT     21H
        MOV     INPUT_CHAR,AL
                ⋮
INPUT_CHAR      DB 0
```

例 9.25   将字符串 HELLO 输出到串行通信口。

```
        MOV     BX,SEG BUFFER          ; DS：BX ＝ addr of string
```

```
            MOV     DS,BX
            MOV     BX,OFFSET BUFFER
            MOV     CX,BUF_LEN          ; CX = length of string
    NEXT:   MOV     DL,[BX]             ; take the next char.
            MOV     AH,4                ; AUX output
            INT     21H                 ; call DOS
            INC     BX                  ; inc pointer
            LOOP    NEXT
            ⋮
    BUFFER      DB      'HELLO'
    BUF_LEN     EQU     $-BUFFER
```

DOS 没有提供读辅助设备状态和检测 I/O 错误(如丢失字符等)的功能,但 ROM 中 BIOS INT 14H 提供了这些功能。

2. BIOS 串行通信口功能

IBM PC 及其兼容机提供了一种有较强的硬件依赖性,但却比较灵活的串行口 I/O 的方法,即通过 INT14H 调用 ROM BIOS 串行通信口例行程序。该例行程序包括将串行口初始化为指定的字节结构和传输速率,检查控制器的状态,读写字符等功能。

表 9.15　串行通信口 BIOS 功能(INT 14H)

AH	功能	调用参数	返回参数
0	初始化串行通信口	AL=初始化参数 DX=通信口号 COM1=0 COM2=1,etc	AH=通信口状态 (AL)=调制解调器状态
1	向串行通信口写字符	AL=所写字符 DX=通信口号 COM1=0 COM2=1,etc	写字符成功: (AH)=0 (AL)=字符 写字符失败: $(AH)_7=1$ $(AH)_{0\sim6}$=通信口状态
2	从串行通信口读字符	DX=通信口号 COM1=0 COM2=1,etc	读成功: $(AH)_7=0$ (AL)= 字符 读失败: $(AH)_7=1$ $(AH)_{0\sim6}=$ 通信口状态
3	取通信口状态	DX=通信口号 COM1=0 COM2=1	(AH)=通信口状态 AL=调制解调器状态

INT14H 的 AH=0 功能把指定的串行通信口初始化为希望的波特率,奇偶性,字长和终止位的位数。这些初始化参数设置在 AL 寄存器中,其各位的含义如图 9.12。

例 9.26　要求 0 号通信口的传输率为 2 400 波特,字长为 8 位,1 位终止位,无奇偶校验。

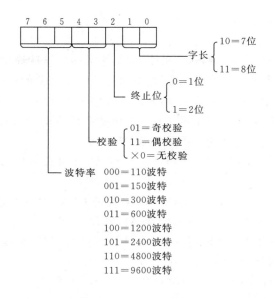

图 9.12 串行通信口初始化参数

```
MOV     AH,0            ; initialize communication
MOV     AL,0A3H         ; 0A3H=10100011B
MOV     DX,0            ; COM1
INT     14H             ; call BIOS
```

返回参数中通信口状态字节各位置 1 的含义如图 9.13。

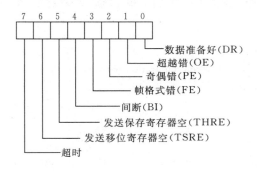

图 9.13 串行通信口状态字节

在接收和发送过程,错误状态位(1,2,3,4 位)一旦被置为 1,则读入的接收数据已不是有效数据,所以在串行通信应用程序中,应检测数据传输是否出错。

奇偶错:通信线上(尤其是用电话线传输时)的噪音引起某些数据位的改变,产生奇偶错。通常检测出奇偶错时,要求正在接收的数据至少应重新发送一段。

超越错:在上一个字符还未被处理机取走,又有字符要传送到数据寄存器里,则会引起超越错。如果处理机处理字符的速度小于串行通信口的波特率,则会产生这种错误。

帧格式错:当接收/发送器未接收到一个字符数据的停止位,则会引起帧格式错。这种错误可能是由于通信线上的噪音引起停止位的丢失。或者是由于接收方和发送方初始化不匹配。

间断:间断有时候并不能算是一个错误,而是为某些特殊的通信环境设置的"空格"状态。当间断位为1时,说明接收的"空格"状态超过了一个完整的数据字传输时间。

PS/2以及所有的 PC 机,AH=04 功能允许程序员将波特率设置为19 200,数据位的长度可以设置为5,6,7或8,而不是像 AH=0 功能那样只能设置成7或8位(INT 14H 的 AH=4 和 AH=5 支持的功能见附录5 BIOS 中断)。

利用 BIOS INT 14H,可以通过 COM 端口与另一台 PC 机传送字符,其过程如下:

(1) 用 INT14H,AH=1,AL=字符,发送一个字符。

(2) 为了接收字符,用 INT14H,AH=3 来获得 COM 端口的状态,其值返回在 AH 寄存器中注意,MODEM 的状态值是返回在 AL 寄存器中。

(3) 检验 COM 端口状态值的第0位,它是数据准备好位,如果该位为1,说明 COM 端口已接收到字符并送到了 8250 UART。

(4) 为了读取接收到的字符,使用 INT 14H AH=2 功能,将字符读到 AL 寄存器。

例 9.27 两台 PC 机通过 COM2 端口进行串行数据通信,编写一个汇编语言程序,要求从一台 PC 机上键盘输入的字符能传送到另一台 PC 机,若按下 ESC 键,则退出程序。在程序中,COM2 端口初始化为4 800波特,8位数据位,无校验,1位终止位;

按照例9.27的通信要求编程时,需要以下几个步骤:

(1) 检测按键,如果按下一键,则获取字符码并将它写到要传输的 COM 端口,同时也要检测是否按动退出键 ESC。

(2) 如无任何键按下,检测 COM 端口的状态,如果已接收到一个字符,则读取并显示在屏幕上。

(3) 转向第(1)步

```
TITLE       SDCOM—serial daia communication between two pcs
            . model    small
            . stack
;--------------------------------------------------------------
            . data
message     db    .Serial communication via COM2,4800,no p,1 stop,8 bit data.',0ah,0dh
            db    .Any key press is sent to other PC.',0ah,0dh
            db    .Press Esc to exit','$'
;--------------------------------------------------------------
            . code
main        proc
            mov    ax, @data
            mov    ds, ax
            mov    ah, 09              ;display string
            mov    dx, offset message
            int    21h
```

```
;Initializing COM2
        mov     ah, 0               ;initialize COM port
        mov     dx, 1               ;COM 2
        mov     al, 0c3h            ;4800, n, 1, 8
        int     14h                 ;call BIOS
;Checking key press and sending key to COM2 to be transferred
again:  mov     ah, 01              ;get keyboard state
        int     16h                 ;for checking key press
        jz      next                ;if ZF=1, there is no key press
        mov     ah, 0               ;there is a key press, get it
        int     16h                 ;call BIOS
;with AH=0 to get the char itself. AL=ASCII char pressed
        cmp     al, 1bh             ;is it ESC key?
        je      exit                ; yes, exit
        mov     ah, 1               ;no, send the char to COM port
        mov     dx, 1               ;COM 2
        int     14h                 ;call BIOS
;Check COM2 port to see there is char. if so get it and display it
next:   mov     ah, 3               ;read COM states to AH
        mov     dx, 1               ;COM 2
        int     14h                 ;call BIOS
        and     ah, 1               ;mask all bits except D0
        cmp     ah, 1               ;check D0 to see if there is a char
        jne     again               ;no char, goto monitor keyboard
        mov     ah, 2               ;yes, read char from COM2
        mov     dx, 1
        int     14h
        mov     dl, al              ;DL get char to be display
        mov     ah, 2               ;display char function
        int     21h                 ;call DOS
        jmp     again               ;keep monitoring keyboard
exit:   mov     ah, 4ch             ;exit to DOS
        int     21h
main    endp
;--------------------------------------------------------------
        end
```

习　　题

9.1 INT 21H 的键盘输入功能 1 和功能 8 有什么区别？

9.2 编写一个程序，接收从键盘输入的 10 个十进制数字，输入回车符则停止输入，然后将这些数字加密后(用 XLAT 指令变换)存入内存缓冲区 BUFFER。加密表为：

　　输入数字：　0,1,2,3,4,5,6,7,8,9

　　密码数字：　7,5,9,1,3,6,8,0,2,4

9.3　对应黑白显示器屏幕上 40 列最下边一个像素的存储单元地址是什么？

9.4　写出把光标置在第 12 行,第 8 列的指令。

9.5　编写指令把 12 行 0 列到 22 行 79 列的屏面清除。

9.6　编写指令使其完成下列要求：

　　(1) 读当前光标位置；

　　(2) 把光标移至屏底一行的开始；

　　(3) 在屏幕的左上角以正常属性显示一个字母 M。

9.7　写一段程序,显示如下格式的信息；

　　Try again, you have n starfighters left.

其中 n 为 CX 寄存器中的 1～9 之间的二进制数。

9.8　从键盘上输入一行字符,如果这行字符比前一次输入的一行字符长度长,则保存该行字符,然后继续输入另一行字符;如果它比前一次输入的行短,则不保存这行字符。按下′$′输入结束,最后将最长的一行字符显示出来。

9.9　编写程序,让屏幕上显示出信息" What is the date(mm / dd / yy)?"并响铃(响铃符为 07),然后从键盘接收数据,并按要求的格式保存在 date 存储区中。

9.10　用户从键盘输入一文件并在屏幕上回显出来。每输入一行(80 个字符),用户检查一遍,如果用户认为无须修改,则键入回车键,此时这行字符存入 BUFFER 缓冲区保存,同时打印机把这行字符打印出来并回车。

9.11　使用 MODE 命令,设置 COM2 端口的通信数据格式为：每字 8 位,无校验,1 位终止位和 1 200bps。

第10章 图形与发声系统的程序设计

编制图形软件是程序设计中一种非常有趣和有价值的工作。在计算机图形设计中，汇编语言具有潜在的优点，因为显示屏幕上的一个图像由几十万个元素组成，处理这些图像元素需要大量的指令。以速度而论，汇编语言程序远比高级语言快得多。最高级的图形技术，例如动画软件，只有以汇编语言设计才能产生更逼真、更有效的动态画面的效果。

配合图形画面各种声响和音乐是不可缺少的，它会使游戏和动画更加有趣。实际上，计算机从来都不是默默无声的，在大多数程序中，声音用来表示出错信息，或表示一个过程的开始和结束，或提醒用户输入信息，这时声音成为一种很有效的人-机通信方式。

本章将主要介绍目前较流行的两种彩色图形适配器 EGA 和 VGA 的基本图形操作以及图形程序设计方法；并在了解 8253/54 计数器/定时器的基础上，讨论如何利用 PC 机的发声系统编写产生各种声音和乐曲的汇编语言程序。

10.1 显示方式

ROM BIOS 显示例程支持多种文本方式和图形方式，每种方式适合于特定的适配器。在不同的显示方式下，屏幕显示的像素分辨率、字符分辨率、颜色数以及视频显示存储器的组织方式都不同。在为 PC 机开发文本和图形软件时，首先要根据硬件配置的情况，选择最能发挥显示器和适配器功能的显示方式；然后调用 BIOS 显示例程或采用直接写入视频存储器的方法显示文本或图形。

10.1.1 显示分辨率

显示分辨率包括字符分辨率和像素分辨率，分别表示显示器在水平和垂直方向上所能显示的字符数和像素(pixel)数。显示分辨率和色彩决定了显示器的显示质量，分辨率越高所显示的字符和图像就越清晰，但是需要更多的处理时间和更大的存储空间，而速度和成本也是不容忽视的问题。

目前较流行的两种彩色图形适配器是增强型图形适配器(enhanced graphics adapter，EGA)和视频图形阵列适配器(video graphics array，VGA)。这两种图形适配器都遵循良好的图形标准，较早期的 MDA 和 CGA，它们的分辨率更高，颜色更丰富；而且 EGA 和 VGA 都提供自己的 BIOS 显示例程，以消除不同硬件实现的差别，增强标准应用软件的移植性和兼容性。

EGA 在增强的彩色显示器上可显示出彩色的文本和图形。为了与先前的适配器相兼容，EGA 具有在单色显示器上和仿真的 CGA 上显示文本的操作方式。EGA 的显示分辨率可达 16 色、640×350，可供选择的颜色数有 64 种。

VGA 是 EGA 的增强版本。它支持 CGA 的显示方式,同时也向下兼容 EGA。VGA 适用于 PS/2(50、60、80 和 90 型)系统的单色和彩色模拟显示器。在文本方式下,分辨率可达 720×400。在图形方式下,VGA 的高分辨率为 640×480,可以选择 256 色中的 16 种颜色同时显示。VGA 还能显示 320×200 分辨率的图形,同时显示 256 种颜色。

表 10.1 列出了 EGA / VGA 的标准图形分辨率。

表 10.1 EGA / VGA 标准图形分辨率

适配器	颜色数	字符分辨率	像素分辨率	字形模式
EGA / VGA	4 色	40×25	320×200	8×8
VGA	256 色	40×25	320×200	8×8
EGA	16 色	40×25	320×350	8×14
VGA	16 色	40×25	360×400	9×16
EGA / VGA	2 色	80×25	640×200	8×8
EGA / VGA	16 色	80×25	640×200	8×8
EGA / VGA	16 色	80×25	640×350	8×14
VGA	2 色	80×30	640×480	8×16
VGA	16 色	80×30	640×480	8×16
VGA	16 色	80×25	720×400	9×16
EGA	单色	80×25	720×350	9×14
VGA	单色	80×25	720×400	9×16

自从 EGA / VGA 图形标准发表以来,各厂商又相继推出了许多显示卡,如 XGA、Super VGA(SVGA)等。它们在 EGA / VGA 标准的基础上提高了分辨率,并提供了更多的颜色选择及更大的显存,还扩充了一些功能;同时也设计了一些非标准的显示分辨率,如 800×600 或 1024×768 等。在一般情况下,使用非标准的显示分辨率及其显示方式会带来软件兼容性和编程复杂性问题。

10.1.2 BIOS 设置显示方式

EGA / VGA 可以在一系列的显示方式中进行选择配置,这些显示方式决定了显示分辨率、可同时显示的颜色数、每个像素所占用的位数、默认的字符集、显存的组织以及显存的起始地址。

显示方式分为两类:文本方式和图形方式。文本方式主要用于字符文本处理,图形方式又称为所有点可寻址(all-point-addressable)方式。在图形方式下,可通过读写屏幕上各个点的映像,显示出单色或彩色图形。EGA / VGA 的文本方式实际上是仿真 MDA 单色显示系统和 CGA 彩色显示系统的操作。

BIOS 显示例程提供了设置各种文本和图形显示方式的功能,程序只要给出调用参数,使用 BIOS INT 10H 即可建立某种显示方式。表 10.2 列出了几种常用的显示方式。

表 10.2　INT 10H 设置显示方式功能

功能号	调用参数		显示方式	适用显示适配器			
AH=00	AL=00	40×25	文本 16 级灰度	CGA	MCGA	EGA	VGA
	01	40×25	文本 16 色或 8 色	CGA	MCGA	EGA	VGA
	02	80×25	文本 16 级灰度	CGA	MCGA	EGA	VGA
	03	80×25	文本 16 色或 8 色	CGA	MCGA	EGA	VGA
	04	320×200	图形 4 色	CGA	MCGA	EGA	VGA
	05	320×200	图形 4 级灰度	CGA	MCGA	EGA	VGA
	06	640×200	图形 黑白显示	CGA	MCGA	EGA	VGA
	07	80×25	文本 黑白显示	MDA		EGA	VGA
	0D	320×200	图形 16 色			EGA	VGA
	0E	640×200	图形 16 色			EGA	VGA
	0F	640×350	图形 黑白显示			EGA	VGA
	10	640×350	图形 4 色或 16 色			EGA	VGA
	11	640×480	图形 黑白显示		MCGA		VGA
	12	640×480	图形 16 色				VGA
	13	320×200	图形 256 色		MCGA		VGA

图形适配器有两种操作方式：文本和图形方式，文本方式是图形适配器的默认方式。利用 BIOS INT 10H 的功能 00 可为当前的执行程序初始化显示方式或在文本方式和图形方式之间切换。显示器一旦设置成图形方式，光标即消失。

设置显示方式的方法如下例：

（1）设置文本方式：

```
MOV     AH,00H          ;Request set mode
MOV     AL,03H          ;Color text
INT     10H             ;Video ROM call
```

（2）设置 VGA 图形方式：

```
MOV     AH,00H          ;Request set mode
MOV     AL,12H          ;Color graphics
INT     10H             ;Video ROM call
```

在为未知的显示器编写程序时，首先应判断 PC 机配置的显示适配器并选择与之相适应的最佳显示方式。BIOS INT 11H 有确定系统配置的功能，这个操作实际上是把设备标志（equip-flag）的值回送给 AX，其中 4 位和 5 位表示显示器的配置及初始的显示方式，如图 10.1 所示。

设备标志字在 ROM BIOS 数据区 0040:0010 存储单元中。当 PC 机第一次加电时，BIOS 的 11H 中断例程读取外部硬件开关状态保留在设备标志字中，通过对

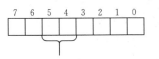

01＝彩色适配板（40×25 彩色）
10＝彩色适配板（80×25 彩色）
11＝黑白适配板（80×25 单色）

图 10.1　设备标志字

其 4 位和 5 位的测试,确定适配器的类型。设备标志字提供的信息是相当原始的,一般只能区分出是单色还是彩色显示。

另一种确定适配器的方法是,先检验 VGA,再检验 EGA,最后确定是 CGA 还是 MDA。具体操作步骤如下:

(1) 确定是否装配 VGA 适配器:

```
    MOV     AH,1AH          ;Request VGA function
    MOV     AL,0            ;and subfunction 0
    INT     10H             ;Video ROM call
    CMP     AL,IAH          ;If AL contains 1AH on return,
    JE      ISVGA           ;system contains an VGA
```

(2) 确定是否装配 EGA 适配器:

```
    MOV     AH,12H          ;Request EGA  function
    MOV     BL,10H          ;Amount of EGA memory
    INT     10H             ;Call video BIOS
    CMP     BL,I0H          ;If BL no longer contains 10H,
    JE      ISEGA           ;system may be exists an EGA
```

因为系统可能同时安装了 EGA 和 MDA、CGA,所以还要确定 EGA 是否在激活状态。BIOS 数据区的 40:87 字节中包含了有关 EGA 的信息,其中第 3 位为 0 时表示 EGA 是激活的,为 1 则说明没有 EGA 被激活,系统中还存在另一个适配器。

```
            ORG     87h
            EGAinfo DB      ?           ;Defined in BIOS data area
;------------------------------------------------------------
    TEST    EGAinfo,8                   ;Check bit 3,
    JE      EGAactive                   ;where 0 means EGA is active
```

(3) 确定 CGA 或 MDA:

BIOS 数据区中的 40:63 处的一个字存有显示控制器(CRTC)的端口地址,如果该地址是 3D4H,则适配器是 CGA,否则适配器为 MDA。

```
            ORG     63h
            VideoPort DW    ?           ;Defined in BIOS data area
;------------------------------------------------------------
            CMP     VideoPort,3D4H      ;3BxH means MDA
            JE      ISCGA               ;3DxH means CGA
    ISMDA:......
```

确定显示适配器后,即可设置能发挥适配器最高性能的显示方式。例如,发现 PC 机配置的是 EGA 卡,可选择显示方式 10H,以允许 640×350 分辨率的 16 色图形功能。如遇到的是 CGA,可设置成方式 6,以显示 640×200 的黑白图形。对 MDA,则设置成方式 7,即 80×25 单色文本显示方式。

单纯为了获取或改变当前显示方式,还有一种方法是调用 INT 10H 显示功能 0FH,该功能将当前显示方式返回在 AL 中:

```
MOV     AH,0FH          ;Finding the current video mode
INT     10H             ;BIOS video call, return the video mode in AL
```

在 BIOS 数据区,存放当前显示方式的字节地址是 00449H,若要改变显示方式,也可利用 DEBUG 直接修改此单元的值。

10.2　视频显示存储器

系统主板和视频显示器之间的通信要通过视频适配器板,适配器板的主要组成是视频控制器和视频显示 RAM。在显示器上要显示的信息(文本或图形数据)都存放在称作视频显示 RAM(VDR)的存储器中,VDR 也称为显示缓冲区。CPU 对视频显示 RAM 和对其他 RAM 一样可以寻址,所以程序可以通过指令对视频 RAM 读取和写入。

为了显示信息,视频显示适配器的控制器(处理器)会连续重复地读取视频 RAM 中的数据,并把它转换成能在屏幕上显示的信号。在屏幕上显示的画面一般需要以 50～70 次/秒的速度更新,视频 RAM 中的一位或几位,可以表示屏幕上一个像素(pixel)的颜色和亮度,所以修改视频 RAM 中的内容时,屏幕上的显示画面也几乎立即跟着改变。

不同的视频系统,用作视频显示 RAM 的空间大小也不一样,这主要取决于视频适配器所支持的显示分辨率以及所选择的显示方式。大部分视频 RAM 可存储多于一个屏幕的数据,因此任何时候显示在屏幕上的数据只是视频 RAM 中的部分内容。

10.2.1　图形存储器映像

EGA/VGA 标准引入了一种新的视频存储器组织方式。我们知道,对 MDA 和 CGA,程序设置或清除屏幕像素,是直接通过对视频显示 RAM 的二维坐标地址写 1 或写 0 来实现的。而在 EGA 和 VGA 的图形方式下,像素的存取是采取一种位映像(bit-mapped)的方式。对视频存储器的一个地址进行读写操作,将会从 4 个并行的位面(bit planes)存取 4 个字节的数据,这 4 个位面的存取由 CPU 根据对相应的锁存寄存器的设置来决定。

1. EGA 视频存储器

对 EGA 视频适配器板,视频 RAM 增大到 256KB,这使得在图形方式下可支持的颜色数和像素数都增加了。虽然 EGA 最多可显示 64 种颜色,但同时在屏幕上可显示的颜色数只有 16 种。EGA 图形存储器定位在 A0000H ～ AFFFFH 的一个独立的 64KB 地址空间中,256KB 的视频 RAM 如何通过 64KB 地址的窗口来存取显示信息呢? 为了解决这个问题,IBM PC 将视频 RAM 组织为 4 个并行的位平面,每个位平面 64KB,以页方式寻址来存取视频 RAM 的全部 256KB。

如图 10.2(a)所示的存储器位面结构中,位面上的每个字节表示显示屏幕上的 8 个像素,每位代表 4 位颜色值中的 1 位,4 个位面上同一地址的 4 位可表示 $2^4=16$ 种颜色。EGA 适配器中,还设置了 16 个调色板寄存器,每个寄存器 8 位,EGA 只用了其中 6 位,用来表示 64 种颜色。为了对 EGA 的调色板寄存器编程,可使用 INT 10H AH=10H 子功能 00H。

```
MOV     AH,10H          ; set palette register
MOV     AL,00H          ; to color correspondence
MOV     BH,color        ; set color
MOV     CL,palette      ; number of palette register (0-0FH)
INT     10H             ; BIOS call
```

EGA 支持的 640×350 像素、16 色显示方式,每个位面需要 28000 个字节对应 640×350=224000 个像素进行寻址。在位面的前 80 个字节中存放的是第一个 640 位的扫描行,紧接着的 80 个字节中存放的是第 2 个扫描行,依此类推。像素的颜色由同一地址而又分别位于 4 个位面的 4 位来组合,如果要改变视频显示器上的一个像素,那就必须对描述该像素的 4 位信息进行修改。因此图形程序一个很重要的工作就是对屏幕上某一位置的像素进行寻址操作,包括位面选择和位选择。

2. VGA 视频存储器

VGA 的分辨率比 EGA 高,VGA 支持的像素数增至 640×480,可同时显示 16 种颜色,这种显示方式(12H)是 VGA 独有的。VGA 最多可同时显示 256 种颜色,每种颜色在调色板中用 18 位表示,红、绿、蓝每种颜色使用 6 位,能对 $2^{18}=262144$ 种颜色进行选择。

VGA 是一种模拟显示器,VGA 电路使用一种数模转换器(digital-to-analog converter,DAC)产生模拟的 RGB 信号,红、绿、蓝每种颜色有一个 6 位的 D/A 转换器,每种颜色允许有 64 种组合,总共能产生 262144(2^{18})种颜色。如果 D/A 转换器由 6 位扩充到 8 位,这三种信号的组合将使调色板的颜色数增加到 $2^{24}=16777216$ 种。

VGA 视频适配器一般都拥有 256KB 的视频 RAM,在 640×480 分辨率,16 色的图形显示方式下,视频 RAM 也和 EGA 一样,由 4 个 64KB 的位面组成,每个像素由 4 位来表示,每个位面 1 位。4 个位面中的每个字节表示相邻的 8 个水平像素,利用 4 个位面的组合可以选择 16 种颜色中的一种。VGA 的图形存储器也定位于 0A0000H ~ 0AFFFH,所以 256KB 视频 RAM 的内容,也必须通过 0A0000H ~ 0AFFFFH 处的 64KB 的窗口,以分区或分页的方式寻址。

VGA 的图形方式 13H 是一种 320×200 的低分辨率显示方式,但它显示的颜色可达 256 种,因而要求一个像素用 8 位来表示。在这种方式下,视频存储器的组织形式与 16 色的不同,视频存储器位面上的一个字节表示一个像素,而不是 8 个像素。每页上的 320×200=64000 个像素,需要有 64000 个字节来表示,用 4 个位面来存储数据,每页实际上只要求 16000 个字节。表示像素 0 的字节位于位面 0,像素 1 的字节位于位面 1,像素 2 的字节位于位面 2,像素 3 的字节位于位面 3,像素 4 的字节跟在像素 0 后面,存于位面 0,以此类推,如图 10.2(b)所示。

VGA 调色板中增加了 256 个颜色寄存器,在调色板寄存器中记录的不是像素的颜色

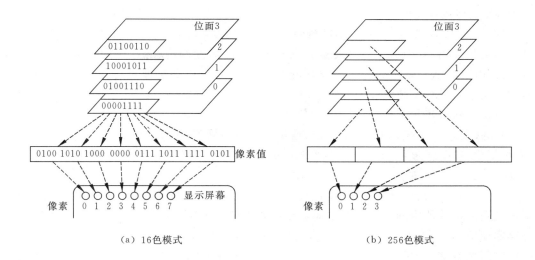

图 10.2 EGA 和 VGA 的位面结构

码,而是访问颜色寄存器的顺序地址,实际的颜色码来自于颜色寄存器。由于在 320×200,256 色的显示方式下,屏幕上的一个像素对应图形存储器中的一个字节,所以图形程序的寻址操作相对比较简单。

3. SVGA 和其他显示适配器

除了前面提到的视频显示的标准外,还有一些标准的适配器被广泛使用,其中有 SVGA 和 XGA,它们都支持 800×600,1024×768,1024×1024 等分辨率。根据所支持的颜色数,适配器板上的视频存储器可达 256K ～ 4MB,有些专用图形适配器板已达 16MB。例如,SVGA 的 800×600 像素,同时显示 256 色的方式,至少需要 800×600＝480000B 的视频存储器。由于采用位面结构,实际上需要 512KB 的存储器,表 10.3 列出了不同分辨率情况下所需要的视频存储器容量。另一个例子是在 800×600 分辨率,$16.7×10^6$ 色的情况下,总共需要 800×600×24＝11520000 位＝1440000B,即 1440KB,组织成位面,实际上是 1.5MB 存储器。

表 10.3 分辨率与视频 RAM 的关系

分辨率	16 色(4 位)	256 色(8 位)	6553 色(16 位)	16777216 色(24 位)
640×480	256K	512K	1M	1M
800×600	256K	512K	1M	1.5M
1024×768	512K	1M	1.5M	2.5M
1280×1024	1M	1.5M	2.5M	4M
1600×1200	1M	2M	4M	6M

10.2.2 数据到颜色的转换

EGA／VGA 中使用了一种颜色编码技术,它包括在 EGA 上使用 2 位的颜色代码,在 VGA 上使用 6 位的颜色代码。

EGA/VGA(除显示方式 13H 外)的每个调色板寄存器中,有 6 位数据作为输出到显示器的颜色码,其中低 3 位(2～0 位)分别表示高亮度的红、绿、蓝三色,高 3 位(5～3 位)表示红、绿、蓝三基色。因为每种颜色由 2 位表示,所以每种颜色可以有 4 种亮度(或饱和度层次)。EGA 调色板中的颜色值通过 6 条控制线输出到显示器,每种颜色 2 条线,这使得显示器可显示多达 64 种不同的颜色。

VGA 要求一个模拟 RGB 显示器,通过数模转换器 D/A 输出红、绿、蓝模拟信号。在 VGA 提供的 13H 显示方式下,每种颜色 6 位,可显示 256K 种颜色,每种颜色有 64 级亮度。

VGA 的调色板中除了 16 个调色板寄存器外,还有 256 个颜色寄存器,如图 10.3(b)所示。图形存储器中的一个字节对应显示屏幕上的一个像素,该字节的 0～3 位对调色板寄存器编址,4～7 位提供了访问颜色寄存器的高 4 位地址。调色板寄存器输出的低 4 位用作访问颜色寄存器的低 4 位地址。从这里可以看出,VGA 的调色板没有必要有两层颜色转换寄存器,只要有颜色寄存器就可以了,VGA 调色板中仍保留调色板寄存器,主要是为了提供和 EGA 的兼容性。图 10.3 给出了 16 色模式和 256 色模式数据到颜色转换的简单示意。

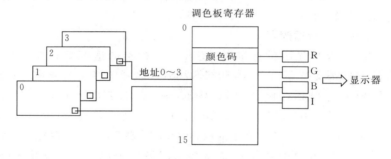

(a) 16 色(EGA)

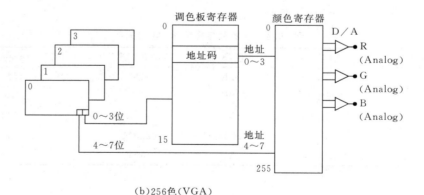

(b) 256 色(VGA)

图 10.3 数据到颜色的转换

从图 10.3 可以看出，在 16 色方式下，EGA 将来自显存位面的 4 位像素值转换为 6 位颜色码，并输出到显示器，完成在 64 种颜色中显示其中 16 种颜色之一的过程。而在 256 色方式下，VGA 将 8 位像素值转换为选择 256 个颜色寄存器的 8 位地址码，再将一个 18 位的颜色码输出到显示器，产生在 256K 颜色中显示 256 种美丽色彩的效应。

10.2.3 直接视频显示

用于 EGA 和 VGA 标准的位面结构的一个重要特征是，在指定的图形方式中，通过设置位面的存储位来确定每个像素的显示状态。例如，在 VGA 的显示方式 12H 中，每个屏幕像素的显示状态取决于该像素映射到具有相同地址的 4 个存储位的值。但是由 80x86 CPU 执行的读写指令是以字节为单位的，而且 80x86 指令集不允许读写一个分离在几个字节中的存储位，因此，为了存取分离在几个位面上的像素值，图形程序必须执行位屏蔽操作。例如，要读或写图 10.2(a)所示的第一行中的 4 号像素时，程序必须首先计算出映射到这个像素在显存中的位地址。对不同的显示方式，所支持的分辨率也不同，因此根据每行像素数和屏幕上的总行数，计算方法也不同。然而在所有位面结构的显示方式下，地址映像操作都需要计算两个值：一个是含有该像素存储位的字节地址，另一个是分离像素位所需要的掩码(mask)。例如，CPU 要读写 4 号屏幕像素，映射到 4 号像素的字节地址是 0A000H，需要分离出这个像素的掩码是 00001000B，结果从 0A0000H 地址所对应的 4 个字节中分离出一个 4 位的值，形成 0111B，这就是显示屏幕上 4 号像素的 IRGB 值。

1. 字节级映像操作

不是所有的图形程序都需要通过位映像的方法逐一计算屏幕上的每个像素值，有的图形应用程序需要对屏幕上的一块区域进行控制。比如在某一区域显示彩色阴影，开窗口，画条块或其他对称的图形，这时只需按字节对像素进行读写操作，而无须细分到位。

显然，使用字节级映像操作要比逐一读写单个像素的速度要快得多。首先，在字节级图形例程中，像素地址的计算是很简单的，如例 10.1 的 COARSE_ADD 例程，根据程序所提供的参数，使用乘法即可获得像素在显存中的字节偏移地址。其次，字节级的图形例程不必进行掩码计算，而只把掩码设置成一种位模式并作用于所有的字节。例如，将掩码设置成 11111111B，则所有像素都由一个视频存储器字节来控制，最终在屏幕区上显示出整条的颜色。如果掩码改变为 00001111B，则会显示出一系列的细条。

例 10.1 根据以水平 8 个像素为一列，以垂直 8 个像素为一行的行列坐标，编写计算读写像素的字节地址的子程序 COARSE_ADD。假定已设定的显示方式为 VGA 12H 方式(16 色，640×480)。

```
;--------------------------------------------------------------
; on entry:
;       CH = pillar number (range 0 to 79) = x coordinate
;       CL = rank number (range 0 to 59) = y coordinate
; Compute byte address in BX from pillar in CH and the rank in CL as follows:
```

```
;         byte address = (CL * 640) + CH
; on exit:
;         BX = byte offset into video buffer
;--------------------------------------------------------------
COARSE_ADD    PROC    NEAR
    PUSH  AX              ; Save  accumulator
    PUSH  DX              ; For word multiply
    PUSH  CX              ; To save CH for addition
    MOV   AX, CX          ; Copy CX in AX
    MOV   AH, 0           ; Clear high-order byte
    MOV   CX, 640         ; Multiplier
    MUL   CX              ; AX * CX results in AX
; The multiplier (640) is the product of 80 bytes per pillar
; times 8 vertical pixels in each rank

    POP   CX              ; Restore CH
    POP   DX              ; and DX
;
    MOV   CL, CH          ; Prepare to add in CH
    MOV   CH, 0
    ADD   AX, CX          ; Add
    MOV   BX, AX          ; Move sum to BX
    POP   AX              ; Restore accumulator
    RET
COARSE_ADD    ENDP
;--------------------------------------------------------------
```

256 色、320×200 方式下的图形程序是另一类字节级映像编程的情况。已经知道,当选择 256 色、320×200 图形方式时,视频 RAM 中的每个字节描述一个单独的屏幕像素,字节的 8 位值就是一个像素的属性值(表示 256 种颜色之一),该图形方式下的显存直接定位于 A0000H～AF9FFH 之间;第一个字节(A000:0000)对应屏幕左上角的第一个像素,最后一个字节(A000:F9FF)对应右下角的最后一个像素。正因为 256 色、320×200 方式下像素与显存字节有一一对应的关系,所以相应的软件编写起来非常容易。例 10.2 的程序通过在屏幕上画一个方框的过程,说明了在 256 色、320×200 方式下计算像素地址的方法。方框的大小和在屏幕上的位置是随意的,通过寄存器把方框的参数传送给画框程序 BOX。

例 10.2　256 色、320×200 图形方式下画方框的例程 BOX。

```
;--------------------------------------------------------------
; poporse : display a box at screen any position
; entry: AL = color (0—255)
;        CX = start column (0—319)
```

```
;            SI = start row (0—199)
;            BP = Box size
;--------------------------------------------------------------------
BOX     PROC    NEAR
        MOV     BX,0A000H           ;video address
        MOV     ES,BX               ;store in ES
        PUSH    AX                  ;save color
        MOV     AX,320              ;calculate start pixel
        MUL     SI
        MOV     DI,AX               ;start address of box
        ADD     DI,CX
        POP     AX
        PUSH    DI                  ;save starting offset
        MOV     CX,BP               ;box size
BOX1:
        REP     STOSB               ;drawing top line
        MOV     CX,BP
        SUB     CX,2                ;update CX
BOX2:
        POP     DI
        ADD     DI,320              ;point to next row
        PUSH    DI
        STOSB                       ;drawing left side
        ADD     DI,BP
        SUB     DI,2
        STOSB                       ;drawing right side
        LOOP    BOX2

        POP     DI
        ADD     DI,320              ;point to last row
        MOV     CX,BP
        REP     STOSB               ;drawing bottom line
        RET                         ;return
BOX     ENDP
;--------------------------------------------------------------------
```

2. 位级映像操作

当 EGA 或 VGA 图形程序读写一个屏幕像素时,必须首先获得与这个像素相对应的字节地址,然后计算位掩码,并用掩码把该像素与映射在同一字节地址中的其他 7 个像素分离出来。对 VGA 12H 方式的一个典型例程是,根据指定像素的 X 坐标(0～639)和 Y 坐标(0～479),计算视频存储器中的字节地址以及用于分离单个像素的位掩码。对于写模式 0 和 2,这个位掩码必须放入图形控制器的位屏蔽寄存器(bit mask register)。

确定位掩码一般有两种方法：

一种方法是通过对一个基本位模式 10000000 右移来获得掩码,移位次数是 X 坐标除以 8 得到的余数。例如,某像素的坐标是(16,0),X 除以 8 的余数为 0,所移位次数为 0,获得的位掩码是 10000000B。如果像素的坐标是(20,0),X 除以 8 余数为 4,则 10000000B 右移 4 位,得到的掩码就是 00001000B。

另一种方法是用这个余数作为索引值,去查找一个包含有 8 个掩码的表。索引 0 对应掩码 10000000B,索引 1 对应掩码 01000000B,以此类推。例 10.3 PIXEL_ADD 程序采用右移基本位模式的方法来完成像素位掩码的计算。

例 10.3 计算像素的字节地址和位掩码的子程序 PIXEL_ADD。

```
;-----------------------------------------------------------
; Address computation from x and y pixel coordinates
; On entry:
;       CX = x coordinate of pixel (range 0 to 639)
;       DX = y coordinate of pixel (range 0 to 479)
; On exit:
;       BX = byte offset into video buffer
;       AH = bit mask for the write operation (using VGA write modes 0 or 2)
;-----------------------------------------------------------
PIXEL_ADD   PROC    NEAR
        PUSH    CX              ; save all entry registers
        PUSH    DX
; Compute address
        PUSH    AX              ; Save accumulator
        PUSH    CX              ; Save x coordinate
        MOV     AX, DX          ; y coordinate to AX
        MOV     CX, 80          ; Multiplier (80 bytes per row)
        MUL     CX              ; AX = y times 80
        MOV     BX, AX          ; Free AX and hold in BX
        POP     AX              ; x coordinate from stack
; Prepare for division
        MOV     CL, 8           ; Divisor
        DIV     CL              ; AX / CL = quotient in AL and
                                ; remainder in AH
; Add in quotient
        MOV     CL, AH          ; Save remainder in CL
        MOV     AH, 0           ; Clear hight-order byte
        ADD     BX, AX          ; Offset into buffer to BX
        POP     AX              ; Restore AX
; Computer bit mask from remainder
        MOV     AH, 10000000B   ; Unit mask for 0 remainder
```

```
        SHR     AH，CL           ; Shift right CL times
; Restore all entry registers
        POP     DX
        POP     CX
        RET
PIXEL_ADD  ENDP
```
;--

10.3 EGA／VGA 图形程序设计

在图形方式下,可以利用 BIOS INT 10H 功能或采用直接存储器映像的方法对屏幕上像素进行读写和处理。调用 BIOS 例程编写的程序通用性和移植性好,但要损失一些效率。而利用存储器映像的方法直接处理显存单元的数据来显示字符和图形,这比调用 BIOS 速度更快,但会损失一些通用性和可移植性,而且要求程序员必须了解视频显示 RAM 的组织方式和表示一个像素的数据结构。在目前 BIOS 提供图形处理功能还相对弱的情况下,许多图形程序都采用了直接视频显示的方法。

10.3.1 读写像素

当 EGA／VGA 为图形方式时,BIOS 仅有两个例程用于读写,即 INT 10H 的 AH＝0CH(写像素),AH＝0DH(读像素),这两个功能可以很方便地读或写一个像素点到视频存储器。调用时,程序员提供颜色、页号、行号和列号。

INT 10H,AH＝0CH:写图形像素

利用该功能,可以将一个点写至像素位置。在合适的视频显示方式下,还可以指定颜色。

调用参数： AH＝0CH
        AL＝像素颜色,取决于显示方式
        BH＝显示页号
        DX＝像素行(0—199 或 349)
        CX＝像素列(0—399 或 639)

返回参数： 无

INT 10H,AH＝0DH:读图形像素

利用该功能,可以读取指定像素位置的颜色。

调用参数： AH＝0DH
        BH＝显示页号
        DX＝像素行(0—199 或 349)
        CX＝像素列(0—399 或 639)

返回参数： AL＝颜色值

使用读写像素功能应注意设置像素的行列坐标应取决于当前的显示方式。另外设置 AL 的 7 位为 1 时,会引起新的像素值与当前像素值进行异或(XOR)操作。

例 10.4 从坐标(50,100)开始到坐标(50,200)结束画一条水平线。

```
        MOV   AX, 0600H      ; scroll the screen
        OV    BH, 07         ; normal attribute
        MOV   CX, 0          ; from (0,0)
        MOV   DX, 184FH      ; to (18h,4fh)
        INT   10H            ; to clear screen
        MOV   AH, 0          ; set mode function
        MOV   AL, 06         ; 640 * 200 in graphics mode
        INT   10H            ; BIOS video call
        MOV   CX, 100        ; start column
        MOV   DX, 50         ; start row
BACK:   MOV   AH, 0CH        ; to draw a line
        MOV   AL, 01         ; pixels = white
        MOV   BH, 0          ; page 0
        INT   10H            ; BIOS call
        INC   CX             ; next dot
        CMP   CX, 200        ; end of line?
        JNZ   BACK           ; no, continue
        RET                  ; end drawing
```

1. 读模式

在 EGA/VGA 图形方式下，CPU 和视频 RAM 之间不是直接传递数据的，而是经过一个叫锁存器(latch)的中间站，每个位面对应一个锁存器。使用 80x86 指令对视频 RAM 的一个地址读数据时，实际上有 4 个字节分别从位面读入这 4 个锁存器，然后在图形控制器的控制和处理下，把数据传送给 CPU。写入时，锁存器中的数据也要经过图形控制器的处理。图 10.4 是 EGA/VGA 图形方式下，CPU 读写时的数据流示意图。在 EGA/VGA 图形方式下设计程序，了解 CPU 和显存之间的数据传递方式也是很重要的。

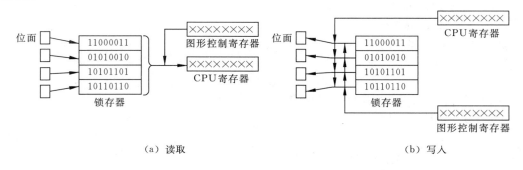

(a) 读取　　　　　　　　　　　　　(b) 写入

图 10.4　EGA/VGA 图形方式下 CPU 读(a)写(b)时的数据流

图形控制器包括九个数据寄存器，它们仅有一个端口地址 3CFH，分别由 0~8 的索引值来选择。访问这些数据寄存器时，首先通过图形地址寄存器(端口地址 3CEH)调入索引值,选择一个寄存器作为当前活动寄存器;然后再通过 3CFH 端口读写数据。图形控制寄存器列于表 10.4 中。

表 10.4 图形控制寄存器

寄存器名称	索 引	端口地址
图形地址寄存器	—	3CEH
置位/重置寄存器	0	3CFH
允许置位/重置寄存器	1	3CFH
颜色比较寄存器	2	3CFH
数据循环/功能选择	3	3CFH
读映像选择寄存器	4	3CFH
模式选择寄存器	5	3CFH
杂用	6	3CFH
颜色无关寄存器	7	3CFH
位屏蔽寄存器	8	3CFH

图形控制器对锁存器中的数据可以用字节方式处理,也就是对每个 8 位的锁存器分别处理;有时也可以用像素方式处理,即把锁存器中的数据看成 8 个像素,分别对每位进行处理。无论使用哪种处理方式,都要选择写入和读取模式。不同的读写模式有不同的操作过程,每种读写模式都有它独特的实用之处,选择正确的读写模式可以提高图形程序的效率。

EGA 和 VGA 提供两种读模式。

读模式 0 是一种默认模式,读模式 0 可以单独读取某个位面的字节,这在位面和主存或磁盘间传输数据时非常有用。选择哪一个位面的字节读入 CPU,取决于读映像选择寄存器(read map select register)。读映像选择寄存器的 0 和 1 位,用来指定哪个位面的锁存器内容读到 CPU。如果读取 4 个位面的内容,则必须对同一地址执行 4 次读操作,在每次读之前,用指令分别设置读映像选择寄存器。

当希望获得多个位面的内容时,使用读模式 1 更方便。在读模式 1 下,CPU 将 8 个像素的值读入锁存器。这 8 个像素值会与颜色比较寄存器做比较,相符合者置为 1,反之置为 0,然后把比较的结果送到 CPU。如颜色比较寄存器为 0011(青色),8 个像素值为 1011,1100,0011,0101,0010,0011,1101,1010,则给 CPU 返回的比较结果为 00100100B,说明有 2 个像素符合,可见读模式 1 在测试某种颜色的像素是否存在时很有用。在比较过程中,颜色无关寄存器(color don't care register)决定哪个锁存器不参加比较。对前面的例子,显然颜色无关寄存器的值是 1111,因为 4 个锁存器都参加了比较。如果颜色无关寄存器设置为 0111,则像素值的最高位都不参加比较,所以 CPU 会得到 10100100B 的结果。如果颜色无关寄存器为 0000,则说明与像素值的 4 位都无关,CPU 会得到一个 0FFH 的值。

例 10.5 在图形方式 12H 下,测试屏幕上第一行中是否有红色像素,测试结果保存于 test_buf 中。

```
;ES to video buffer, Mode 12H is 640 by 480 pixels in 16 colors
        MOV     AX, 0A000H              ; VGA graphics mode start address
        MOV     ES, AX                  ; in video buffer to ES
;graphics controller set to Read Mode 1
        MOV     DX, 3CEH                ; graphic controller address register
        MOV     AL, 5                   ; select Read Type register
        OUT     DX, AL                  ; Activate
        INC     DX                      ; control register is at 3CFH
        MOV     AL, 00001000B           ; set bit 3 for Read Mode 1
        OUT     DX, AL
;set color compare register
        MOV     DX, 3CEH                ; graphic controller
        MOV     AL, 2                   ; select color compare
        OUT     DX, AL
;
        INC     DX                      ; write to color compare register
        MOV     AL, 00000100B           ; red color
        OUT     DX, AL                  ; color loaded
;set color Don't care register
        MOV     DX, 3CEH                ; graphic controller
        MOV     AL, 7                   ; select color Don't Care
        OUT     DX, AL
;
        INC     DX                      ; write to color Don't Care
        MOV     AL, 00001111B           ; all of 4 bit planes
        OUT     DX, AL
; reset address in BX
        MOV     BX, 0                   ; byte offset into video buffer
        MOV     SI, 0                   ; pointer of test_buf
;read pixel into AL
READP:
        MOV     AL, ES:[BX]             ; read byte in video buffer
        MOV     TEST_BUF[SI], AL        ; result in test_buf
        INC     BX
        INC     SI
        CMP     BX, 80                  ; end of row
        JNZ     READP                   ; read next video byte
        RET                             ; end of first row
```

2. 写模式

EGA 有 3 种写模式,VGA 有 4 种写模式,选择写模式通过设置图形控制器的模式选择寄存器来实现。

写模式 0 是 EGA 和 VGA 的默认方式。在写模式 0 中,所写入的 CPU 数据可以更新任何一个或是全部的位面,同时,还可以与一个事先定义好的值进行逻辑运算,以更新锁存器中的 8 个像素或其中任一个像素。这种同时以字节方式和像素方式更新锁存器值的操作,要用到允许置位/重置寄存器、数据循环/功能选择寄存器和位屏蔽寄存器。

位屏蔽寄存器决定了新的像素值产生的方法。当位屏蔽寄存器的某位设为 0 时,相对应的像素值直接由锁存器写入显存;位屏蔽寄存器为 1 的位所对应的像素值由锁存器中的像素值与 CPU 数据或置位/重置寄存器中相应位合并之后产生。

数据循环/功能选择寄存器分为两个字段,数据循环字段(0～2 位)决定在写操作之前,CPU 数据向右循环的位数;功能选择字段(3～4 位)决定写入显存的数据和 CPU 数据进行何种位运算(00:代换,01:AND,10:OR,11:XOR),这些操作在编写动画程序时经常需要。

写模式 1 能把先前读入锁存器的数据直接复制到位面中。这个写入模式在把显存中一个区的数据移动到另一个区时特别有用。例如,在 VGA 图形方式 12H 中,每个显示位面有 64KB RAM,而屏幕像素只占用了 640×480＝38400B,其余 27135B 未用,程序员就可利用写模式 1 将图像或数据写入这个区域。

写模式 2 是写模式 0 的简化方式,它也允许把像素设置为任何希望的颜色,但它比写模式 0 的执行速度快。因为在写模式 2 中,像素颜色只由 CPU 数据来确定,而不使用位屏蔽寄存器或置位/重置寄存器来控制。

写模式 3 是 VGA 系统独有的,在 EGA 中没有等同的模式。在写模式 3 下,像素值是由锁存器中的像素值和置位/重置寄存器的值合并后产生的,而数据循环/功能选择寄存器也设置了合并时所要做的逻辑运算和移位操作,位屏蔽寄存器也是决定位面中的哪一个像素要被更新。写模式 3 与写模式 0、2 的区别是,按图形控制器各寄存器设置所进行的合并操作的程序不同。

例 10.6 所示的通用子程序是通过设置图形控制器的模式选择寄存器来选择写模式的实例。例 10.7 是使用写模式 2 的一种快速输出程序。在介绍动画显示技术一节中,例 10.11 例程分别将数据循环寄存器设置为正常方式和异或方式。

例 10.6 选择写模式的通用例程 SET_WRITE_MODE。

```
;------------------------------------------------------------
; Set the Graphics Controller's Graphics Mode register to the desired write mode
; on entry:
;        AL = write mode requested
; Also set default bit mask
;------------------------------------------------------------
SET_WRITE_MODE      PROC     NEAR
        PUSH     AX                  ; Save mode
        MOV      DX, 3CEH            ; Graphics Controller Address register
        MOV      AL, 5               ; offset of the Mode register
        OUT      DX, AL              ; Select this regiser
        JMP      SHORT $+2
```

```
        MOV     DX, 3CFH        ; Point to data register
        POP     AX              ; Recover mode in AL
        OUT     DX, AL          ; Selected
        JMP     SHORT $+2
; Set Bit Mask register to default setting
        MOV     DX, 3CEH        ; Graphics controller latch
        MOV     AL, 8
        OUT     DX, AL          ; Select data register 8
        JMP     SHORT $+2
        MOV     DX, 3CFH        ; To 3CFH
        MOV     AL, 0FFH        ; Default mask
        OUT     DX, AL          ; Load bit mask
        JMP     SHORT $+2
        RET
SET_WRITE_MODE  ENDP
;------------------------------------------------------------
```

例10.7 写像素例程 WRITE_PIX

```
;------------------------------------------------------------
; VGA mode 12H device driver for writing an individual pixel
; or a pixel pattern to the graphics screen
; on entry:
;       ES = 0A000H
;       BX = byte offset into the video buffer
;       AL = pixel color in IRGB format
;       AH = bit pattern to set
; This routine assumes that write mode 2 has been set
;------------------------------------------------------------
WRITE_PIX   PROC    NEAR
        PUSH    DX              ; Save outer loop counter
        PUSH    AX              ; Color byte
        PUSH    AX              ; Twice
; Set Bit Mask requster according to mask in AH
        MOV     DX, 3CEH        ; Graphics controller latch
        MOV     AL, 8
        OUT     DX, AL          ; Select data register 8
        JMP     SHORT $+2
        INC     DX              ; To 3CFH
        POP     AX              ; AX once from stack
        MOV     AL, AH          ; Bit pattern
        OUT     DX, AL          ; Load bit mask
        JMP     SHORT $+2
```

```
        ; Write color code
            MOV     AL, ES:[BX]         ; Dummy read to load latch registers
            POP     AX                  ; Restore color code
            MOV     ES:[BX], AL         ; Write the pixel with the color code in AL
            POP     DX                  ; Restore outer loop counter
            RET
WRITE_PIX   ENDP
;************************************************************
```

例 10.7 是一个快速写像素例程,它不重置默认的读或写模式,也不改变设置的位屏蔽寄存器。该例程首先通过图形地址寄存器(0CEH)以及索引值 8,实现对位屏蔽寄存器的选择,然后给位屏蔽寄存器(0CFH)设置指定的位模式。位模式决定在把一个字节写入显存时要修改的位。如果位屏蔽寄存器中的位模式为 0FFH,则该字节地址对应的 8 个像素一起被更新。

10.3.2 图形方式下的文本显示

大多数图形程序都需要以某种形式显示文本,在 EGA／VGA 标准下,有几种生成文本字符的方法,最简单的一种方法是使用 BIOS 字符显示功能,而在很多情况下需要扩充编码。

INT 10H 的功能 9 是 BIOS 提供的惟一能用于图形方式的字符显示功能。另外 INT 10H 的功能 2 也能在图形方式下设置字符显示位置。对程序员来说,应了解在图形方式下,屏幕可显示字符的总行数以及每行的字符数。

许多图形程序还通过使用 ASCII 码扩展字符集中的图形符号来组成大号的显示字符。例如,利用 ASCII 码为 0DBH,0DCH,0DDH,0DEH,0DFH 等方块符来设计大号字母或文章中的方块标记。采用这种方法的字符或串显示例程,在数据段将图形文本定义为一系列的数据编码。例 10.8 程序显示的词"GALLERY"如图 10.5 所示。由此形成的数据编码可定义在数据段的 GALL_MS 数组中。

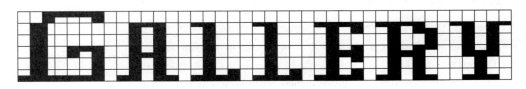

图 10.5 图形文本

```
; Graphics block message for the words shooting GALLERY
; 00 → end of massage, FF → end of screen line
GALL_MS     DB      2                   ; Start row
            DB      2                   ; Start column
            DB      10000011            ; Color attribute
```

                DB   'Shooting',0FFH,0FFH
                DB   7 DUP (0DCH),0FFH
; Graphics encoding of the word Gallery using IBM character set
                DB   20H,20H,0DBH,3 DUP(0DFH),0DBH,0FFH
;

                DB   20H,0DBH,0DDH,6 DUP(20H),0DBH,0DFH,0DBH,20H,20H
                DB   0DFH,0DBH,3 DUP(20H),0DFH,0DBH,3 DUP(20H),0DFH
                DB   0DBH,0DFH,0DBH,20H,0DFH,0DBH,0DFH,0DBH,0DCH,20H
                DB   0DFH,0DBH,20H,0DBH,0DFH,0FFH
;

                DB   20H,0DBH,0DDH,6 DUP(20H),0DBH,20H,0DBH,3 DUP(20H)
                DB   0DBH,4 DUP(20H),0DBH,4 DUP(20H),0DBH,4 DUP(20H)
                DB   0DBH,20H,20H,0DBH,20H,20H,0DBH,20H,0DBH,0FFH
;

                DB   20H,0DBH,0DDH,20H,0DBH,0DFH,0DBH,20H,20H,0DBH
                DB   0DCH,0DBH,3 DUP(20H),0DBH,20H,20H,0DBH
                DB   4 DUP(20H),0DBH,0DCH,0DBH,20H,20H,0DBH,0DCH
                DB   0DBH,0DFH,20H,20H,0DBH,0DCH,0DBH,0FFH
;

                DB   20H,0DBH,0DDH,3 DUP(20H),0DBH,20H,20H,0DBH,20H
                DB   0DBH,3 DUP(20H),0DBH,4 DUP(20H),0DBH, 4 DUP(20H)
                DB   0DBH,20H,0DFH,2 DUP(20H),0DBH,20H,0DBH,4 DUP(20H)
                DB   0DBH,0FFH
;

                DB   0DCH,0DBH,4 DUP(0DCH),0DBH,20H,0DCH,0DBH,20H,0DBH
                DB   0DCH,20H,0DCH,0DBH,0DCH,0DBH,20H,0DCH,0DBH,0DCH
                DB   0DBH,20H,0DCH,0DBH,0DCH,0DBH,20H,0DCH,0DBH,20H
                DB   0DBH,0DCH,20H,20H,0DCH,0DBH,0DCH
                DB   00

例 10.8 图形方式下的文本显示例程 GRAPHIC_TEXT。

;***
; Text display procedures: display a message on the graphics screen
; On entry:
;   DS:DI→Address of graphics display block (e.g., GALL_MS)
;---
GRAPHIC_TEXT PROC   NEAR
        MOV   DH,[DI]   ; Get row into DH
        INC   DI   ; Bump pointer
        MOV   DL,[DI]   ; And column into DL
        MOV   START_COL,DL   ; Store start column
        MOV   AH,2   ; set cursor position

```
        MOV         BH, 0                   ; page 0
        INT         10H                     ; BIOS video call
        INC         DI                      ; Bump pointer to attribute
        MOV         BL, [DI]                ; Get color code into BL
CHAR_WRITE:
        INC         DI                      ; Bump to message start
        MOV         AL, [DI]                ; Get character
        CMP         AL, 0FFH                ; End of line?
        JE          BUMP_ROW                ; Next row
        CMP         AL, 0                   ; Test for terminator
        JZ          END_TEXT                ; Exit routine
        CALL        SHOW_CHAR
        JMP         CHAR_WRITE
END_TEXT:
        RET
BUMP_ROW:
        INC         DH                      ; Row control register
        MOV         DL, START_COL           ; Column control to start column
        MOV         AH, 2                   ; set cursor position
        MOV         BH, 0                   ; page 0
        INT         10H                     ; BIOS video call
        JMP         CHAR_WRITE
GRAPHIC_TEXT    ENDP

;----------------------------------------------------------------
; Display character in AL and using the color code in BL
SHOW_CHAR  PROC   NEAR
        MOV         AH, 9                   ; BIOS service request number
        MOV         BH, 0                   ; Page
        MOV         CX, 1                   ; No repeat
        INT         10H
; Bump cursor
        INC         DL
        MOV         AH, 2                   ; set cursor position
        MOV         BH, 0                   ; page 0
        INT         10H                     ; BIOS video call
        RET
SHOW_CHAR  ENDP
;***************************************************************
```

当视频 BIOS ROM 中的字符集不能满足要求,使用方块符组成的字符或图形也不能达到满意的效果时,程序员可以自行设计位映像的字体和图形。对位映像的图形或字体进行编码,没有技术上的困难,只是一件耗时的工作。

10.3.3 彩色绘图程序

许多绘图程序的技术,都是从处理视频缓冲区(显存)中的像素做起的。一旦计算出一个指定像素在显存中的地址,包括字节偏移量和位偏移量,那么读写一个像素值就是很简单的了,只要使用汇编指令 MOV,MOVS,STOS 等传输像素值即可。

由于很多绘图程序需要处理的像素数量较大,因此高效快速是绘图程序很重要的一个指标,这也是绘图程序大多采用直接视频显示,而不用 BIOS 视频例程的原因。即使这样,编写简单而又高效的图形程序仍有许多设计技巧值得探讨。例如,选择一种快速的读写模式,利用图形控制器提供的硬件支持,完成某些像素运算或快速位面更新等。这里介绍例 10.9 和例 10.10 两个实例中涉及到编写图形程序的一些基本方法。

例 10.9 在 VGA 640×480 方式下,编写填充指定的屏幕区域的例程 COARSE_FILL。

本例在按指定的颜色填充某块屏幕区域时,指定屏幕位置的 XY 坐标是以 8 个水平像素(a pillar)和 8 个垂直像素(a rank)为单位的。因此,采用字节级的处理方法,并为位屏蔽寄存器提供 0FFH 的位模式,使调用的快速写像素例程一次就填充 8 个像素。

```
;******************************************************************
; Fill a screen area on a byte boundary base
; On entry:
;         CH = pillar address of start point (range 0 to 79)
;         CL = rank address of start point (range 0 to 59)
;         DL = total ranks to fill (1 to 60)
;         DH = total pillars to fill (1 to 80)
;         AL = IRGB color to use in fill area
;------------------------------------------------------------------
COARSE_FILL PROC NEAR
        PUSH    CX              ; save entry requsters
        PUSH    DX
        PUSH    AX
;
        MOV     AH, 0FFH        ; Bit pattern for all bits set
        PUSH    AX
        CALL    COARSE_ADD      ; in section 10.2.3.1
; Byte offset into video ROM is now set in BX
; Next set up column counter in CX and row counter in DX
        MOV     CL, DH          ; Horizontal fill
        MOV     CH, 0           ; CX = horizontal counter
        PUSH    CX              ; Save temporarily
```

```
; Row counter = DL * 8
    MOV         AX, DX
    MOV         AH, 0           ; Clear hight-order byte
    MOV         CL, 8           ; Multiplier
    MUL         CL              ; AX = DL * 8
; Set up outer counter in DX
    MOV         DX, AX
    POP         CX              ; Inner loop counter
    POP         AX              ; AL = color
                                ; AH = 11111111B(all bits set \)
OUTER:
    PUSH        CX              ; Save inner loop counter
    PUSH        BX              ; Save start address
INNER:
    CALL        WRITE_PIX       ; in section 10.3.1.3
    INC         BX              ; Index to next 8-pixel block
    LOOP        INNER
; Index to next column
    POP         BX              ; Restore start of this row
    ADD         BX, 80          ; Index to next row start
    POP         CX              ; Restore inner loop counter
    DEC         DX              ; Decrement total rows
    JNZ         OUTER           ; Continue if not end of count
; Restore input registers
    POP         AX
    POP         DX
    POP         CX
    RET
COARSE_FILL ENDP
;******************************************************************
```

前面几节已介绍了屏幕像素与视频显示存储器的映像关系,并编写了一个计算像素字节地址及位掩码的子程序 PIXEL_ADD,同时还介绍了利用图形控制器读写像素的方法及写像素例程 WRITE_PIX。现在只要将显示图形设计成位映像的图样,并对图样的位模式及颜色进行编码,这些编码作为数据定义在图形程序的数据段。

在编写位映像图形显示例程 VARI_PATTERN(例 10.10)之前,应先在数据段定义显示图形的数据变量。图 10.6 给出了"十"字符和一个"鸭子"的位映像图,右边的数据是位模式的编码,图样中,"0"表示该位保持背景色,"w"和"r"分别表示像素显示的颜色为白色(0FH)和红色(0CH)。

```
00000www   |www00000    ————   0 7 E 0 H
000w000w   |000w0000    ————   1 1 1 0 H
00w0000w   |0000w000    ————   2 1 0 B H
0w00000w   |00000w00    ————   4 1 0 4 H
w000000r   |000000w0    ————   8 1 0 2 H
w000000r   |000000w0    ————   8 1 0 2 H
w000000r   |000000w0    ————   8 1 0 2 H
wwwwrrrr   |rrrwwww0    ————   F F F E H
w000000r   |000000w0    ————   8 1 0 2 H
w000000r   |000000w0    ————   8 1 0 2 H
w000000r   |000000w0    ————   8 1 0 2 H
0w00000w   |00000w00    ————   4 1 0 4 H
00w0000w   |0000w000    ————   2 1 0 8 H
000w000w   |000w0000    ————   1 1 1 0 H
0000wwww   |www00000    ————   0 7 E 0 H

0000www0   |00000000    ————   0 E 0 0 H
000wwwww   |000000ww    ————   0 F 0 3 H
00wwwwww   |00000www    ————   3 F 0 7 H
0wwwwwww   |0000wwww    ————   7 F 0 F H
wwwwwwww   |wwwwwwww    ————   F F F F H
000wwwww   |wwwwwww0    ————   1 F F E H
00wwwwww   |wwwwwww0    ————   3 F F E H
0wwwwwww   |wwwwwwww    ————   7 F F F H
0wwwwwww   |wwwwwwww    ————   7 F F F H
```

w:白色(颜色码:0FH)   r:红色(颜色码:0CH)   0:背景色

图 10.6 "十"字符和"鸭子"位模式图

调用例程 VARI_PATTERN 显示"十"字符和"鸭子"图形的数据定义如下,要注意的是,用来描述图形的数据结构必须符合显示例程的要求。

```
;--------------------------------------------------------------
; Graphics block control area for cross—hair
CROSS_X    DW    320            ; Present x coordinate, range= 0 to 630
CROSS_Y    DW    360            ; Present y coordinate, range=360 to 460
           DB    15             ; Horizontal rows in block
           DB    2              ; Number of bytes per row
; Symbol encoding of cross—hair symbol
           DB    07H,0E0H,11H,10H,21H,08H,41H,04H,81H,02H
           DB    81H,02H,81H,02H,0FFH,0FFH,81H,02H,81H,02H
           DB    81H,02H,41H,04H,21H,08H,11H,10H,07H,0E0H
           DW    0000H                ; Safety padding
; Color encoding of cross-hair symbol
CR_COL     DB    0,0,0,0,0,0FH,0FH,0FH,0FH,0FH,0FH,0,0,0,0,0
           DB    0,0,0,0FH,0,0,0,0FH,0,0,0,0FH,0,0,0,0
           DB    0,0,0FH,0,0,0,0,0FH,0,0,0,0,0FH,0,0,0
           DB    0,0FH,0,0,0,0,0,0FH,0,0,0,0,0,0FH,0,0
           DB    0FH,0,0,0,0,0,0,0CH,0,0,0,0,0,0,0FH,0
```

```
        DB    0FH,0,0,0,0,0,0,0CH,0,0,0,0,0,0,0FH,0
        DB    0FH,0,0,0,0,0,0,0CH,0,0,0,0,0,0,0FH,0
        DB    0FH,0,0,0,0,0,0,0CH,0,0,0,0,0,0,0FH,0
        DB    4 DUP(0FH),0CH,0CH,0CH,0,0CH,0CH,0CH,4 DUP(0FH),0
        DB    0FH,0,0,0,0,0,0,0CH,0,0,0,0,0,0,0FH,0
        DB    0FH,0,0,0,0,0,0,0CH,0,0,0,0,0,0,0FH,0
        DB    0FH,0,0,0,0,0,0,0CH,0,0,0,0,0,0,0FH,0
        DB    0,0FH,0,0,0,0,0,0FH,0,0,0,0,0,0FH,0,0
        DB    0,0,0FH,0,0,0,0,0FH,0,0,0,0,0FH,0,0,0
        DB    0,0,0,0FH,0,0,0,0FH,0,0,0,0FH,0,0,0,0
        DB    0,0,0,0,0,0FH,0FH,0FH,0FH,0FH,0FH,0,0,0,0,0
;------------------------------------------------------------------
; Block control area for duck
DUCK_X      DW    620                ; Present x coordinate
DUCK_Y      DW    410                ; y coordinate
            DB    9                  ; Horizontal rows in block
            DB    2                  ; Number of bytes per row
; Symbol encoding for duck
            DB    0EH,00H,1FH,03H,3FH,07H,7FH,0FH,0FFH,0FFH
            DB    1FH,0FEH,3FH,0FEH,7FH,0FFH,7FH,0FFH
            DW    0000H
; Color encoding of cross-hair symbol
DUCK_COLOR  DB    0,0,0,0,0FH,0FH,0FH,0,0,0,0,0,0,0,0,0
            DB    0,0,0,0FH,0FH,0FH,0FH,0FH,0,0,0,0,0,0FH,0FH
            DB    0,0,0FH,0FH,0FH,0FH,0FH,0FH,0,0,0,0,0FH,0FH,0FH
            DB    0,0FH,0FH,0FH,0FH,0FH,0FH,0FH,0,0,0,0,20 DUP (0FH)
            DB    0,0,0,12 DUP(0FH),0,0,0,13 DUP(0FH),0,0,15 DUP (0FH)
            DB    0,15 DUP (0FH)
            DW    0000H              ; Padding
X_COORD     DW    0000H              ; Storage for x coordinate
BYTES       DB    0H                 ; Number of bytes per block row
COUNT_8     DB    8                  ; Operational bit counter
;------------------------------------------------------------------
```

例10.10 位映像图形显示例程 VARI_PATTERN。

```
;================================
; Display an encoded graphics block
; on entry:
;        SI → Start of control area of graphics block to be displayed
;        BX → Start of block holding color codes
; Register setup:
;        CX = x coordinate of block start
```

```
;               DX = y coordinate of block start
;               BL = number of rows in block
;               BH = number of bytes per block row
;               DI → start of encoded graphics block
;               SI → start of encoded color codes
;===================================
VARI_PATTERN    PROC    NEAR
        CALL    DATA_XOR            ; Set the Data Rotate register to the XOR mode
        PUSH    BX                  ; Save pointer in stack
        MOV     CX, WORD PTR [SI]   ; x coordinate
        MOV     X_COORD, CX         ; store in variable
        ADD     SI, 2               ; Bump pointer
        MOV     DX, WORD PTR [SI]   ; Y coordinate
        ADD     SI, 2               ; Bump pointer
        MOV     BL, BYTE PTR [SI]   ; Number of rows
        INC     SI                  ; Bump pointer
        MOV     BH, BYTE PTR [SI]   ; Bytes per block
        MOV     BYTES, BH           ; Store in variable
        INC     SI                  ; Bump pointer
        XCHG    SI, DI              ; Buffer start to DI
        POP     SI                  ; Color code block pointer
        MOV     COUNT_8, 8          ; Prime bit counter
DISPLAY_BYTE:
        MOV     AH, [DI]            ; High-order nibble to AH
TEST_BIT:
        TEST    AH, 10000000B       ; Is high-order bit set?
        JZ      NEXT_BIT            ; Bit not set
        MOV     AL, [SI]            ; Get color code
; Set the pixel
        PUSH    AX                  ; Save entry reqisters
        PUSH    BX
        CALL    PIXEL_ADD           ; see 10.2
        CALL    WRITE_PIX           ; see 10.3
        POP     BX                  ; Restore registers
        POP     AX
NEXT_BIT:
        SAL     AH, 1               ; shift AH to test next bit
        INC     CX                  ; Bump x coordinate counter
        INC     SI                  ; Bump color table pointer
        DEC     COUNT_8             ; Bit counter
        JZ      NEXT_BYTE           ; Exit if counter rewound
        JMP     TEST_BIT            ; Continue
```

```
; Index to next byte in row, if not at end of row
NEXT_BYTE:
        DEC     BH                      ; Bytes per row counter
        JZ      NEXT_ROW                ; End of graphics row
BYTE_ENTRY:
        INC     DI                      ; Bump graphics code pointer
        MOV     COUNT_8, 8              ; Reset bits counter
        JMP     DISPLAY_BYTE
; Index to next row
NEXT_ROW:
; Test for last graphic row
        DEC     BL                      ; Row counter
        JZ      GRAPH_END               ; Done, exit
        MOV     BH, BYTES               ; Reset bytes counter
        INC     DX                      ; Bump y coordinate control
        MOV     CX, X_COORD             ; Reset x coordinate control
        JMP     BYTE_ENTRY
GRAPH_END:
        CALL    DATA_NORMAL             ; Set the data rotate register to the normal mode
        RET
VARI_PATTERN ENDP
;********************************************************************
```

10.3.4 动画显示技术

计算机动画是利用计算机图形显示技术来模仿物体活动的效果,一般分为两种类型:逐帧(frame-by-frame)动画和实时(real-time)动画。逐帧动画技术的最重要的用途是设计图形系列以建立用不同介质表现的动画图像。例如,利用图形系统来制作卡通电影的图片,当把这些图片以适当的速度投影时,就会产生运动的影像。设计这种逐帧动画的图片,对计算机系统的性能没有严格的要求。

对实时动画来说,因为直接在终端上显示动画程序执行的结果,所以计算机系统的性能的优劣直接影响动画生成的速度和图像的质量,此时计算机系统的性能就显得非常重要了。

屏幕物体的动画效果,经常通过几何变换来产生,最简单的几何变换有平移、旋转和比例变换,复杂的动画通过组合两个或更多的变换来完成。在所有情况下,变换都是以一个新的图像代替先前的图像来实现。在简单的平移变换中,如果在连续递减的 X 坐标上不断地重画物体,就会在屏幕上出现物体从右向左水平移动的画面。在动画软件中,不仅要能画出一系列连续的图像,而且要能从屏幕上擦除先前的图像,否则,运动的物体就会在屏幕上留下痕迹。

擦除和重画屏幕物体的周期有几种实现的方法,最直接的一种方法是在显示图形之前,把图形将要占据的屏幕部分的背景图像保存下来,当要擦除图形时,再把保存的背景图像重新显示出来。

另一种擦除屏幕图像的方法是基于 XOR 操作的方法,有时采用这种方法能获得满意的效果。我们知道,对一个数据用同一个数值 XOR 两次就能恢复原来的内容,例子如下:

```
         10000001B
XOR      10110011B
         ─────────
         00110010B
XOR      10110011B
         ─────────
         10000001B
```

EGA 和 VGA 系统应用 XOR 的方法非常方便,因为汇编语言能对图形控制器的数据循环寄存器编程,选择以正常方式写数据,或用一个锁存器中的数据和 CPU 数据实行 AND,OR 或 XOR 操作。例 10.11 的例程 DATA_XOR 和 DATA_NORMAL 就是设置这两种写方式的程序。

注意:使用 BIOS INT 10H 的功能 9 在图形屏幕上显示文本时,总是把数据循环寄存器设置为正常方式。

在动画程序中,XOR 操作提供了一种连续显示图形和擦除图形的方法,XOR 方法的优点是简单、快速,而且当多个显示物体在同一屏幕位置上重叠时特别有用。XOR 方法的缺点是显示物体的颜色有时要受到背景色的影响,但在动画显示中,有时它反而会产生一种很有趣的效果。

例 10.11　分别设置数据循环寄存器为正常方式(DATA_NORMAL)和异或方式(DATA_XOR)

```
;******************************************************************
DATA_NORMAL    PROC    NEAR
; Set the Graphics controller data rotate register to the normal mode
    MOV    DX, 03CEH          ; Graphics controller port address
    MOV    AL, 3              ; Select Data Rotate register
    OUT    DX, AL
    JMP    SHORT $+2
    INC    DX                 ; 03CFH register
    MOV    AL, 00000000B      ; Reset bits 3 and 4 for normal
    OUT    DX, AL
    JMP    SHORT $+2
    RET
DATA_NORMAL    ENDP
;******************************************************************
```

```
DATA_XOR    PROC    NEAR
; Set the Graphics controller Data Rotate register to the XOR mode
    MOV     DX, 03CEH           ; Graphics controller port address
    MOV     AL, 3               ; Select Data Rotate register
    OUT     DX, AL
    JMP     SHORT $+2
    INC     DX                  ; 03CFH reqister
    MOV     AL, 00011000B       ; Set bits 3 and 4 for XOR
    OUT     DX, AL
    JMP     SHORT $+2
    RET
DATA_XOR    ENDP
;********************************************************************
```

计算机动画的显示操作和擦除操作是在一定速率下进行的，理想的重画速率应不小于每秒 24 幅图像，每秒 15 幅图像是动画显示的临界速率，小于这个临界速率，画面就会产生颠簸和抖动现象。

对许多图形应用程序来说，获得定时控制的最好方法是在垂直回扫周期的开始产生一个中断，这可以通过对 EGA 和 VGA 的编程来检测 CRT 垂直回扫周期的起始点。由于 EGA 和 VGA 的屏幕刷新速率是每秒 70 周期，程序以每秒 70 次的速度完成图像或数据的处理，就足以产生非常平滑的动画效果。在图形比较简单，动画质量也要求不高的情况下，如显示前面介绍的"十字符"和"鸭子"图形时，只要利用定时器中断（每秒 18.2 次）的周期就可以完成动画的处理了。有兴趣编写动画程序的读者可以查阅相关资料。

已经知道计算机动画是由一定速率的显示操作和擦除操作完成的，有时这个技术也被称作定时脉冲动画（timed-pulse animation）。在定时脉冲动画中，理想的重画速率应不小于每秒 24 幅图像，每秒 15 幅图像是动画显示的临界速率，小于这个临界速率，画面就会产生颠簸和抖动现象。

对许多图形应用程序来说，获得定时脉冲控制的最好方法是在垂直回扫周期的开始产生一个中断，这可以通过对 EGA 和 VGA 的编程来实现。EGA 和 VGA 的屏幕刷新速率是每秒 70 周期，这足以能产生非常平滑的动画，因为在这个定时脉冲区间的很短的时间里就可完成任何图像或数据的处理。

利用垂直回扫中断作为定时脉冲的优点是自动避免了屏幕干扰。采用这种方法，使中断例程在 CRT 控制器的垂直回扫周期开始时接收控制，在这 1/70 秒里能完成多少处理工作取决于系统硬件，主要是 CPU 的型号和速度以及存储器的存取周期。

在图形比较简单，动画质量也要求不高的情况下，只要使用时钟中断（每秒 18.2 次）就可以实现简单动画的显示与擦除。有兴趣的读者可以根据前几节的介绍，编写一个"鸭子"在"河里"游动的动画程序。

10.4 通用发声程序

在演示游戏程序时，音乐和声响的效果会使表演更加丰富多彩。那么计算机是如何产生声音和乐曲的呢？原来在计算机中有一个可编程时间间隔定时器 8253/54 (programmable interval timer,PIT)，它能根据程序提供的计数值和工作模式，产生各种形状和各种频率的计数/定时脉冲，提供给系统的各个部件使用。例如，提供计时信号给系统日时钟，提供刷新定时信号给动态存储器，对分时系统产生时间片等。它还可以产生不同频率的脉冲作为扬声器的声源。本节先介绍计算机发声的原理，然后在 10.5 节介绍乐曲的编程方法。

10.4.1 可编程时间间隔定时器 8253/54

在 8253/54 定时器内部有 3 个独立工作的计数器：Counter0，Counter1 和 Counter2，每个计数器都分配有一个端口地址，分别为 40H、41H 和 42H。8253/54 内部还有一个公用的控制寄存器，端口地址为 43H。端口地址输入到 8253/54 的 CS，A1，A0 端，分别对 3 个计数器和控制器寻址。

对 8253/54 编程时，先要设定控制字，以选择计数器，确定工作模式和计数值的格式。每个计数器由三个引脚与外部联系，如图 10.7 所示。CLK 为时钟输入端，GATE 为门控信号输入端，OUT 为计数/定时信号输出端。每个计数器中包含一个 16 位的计数寄存器，这个计数器是以倒计数的方式计数的，也就是说，从计数初值开始逐次减 1，直到减为 0 为止。

8253/54 的三个计数器是是分别编程的，在对任一个计数器编程时，必须首先将控制字节写入控制寄存器。控制字的作用是告诉 8253/54 选择那一个计数器工作，要求输出什么样的脉冲波形。另外，对 8253/54 的初始化工作还包括，向选定的计数器送入一个计数初值，因为这个计数值可以是 8 位的，也可以是 16 位的，而 8253/54 的数据总线是 8 位的，所以要用两条输出指令来写入初值。

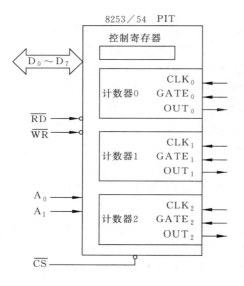

图 10.7　8253/54 的编程结构

图 10.8 所示的是控制字格式，在 8253/54 初始化时，这个字节被送入控制寄存器。控制字节各位的含义如下：

D0——选择计数值的格式。D0 = 0,计数值是二进制格式；D0 = 1,计数值是 BCD 码格式。这个计数值在编程时作为与 CLK 输入频率相除的除数，最小值是 0001,最大值是 2^{16}（二进制）或 10^4（BCD）。为了得到这个最大数，计数器装入的是 0000,它作为 0FFFFH+1 来看待，相当于 10000H。

D3,D2 和 D1——选择操作模式。D3 D2 D1 = 000 ～ 010 分别表示 Mode0 ～ Mode5。这 6 种操作模式决定了输出脉冲的形状。

D5 和 D4——读写指示位。有 4 种选择：① 计数器锁存操作，锁定当前计数值，以便读出；② 只读/写高字节（MSB）；③ 只读/写低字节（LSB）；④ 先读/写 LSB，紧接着读/写 MSB。

D7 和 D6——选择计数器。确定控制字是对哪一个计数器进行初始化。D7 D6 = 00 ～ 11，分别选择 Mode0，Mode1 和 Mode2。

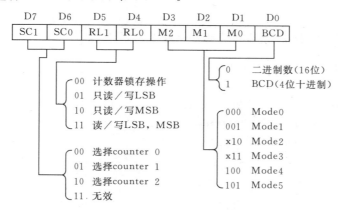

图 10.8　8253/54 的控制字格式

对 8353/54 设置初始值时，一定要符合控制字的格式规定，是二进制数还是 BCD 码表示的数，是只写高（低）字节，还是高低字节都写。控制字一旦做出规定，程序中给出的计数值就要和控制字的要求一致。

例 10.12　写出 8253/54 的初始化程序段。

① 将计数器 0 设定为模式 3，计数初值为 4282（BCD）

```
    MOV    AL,00110111B      ;counter0,mode3,BCD
    OUT    43H,AL            ;Send it to control register
    MOV    AX,4282H          ;load the divisor
    OUT    40H,AL            ;send the low byte to counter0
    MOV    AL,AH             ;then the high byte
    OUT    40H,AL            ;to counter0
```

② 将计数器 2 设定为模式 3，计数初值为 65536

```
    MOV    AL,10110110B      ;counter2,mode3,binary
    OUT    43H,AL            ;Send it to control register
    SUB    AL,AL             ;load the divisor 0000
    OUT    42H,AL            ;send the low byte
    OUT    42H,AL            ;and then the high byte to counter2
```

在 IBM PC 机中，8253/54 的三个时钟端 CLK0、CLK1 和 CLK2 的输入频率都是 1.1931817MHz，计数器 0 和计数器 1 的门控 GATE0 和 GATE1 接 +5V，计数器 2 的 GATE2 与 8255 的端口 PB0 相连。知道了每个计数器的输入，下面分别介绍它们的编程和应用。

计数器 0 作为定时器为系统日时钟提供计时基准,计数器 0 的输出端 OUT 与中断控制器 8259A 的中断请求端 IRQ0 相连,为 IRQ0 提供每秒 18.2 次的中断信号,也就是说,OUT0 的输出频率应当是 18.2Hz,这正是 CLK0 的输入频率 1.1931817 MHz 与 2^{16} 相除的结果。计数器 0 的操作模式选择 Mode3。根据这些要求,控制字应当是 00110110B＝36H。

在 IBM PC BIOS 中,计数器 0 的初始化程序如下:

```
            22      TIMER   EQU    40H
...   ...   ...     ...     ...    ...
E277  B036  695     MOV     AL,    36H              ;SET TIM0,LSB,MSB,MODE3
E279  E643  696     OUT     TIMER+3,AL              ;WRITE TIMER MODE REG
E27B  B000  697     MOV     AL,0
E27D  E640  698     OUT     TIMER,AL                ;WRITE LSB TO TIMER0 REG
...   ...   ...     ...     ...    ...
E284  E640  704     OUT     TIMER,AL                ;WRITE MSB TO TIMER0 REG
```

计数器 1 作为定时器使用,其输出脉冲用作 DRAM 刷新的定时信号。在 IBM PC 中,刷新 DRAM 的任务由 8237 DMA 来完成。DRAM 要求每隔 15μs 刷新一次,这样,OUT1 输出脉冲的频率是 66.2KHz。因为 CLK1 的输入频率为 1.19318MHz,所以计数初值应是 18(1.19318MHz/18＝66.2 KHz)。在操作模式 2 下,OUT1 连续输出周期为 15μs 的定时信号,这个定时信号就作为 DRAM 的刷新请求信号。

IBM BIOS 中,有关计数器 1 的初始化程序如下:

```
        MOV     AL,54H      ; set counter1,LSB only,mode2,binary
        OUT     43H,AL      ; the control word to control register
        MOV     AL,18       ; 18 decimal,the divisor
        OUT     41H,AL      ; to counter1
```

计数器 2 用来控制扬声器发声。在 IBM BIOS 中有个 BEEP 子程序,它在模式 3 下,能产生频率为 896Hz 的声音,装入计数器 2 的计数初值为 533H(1.19318MHz / 896 Hz＝1331＝533H),这样,得到的控制字为 10110110B ＝ 0B6H(Counter2,LSB 和 MSB,Mode3,二进制格式)。

BIOS 中计数器 2 的初始化程序如下:

```
        MOV     AL,0B6H     ; set counter2,LSB, MSB,mode3,binary
        OUT     43H,AL      ; to control register
        MOV     AL,33H      ; low byte
        OUT     42H,AL
        MOV     AL,05       ; high byte
        OUT     42H,AL
```

10.4.2 扬声器驱动方式

PC 机上的大多数 I/O 都是由主板上的 8255（或 8255A）可编程序外围接口芯片（PPI）管理的。PPI 包括三个 8 位寄存器，两个用于输入功能，一个用于输出功能。输入寄存器分配的 I/O 端口号为 60H 和 62H，输出寄存器分配的 I/O 端口号为 61H。由 PPI 输出寄存器中的 0 和 1 两位来控制扬声器的驱动方式（见图 10.9 扬声器驱动系统）。

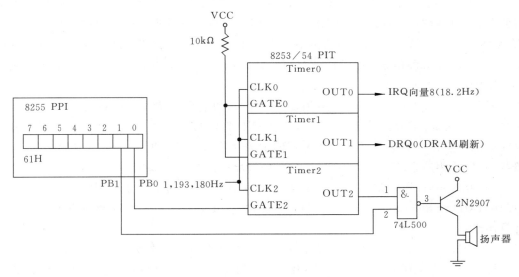

图 10.9 扬声器驱动系统

在第 8 章中例 8.1 曾介绍过一个 Sound 程序，它是直接控制输出端口 61H 的 PB1 交替为 1 或为 0，以产生一个脉冲电流，经过放大器后控制扬声器交替地接通与断开，从而推动扬声器的纸盆振动，发出不同音高和音长的声音。这种控制发声的方式称为位触发方式。

另一种产生声音的方式是利用 8253/54 定时器来驱动扬声器。连接到扬声器上的是定时器 2，从图 10.9 中可以看到，GATE2 与端口 61H 的 PB0 相连，当 PB0＝1 时，GATE2 获得高电平，使定时器 2 可以在模式 3 下工作。定时器 2 的 OUT2 与端口 61H 的 PB1 通过一个与门与扬声器的驱动电路相连。当 PB1＝1 时，允许 OUT2 的输出信号到达扬声器电路。下面是控制扬声器的代码：

```
IN      AL,61H           ; get the current setting of port B
MOV     AH,AL            ; save it
OR      AL,00000011B     ; make PB0=1 and PB1=1
OUT     61H,AL           ; turn the speaker on
……                       ; how long the beep sound goes here
MOV     AL,AH            ; get the original setting of port B
OUT     61H,AL           ; turn off the speaker
```

对 IBM PC 及兼容机来说,无论是 8086,80486 还是 Pentium,驱动扬声器的过程都是相同的。至于音调发出的时间量(音长),则取决于它延迟的时间,这个延迟时间可以在 PC 机的 80x86 主处理器的帮助下实现。

10.4.3 通用发声程序

BIOS 中的 BEEP 子程序能根据 BX 中给出的计数值控制 8253/54 定时器,产生频率为 896Hz 的声音。这个子程序在 IBM PC 技术手册的 TEST4 部分,它的用途是当"加电自测"系统发现硬件错误后,由 ERROR_BEEP 处理程序调用 BEEP 子程序,使扬声器发出"嘟嘟"的信号。BX 中的计数值由 ERROR_BEEP 程序设置为 533H,然后 BEEP 程序又将 8255 的输出寄存器(I/O 端口 61H)的 PB0 和 PB1 置 1,使扬声器接通并发出音频为 896Hz 的声音。

实际上 BEEP 是一个很好的通用发声程序,可以利用并修改 BEEP,使其产生任一频率的声音。为此,需要做两点修改。首先,BEEP 程序只能产生 896Hz 的声音,通用发声程序应能产生任一频率的声音。其次,BEEP 产生声音的持续时间(音长)只能是 0.5s 的倍数,我们希望声音的持续时间更易于调整,例如可以是 10ms 的倍数。

从 10.4.1 小节中知道,给计数器 2 装入计数值 533H 就能产生 896Hz 的声音 (1.1931MHz÷896Hz=1331=533H),同样,产生其他频率声音的计数值也可以用下式计算出来:

$$1193100 \div 给定频率 = 12348CH \div 给定频率$$

假定发声频率存放在 DI 寄存器中,下面的指令使 AX 中得到送往定时器 2 的计数值:

```
MOV    DX, 12H
MOV    AX, 348CH
DIV    DI
```

利用 CPU 来产生延迟时间是最常用的一种方法,但对 8088/86,80286,386,486 以及 Pentium 来说,它们产生的延迟时间都是不一样的。对基于 8088/86 的 PC/XT,PS/2 等 PC 机,可用下面的程序来产生时间延迟:

```
        MOV    CX, N
DELAY:  LOOP   DELAY
```

在 8088/86 CPU 中,执行一条 LOOP 指令需要 17 个时钟周期,因此,执行上面的指令大约需要 N×17×T 时钟周期。例如,N=2800,系统频率为 4.7MHz(时钟周期 T 约为 212ns),那么得到的延迟时间为 10ms(2800×17×212ns)。

在 BIOS 中,利用上述方法来产生 250ms 延迟时间的程序如下:

```
        SUB    CX, CX
G7:     LOOP   G7            ; 65536×212ns×17=236ms
```

如果要产生与10ms成倍数的延迟时间,可在BX寄存器中放入倍数。例如,要产生1s的持续时间,则在BX中放入100,以控制LOOP指令执行100×2800次,也就是10ms的100倍。指令如下:

```
            MOV     BX,100
WAIT:       MOV     CX,2800
DELAY:      LOOP    DELAY
            DEC     BX
            JNZ     WAIT
```

下面在修改后的BEEP程序的基础上,为8088/86编写一个任一频率(由DI指定)和任一持续时间(由CX和BX指定)的通用发声程序。此程序利用定时器产生声音比例8.1的Sound程序稍微复杂一些,它包括三个步骤:

(1) 在8253/54中的43端口送一个控制字0B6H(10110110B),该控制字对定时器2进行初始化,使定时器2准备接收计数初值。

(2) 在8253/54中的42H端口(Timer2)装入一个16位的计数值(533H×896/频率),以建立将要产生的声音频率。

(3) 把输出端口61H的0、1两位置1,发出声音。

例10.13 编写通用发声程序GENSOUND,它能利用定时器发出指定频率的声音。

```
TITLE GENSOUND —— The speaker beeper
; Produces a tone of a specified frequency
;    and duration public gensound
;--------------------------------------------------------------
public      gensound
cseg        segment para 'code'
            assume      cs:cseg
gensound    proc        far
            push        ax              ;save registers
            push        bx
            push        cx
            push        dx
            push        di
            mov         al,0b6h         ;write timer mode reg.
            out         43h,al
            mov         dx,12h          ;timer divisor
            mov         ax,348Ch        ;1193100Hz /freq
            div         di              ;value of freq
            out         42h,al          ;write timer2 count low byte
            mov         al,ah
            out         42h,al          ;write timer2 count high byte
```

```
            in      al,61h              ;get current port setting
            mov     ah,al               ;and save it in ah
            or      al,3                ;turn speaker on
            out     61h,al
wait1:      mov     cx,2800             ;wait for specified interval
delay:      loop    delay
            dec     bx
            jnz     wait1
            mov     al,ah               ;recover value of port
            out     61h,al
            pop     di                  ;recover the rigister
            pop     dx
            pop     cx
            pop     bx
            pop     ax
            ret                         ;exit
gensound    endp
cseg        ends
;------------------------------------------------------------
            end
```

GENSOUND 程序能产生 19～65535Hz 的声音,这个频率的下限 19Hz 是使除法不产生溢出的最小的 DI 值((DX)=12H=18 d<19)。其上限 65 535Hz 实际上是多余的,因为人们最高能听到的音频约为 20000Hz。

注意:GENSOUND 程序产生的声音不仅与输入频率有关,而且与 CPU 有关。如果 80x86 的工作频率为 8MHz(如 IBM PS/2-25 型),则 T=125ns(1/8MHz=125ns),那么上面程序产生的时间延迟就要短的多。在 80286 中,LOOP 指令只需 8 个执行周期,而不是 17 个执行周期,这也会使延迟时间缩短很多。因此,从 PC/AT 开始,对所有的 80286,80386,80486 和 Intel Pentium 计算机,IBM 都提供一种利用硬件产生时间延迟的方法,这种方法不仅与频率无关,也与 CPU 无关。在 10.4.4 小节来介绍这种方法。

10.4.4 80x86 PC 的时间延迟

80x86 的各种处理器采用 6～66MHz 的工作频率,LOOP 指令的执行时间在这些处理器上也不相同。为了建立一个与处理器无关的时间延迟,IBM 采用了一种利用硬件产生时间延迟的方法,即通过监控端口 61H 的 PB4,使 PB4 每 15.08μs 触发一次,以产生一个固定不变的时间基准。在 IBM PC AT BIOS 中的 WAITF 子程序,就是一个产生 N×

15.08μs 时间延迟的程序。调用 WAITF 子程序时,CX 寄存器必须装入 15.08μs 的倍数 N。

```
        ; (CX) = Count of 15.08μs
WAITF   PROC   NEAR
        PUSH   AX
WAITF1:
        IN     AL,61H
        AND    AL,10H          ; check PB4
        CMP    AL,AH           ; did it just change
        JE     WAITF1          ; wait for change
        MOV    AH,AL           ; save the new PB4 status
        LOOP   WAITF1          ; continue until CX becomes 0
        POP    AX
        RET
WAITF   ENDP
```

必须注意:在 80x86 各种型号的 PC 机中,端口 B(端口地址 61H)既作为输入寄存器又作为输出寄存器,而在 8088/86 的 PC 机中,端口 B 只作输出寄存器用。

利用 WAITF 子程序能获得任意的延迟时间,而再不必考虑 CPU 的型号和工作频率。例如,为了产生 0.5s 的延迟,先设置(CX)=33144(33144×15.08μs=0.5sec),然后,调用 WAITF 子程序。

```
        MOV    CX,33144
        CALL   WAITF
```

例 10.14 调用 WAITF 子程序,产生 1.5s 的延迟时间。

因为 1.5s 的延迟时间需要设置计数值 99436(1.5 sec /15.08μs = 99436),而 16 位寄存器表示的最大数为 65536,所以可以连续调用 3 次产生 0.5s 延迟的程序。

```
        MOV    BL,03
BACK:   MOV    CX,33144        ; 0.5 second delay
        CALL   WAITF
        DEC    BL
        JNZ    BACK
```

依此原理可以进一步修改通用发声程序 GENSOUND,使其不再依赖 CPU 的工作频率。GENSOUND 程序中的 10ms 延迟时间是通过执行循环指令来获得的,现在可以用 WAITF 子程序来取代它,计数值设置为 663(10ms /15.08μs = 663.13)。为了区别起见,修改后的通用子程序改名为 SOUNDF,延迟时间仍为 10ms。

```
SOUNDF  PROC  FAR
        ...   ...           ; same as gensound
WAIT1:  MOV   CX,663         ; 663×15.08μs=10ms
        CALL  WAITF
        ...   ...           ; same as gensound
```

10.5 乐曲程序

10.5.1 音调与频率和时间的关系

利用计算机控制发声的原理，可以编写演奏乐曲的程序。乐曲是按照一定的高低、长短和强弱关系组成的音调。在一首乐曲中，每个音符的音高和音长与频率和节拍有关。图 10.10 画出了两个音阶（一个音阶是 8 个音符）的钢琴键和每个键的音名及其频率（Hz）。低音阶从低 C(130.8Hz)到中 C(261.7Hz)，高音阶从中 C 到高 C(523.3Hz)。白色键（A 到 G）演奏本位音符，黑色键演奏升降音符，黑键比它旁边的白键高半个音或低半个音。

组成乐曲的每个音符的频率值和持续时间是乐曲程序发声所需要的两个数据。音符的频率可以从图 10.10 中查到，但实际上送入计数器 2 的是输入频率 1.1931MHz 与音符频率相除的值，通用发声程序 SOUNDF 的前半部分就是完成这个计算，并将计数值送入计数器 2 的功能。

```
        ;(DI) = the desired output frequency
        MOV   AL,0B6H         ; write the control word
        OUT   43H,AL          ; into control register
        MOV   DX,12H          ; 1193100Hz = 12384CH
        MOV   AX,384CH
        DIV   DI              ;(AX) = value loaded into counter2
        OUT   42H,AL          ; write counter2 count low byte
        MOV   AL,AH
        OUT   42H,AL          ; write counter2 count high byte
        ...   ...
```

音符的持续时间是根据乐曲的速度及每个音符的节拍数来确定的。在 4/4（四四拍）中，四分音符为 1 拍，每小节 4 拍，全音符持续 4 拍，二分音符持续 2 拍，四分音符持续 1 拍，八分音符持续半拍等。如果给全音符分配 1s(100×10ms)的时间，则二分音符的持续时间为 0.5s(50×10ms)，四分音符的持续时间为 0.25s(25×10ms)，八分音符的持续时间为 0.125s(12.5×10ms)。

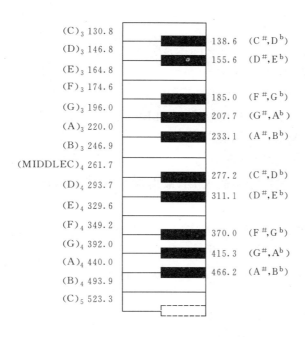

图 10.10 两个音阶的琴键

10.5.2 演奏乐曲的程序

知道了音调与频率和时间的关系,就可以按照乐曲的曲谱将每个音符的频率和持续时间定义成两个数据表;然后编写程序依次取出表中的频率值和时间值,调用 SOUNDF 程序发出各种声音。

下面按照图 10.11 的乐谱编写一个演奏程序。

图 10.11 乐谱

编写乐曲程序可分为四个步骤:

(1) 为演奏的乐曲定义一个频率表和一个节拍时间表。频率表中的数据可以从图 10.10 中查到,节拍时间表中的数据是延迟时间 10ms 的倍数。

```
MUS_FREQ  DW  330, 294, 262, 294, 3 DUP(330)         ;1,2 小节
          DW  294, 294, 294, 330, 392, 392            ;3,4 小节
          DW  330, 294, 262, 294, 4 DUP(330)         ;5,6 小节
```

```
            DW    294,294,330,294,262,0FFFFH           ;7,8 小节
MUS_TIME    DW  6 DUP(25),50                           ;1,2 小节
            DW  2 DUP(25,25,50)                        ;3,4 小节
            DW  12 DUP(25),100                         ;5～8 小节
```

(2) 分别将两个表的偏移地址放入 SI 和 BP

```
            LEA    SI,MUS_FREQ
            LEA    BP,DS:MUS_TIME
```

(3) 从表中取出音符的频率放入 DI,取出音符的持续时间(实际上是 10ms 的倍数)放入 BX。

```
            MOV    DI,[SI]
            MOV    BX,DS:[BP]
```

频率表中的最后一个数据 0FFFFH 作为乐曲的结束符,也可用 0 或其他特定值来代替。

(4) 调用 SOUNDF 子程序发出音调。

例 10.15 演奏乐曲的程序 MUSIC。

```
TITLE MUSIC --A music of 'Mary had a little lamb'
; assemble with: masm music
; link with: link music + soundf
;------------------------------------------------------------
extrn     soundf:far
stack     segment para stack 'stack'
          db    64 dup('stack...')
stack     ends
;------------------------------------------------------------
dseg      segment para 'data'
mus_freq  dw    330,294,262,294,3 dup(330)       ;bar 1&2
          dw    3 dup(294),330,392,392           ;bar 3&4
          dw    330,294,262,294,4 dup(330)       ;bar 5&6
          dw    294,294,330,294,262,-1           ;bar 7&8
mus_time  dw    6 dup(25),50                     ;bar 1&2
          dw    2 dup(25,25,50)                  ;bar 3&4
          dw    12 dup(25),100                   ;bar 5&6
dseg      ends
;------------------------------------------------------------
cseg      segment para 'code'
          assume cs:cseg,ss:stack,ds:dseg
music     proc far
          mov    ax,dseg                         ;initialize DS
```

```
            mov     ds,ax
            lea     si,mus_freg;    puts the freg table offset in SI
            lea     bp,ds:mus_time        ;puts the time table offset in BP
    freq:
            mov     di,[si]               ;read next frequency
            cmp     di,-1                 ;end of tone?
            je      end_mus               ;if yes,exit
            mov     bx,ds:[bp]            ;else,fetch the duration
            call    soundf                ;play the note
            add     si,2                  ;update the table pointer
            add     bp,2
            jmp     freq                  ;go process next note
    end_mus:
            mov     ax,4c00h              ;return to DOS
            int     21h
    music   endp
    cseg    ends
;------------------------------------------------------------
            end     music
```

这个程序是比较简单的,如果想演奏另一个乐曲,只需把 mus_freq 和 mus_time 两个表中的数据换成另一个乐曲的频率和节拍时间就可以了。例如可把《太湖船》的曲调写入数据段,则此程序运行的结果是演奏了一首民乐《太湖船》。

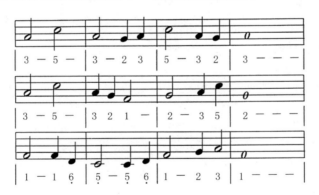

图 10.12 《太湖船》乐谱

```
    DSEG         SEGMENT PARA 'DATA'
    MUS_FREQ  DW  330,392,330,294,330,392,330,294,330
              DW  330,392,330,294,262,294,330,392,294
              DW  262,262,220,196,196,220,262,294,330,262
              DW  -1
    MUS_TIME  DW  3 DUP(50),25,25,50,25,25,100
              DW  2 DUP(50,50,25,25),100
              DW  3 DUP(50,25,25),100
    DSEG      ENDS
```

10.5.3 键盘控制发声程序

音符和频率之间有一定的对应关系,如果计算机键盘上的某些键(例如数字键或字母键)和音符、频率也形成一种对应关系,则可通过键盘控制扬声器发出各种音符声音,这时计算机键盘就变成了钢琴键盘,就可以用它弹奏出简单的音乐。钢琴有88个音符,为了简单起见,下面通过编写一个八度音程的钢琴程序来了解键盘控制发声的原理。为了弹奏方便,让数字键1~8对应一个音阶的八个音符。

弹奏时,对应乐谱上的C音符(简谱中为1),按下数字键"1"(ASCII码为31H),程序的工作就是要接收这个键,并将频率表中和它对应的频率值262Hz送给SOUNDF程序,以发出C的音调。按下数字键"2"(ASCII码为32H),程序就将294Hz的频率值送给SOUNDF程序,从而发出D的音调。按下数字键"8",将发出比第一个C高八度的音调,音符的频率表作为数据定义在数据段中。图10.13是一个八度的钢琴程序PIANO的流程图。

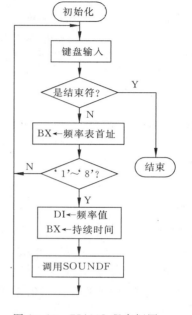

图10.13 PIANO程序框图

例10.16 八度音程的钢琴程序PIANO。

```
TITLE PIANO ————Use timer2 to run speaker
; Keyboard number keys play notes of the scale
; assemble with: masm piano
; link with: link piano + soundf + waitf
;————————————————————————————————————
extrn    soundf :far
stack    segment   para    'stack'
         db    64 dup('stack...')
stack    ends
;————————————————————————————————————
dseg     segment   para    'data'
table    dw    262         ; C
         dw    294         ; D
         dw    330         ; E
         dw    349         ; F
         dw    392         ; G
         dw    440         ; A
         dw    494         ; B
         dw    523         ; C
```

```
dseg       ends
;----------------------------------------------------------------
cseg       segment    para   'code'
           assume     cs:cseg, ds:dseg, ss:stack
main       proc       far
           mov        ax,dseg              ;initialize DS
           mov        ds,ax
new_note:
           mov        ah,0                 ;read the keyboard
           int        16h
           cmp        al,0dh               ;is it the return key?
           je         exit                 ;yes,return
           mov        bx,offset table      ;freq table address
           cmp        al,'1'               ;is '1'—'8'
           jb         new_note             ;no,read the next key
           cmp        al,'8'
           ja         new_note
           and        ax,0fh               ;mask off high 4 bits
           shl        ax,1                 ; * by 2
           sub        ax,2                 ;convert the ASCII to
           mov        si,ax                ; an index in SI
           mov        di,[bx][si]          ;look_up a frequency
           mov        bx,10                ;for 0.1 seconds
           call       soundf
           jmp        new_note
exit:
           mov        ax,4c00h             ;return to DOS
           int        21h
main       endp
cseg       ends
;----------------------------------------------------------------
           end        main
```

从键盘输入一个键这里用的是 BIOS 16H 中断，AL 中回送的是键的 ASCII 码，AH 中回送的是键的扫描码。因为此程序没有用到功能键或控制键，所以也可以用 DOS 的键盘输入功能。

```
MOV   AH,07       ;DOS Kbd function,no echo
INT   21H         ;call DOS
```

习　题

10.1 写出指令,选择显示方式 10H,并将背景设为绿色。

10.2 如何使用 INT 10H 的功能调用改变显示方式?

10.3 VGA 独有的一种显示方式是什么?

10.4 对于 EGA 和 VGA 显示适配器,使用显示方式 13H 时,显示数据存在哪里?

10.5 对于 VGA 的显示方式 13H,存放一屏信息需要多少字节的显存?

10.6 利用 BIOS 功能编写图形程序:设置图形方式 10H,选择背景色为蓝色,然后每行(水平方向)显示一种颜色,每 4 行重复一次,一直到整个屏幕都显示出彩条。

10.7 修改 10.6 题的程序,使整个屏幕都显示出纵向的彩条。

10.8 按动键盘上的光标控制键,在屏幕的上下左右任一方向上绘图,每画一点之前,由数字键 0~3 指定该点的颜色值,按动 ESC 键,绘图结束,返回 DOS。

10.9 位屏蔽寄存器的作用是什么?在 16 色,640×480 显示方式中如何使用位屏蔽寄存器?

10.10 读映像选择寄存器的作用是什么?如果 4 个位面的内容都需要读取,读映像选择寄存器应如何设置?

10.11 编写程序使一只鸟飞过屏幕。飞鸟的动作可由小写字母 V(ASCII 码 76H)变为破折号(ASCII 码 0C4H)来模仿,这两个字符先后交替在两列显示。鸟的开始位置是 0~20 行,每个字符显示 0.5s,(可用指令循环来模拟),然后消失。

10.12 用图形文本的方法设计"Name = XXX"(X 为你自己姓名的缩写),并将其数据编码定义在一个数组中。

10.13 游戏程序常常用随机数来控制其图形在屏幕上移动。请编写一程序,用随机数来控制笑脸符(ASCII 码 02)显示的位置。笑脸符每次显示的列号总是递增 1,而行的位置可能是前次的上一行、下一行或同一行,这根据随机数是 0,1 或 2 来决定,当行号变为 24 或列号变为 79 时显示结束。笑脸在每个位置上显示 1/4s。(提示:随机数的基数可读定时器 0 来获得,再经过一定的算法取得随机数)。

10.14 分配给 PC 机主板上的 8253/54 定时器的端口地址是什么?

10.15 8253/54 定时器的三个计数器,哪一个用于扬声器?它的端口地址是什么?

10.16 下面的代码是利用监控端口 61H 的 PB4 来产生延迟时间的,它适用于所有的 286,386,Pentium PC 及兼容机。请指出该程序的延迟时间是多少?

```
        MOV   DL,200
BACK:   MOV   CX,16572
WAIT:   IN    AL,61H
        AND   AL,10H
        CMP   AL,AH
        JE    WAIT
        MOV   AH,A;
        LOOP  WAIT
        DEC   DL
        JNZ   BACK
```

10.17 在 PC 机上编写乐曲程序"Happy Birthday",乐曲的音符及音频如下:

歌词	音符	音频	节拍
hap	C	262	1/2
py	C	262	1/2
birth	D	294	1
day	C	262	1
to	F	349	1
you	E	330	2
hap	C	262	1/2
py	C	262	1/2
birth	D	294	1
day	C	262	1
to	G	392	1
you	F	349	2
hap	C	262	1/2
py	C	262	1/2
birth	D	523	1
day	A	440	1
dear	F	349	1
so	E	330	1
so	D	294	3
hap	Bb	466	1/2
py	Bb	466	1/2
birth	A	440	1
day	C	262	1
to	G	392	1
you	F	349	2

10.18 编写用键盘选择计算机演奏歌曲的程序。首先在屏幕上显示出歌曲名单如下：

    A   music 1

    B   music 2

    C   music 3

当从键盘上输入歌曲的序号 A,B 或 C 时,计算机则演奏所选择的歌曲,当在键盘上按下 0 键时,演奏结束。

第 11 章　磁盘文件存取技术

外部设备一般分为两类,一类为字符设备,另一类为大容量存储设备。前面几章介绍的键盘、显示器、打印机、串行通信口等都是字符设备。大容量存储设备包括软磁盘、硬磁盘、磁带和光盘(optical disk)。目前微型机普遍使用的是温彻斯特(winchester)硬磁盘和3.5英寸高密度软磁盘。这两种磁盘容量大,速度快,一直是微型机理想的外存储器。近年来,光盘以其高记录密度与高可靠性,已成为一种很有竞争力的外存储器。

磁盘操作系统(DOS)提供了一组 DOS 磁盘存取功能,可以很方便地引用这组功能调用从磁盘上读取某个文件或把一个文件写入磁盘中去。所谓文件就是存放在磁盘上的程序或数据。DOS 1.00 和 1.10 为读写磁盘文件提供了三种方式:顺序存取方式,随机存取方式和随机分块存取方式。DOS 2.00 以上的版本除了兼容这三种存取方式而外,还提供了一个更新的方法:文件代号式磁盘存取。DOS 3.0 以上的版本增加了文件共享等网络磁盘的基本功能。包括在低一级的 DOS 功能中,中断 25H 和 26H 提供了按磁盘扇区号来绝对寻址的方法。最低级的是 BIOS 中断 13H,此功能要求指定扇区号及磁道号来进行读写操作。

本章将介绍一些有关磁盘存取的基本概念,并主要讨论文件代号式磁盘存取方式,以及 BIOS 提供的磁盘存取方式,读者可根据具体的程序运行环境有选择地学习本章的内容。

11.1　磁盘的记录方式

软盘与硬盘的存储原理与记录方式是相同的,但在结构上存在一些差别:硬盘转速高,存取速度快,存储容量大;软盘转速低,存取速度慢,存储容量相对小;硬盘有固定磁头、固定盘、盘组等结构;软盘都是活动头,可换盘片结构;硬盘是靠浮动磁头读写,磁头不接触盘片;软盘磁头是接触式读写;硬盘系统及硬盘片价格都比较高,大部分盘片不能互换;软盘造价低,盘片使用灵活,且具有互换性;硬盘要求有超净措施,软盘则对环境的要求不太严格。可以说,硬盘是计算机系统中不可缺少的外存设备,软盘是用户最灵活的记录设备。为了存取磁盘上文件信息,先介绍磁盘的记录方式和一些术语。

11.1.1　磁盘记录信息的地址

1. 磁道和扇面

硬磁盘的盘片一般以铝合金为基体,软盘是用聚酯塑料为基片,它们表面涂有一层磁性材料。硬盘的多个盘片固定在同一根主轴上,称为盘组。为了正确存储、检索信息,必须将盘片划分成磁道和扇区,它们是磁盘记录信息的地址。

早期的 PC 机使用单面双密度盘,后来用双面双密度盘,现在的 286,386 和 Pentium 机使用双面双倍密度盘。无论哪一种盘,信息都记录在盘面的一个个同心圆上,这些同心圆称为磁道。盘片的每面都有一个磁头,存取数据时磁头沿盘面径向移动,从而对每个磁道进行读写。磁道从外圈到里圈依次编号为 00,01,…,每个磁道被格式化为 512 字节的扇面,扇面的编号从 1 开始。当磁头定位到目标磁道,所要求的扇区也旋转到磁头位置时,在磁盘控制器的控制下,磁头完成一次读写操作。

2. 柱面

柱面是由每个盘面上同一编号的磁道组成的。例如,0 号柱面是每个盘面上 0 号磁道的集合,1 号柱面是所有 1 号磁道的集合,等等。

当写文件时,磁盘控制器写满一个柱面上的所有磁道,再移动到下一个柱面进行写操作。例如,系统先写满软盘的柱面 0(盘面 1 和 2 磁道 0 上的所有扇面),再在盘面 1 柱面 1 的磁道 1 上写信息。

3. 磁盘控制器

磁盘控制器的功能是解释来自主机的命令,并向磁盘驱动器发出各种控制信号;同时还要监测驱动器的状态,按规定的数据格式向驱动器读写数据。例如,处理器发出一个从指定的柱面-磁头-扇面读数据的请求信号,控制器的任务是把磁头移动到所要求的柱面,并选择磁头将其定位在目标磁道上,当磁盘上的地址信息和目标地址相符时,从扇区读出数据送往内存。

4. 簇

将几个扇区组成一组称为簇。系统将簇作为磁盘存储空间的单位,簇的大小总是 2^n 个扇区,如 1 扇区、2 扇区、4 扇区或 8 扇区。在每簇 1 个扇区的磁盘中,扇区和簇是相同的。在典型的硬盘中,每簇 4 个扇区,其磁盘组织如下:

假如有一个 100 字节的文件存储在每簇 4 扇区的磁盘中,那么这个文件占据了 $4 \times 512 = 2048$ 字节,虽然它只有一个扇区的数据。

5. 磁盘容量

表 11.1 是常用软盘的存储容量:

表 11.1 常用软盘的存储容量

磁盘.容量	磁道数/面	扇区数/道	字节数/扇	双面总字节	扇区数/簇
5.25″ 360KB	40	9	512	368 640	2
5.25″ 1.2MB	80	15	512	1 228 800	1
3.5″ 720KB	80	9	512	737 280	2
3.5″ 1.44MB	80	18	512	1 474 560	1
3.5″ 2.88MB	80	36	512	2 949 120	—

11.1.2 磁盘系统区和数据区

软盘和硬盘的组织根据它们的容量是不同的。硬盘和某些软盘被格式化为"自引导",也就是说,当接通电源或用户按下 Ctrl + Alt + Del 键时,机器可以自行启动。磁盘一般由系统区和数据区组成。

1. 系统区

磁盘的第一个区域是系统区,从 0 面,0 磁道,1 扇区开始。系统区一般用来存放系统存储和维护的信息。例如,磁盘上每个文件的起始地址就存放在系统区。系统区由三部分组成:

① 引导记录(帮助系统将操作系统从磁盘装入内存);
② 文件分配表 FAT(为文件分配磁盘空间);
③ 目录(包括文件名、磁盘地址和文件的状态)。

系统区和数据区的组织如下:

举例来说,720KB 软盘上的系统区在 0 柱面,0 盘面,数据区从 0 柱面,1 盘面的 6 扇区开始。这种先在同一柱面的正反两个盘面上存储信息,再在下一柱面上存储信息的方法,既适用于软盘也适用于硬盘。

2. 数据区

在可引导磁盘的数据区,装有两个系统文件:IO.SYS 和 MSDOS.SYS(MS-DOS)或 IBMBIO.COM 和 IBMDOS.COM(IBM PC-DOS)。当用 FORMAT/S 来格式化磁盘时,DOS 把这几个系统文件复制到数据区的第一个扇区,系统文件之后紧接着是用户文件。如果没有系统文件,数据区的第一个扇区就用来保存用户文件。

11.1.3 磁盘目录及文件分配表

1. 磁盘目录

每个磁盘上都有一个根目录,它是寻找磁盘文件最重要的一张表。根目录下可以包含子目录,子目录下又可以有文件名和它的子目录,这样形成了一个分层次的树形结构,顶端是根目录,如图 11.1 所示。

目录中的每一个目录名和文件名都限定在它上层的目录下,这称为路径。例如,ASM 目录下的子目录 PROGS 中的 PRG1.ASM 文件的路径是:

C:\ASM\PROGS\PROG1.ASM

对每一个文件,DOS 都建立一个 32 字节的目录项,以记录文件名、建立文件的日期、

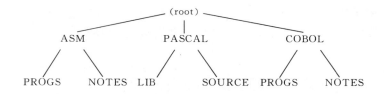

图 11.1 树形结构的目录

文件的大小以及文件起始簇的地址。目录项的格式如表 11.2。

表 11.2 目录项的格式

字 节	域 名	格 式
00～07	文件名	ASCII
08～0A	扩展名	ASCII
0B	属性	8 位二进制
0C～15	DOS 保留	
16～17	时间	16 位二进制
18～19	日期	16 位二进制
1A～1B	起始簇号	16 位二进制
1C～1F	文件大小	16 位二进制

2. 文件分配表

文件分配表(file allocation table,FAT)的功能是为文件分配磁盘空间,磁盘上的每一簇在 FAT 中都有一个入口项。当建一个新文件或修改一个已有文件时,系统根据磁盘文件的地址修改相应的 FAT 入口项。对每簇 4 个扇区的磁盘,同一编号的 FAT 入口项可引用 4 次,因此,簇的使用减少了 FAT 入口项的数目,使系统能对更大存储空间的磁盘寻址。

FAT 的第一个入口项是一个字节的介质说明项,指出磁盘的类型:

F0H　3.5″,双面,18 扇区/磁道(1.44MB)和 3.5″,双面,36 扇区/磁道(2.88MB)

F8H　硬盘

F9H　3.5″,双面,9 扇区/磁道(720KB)和 5.25″,双面,15 扇区/磁道(1.2MB)

FCH　5.25″,单面,9 扇区/磁道(180KB)

FDH　5.25″,双面,9 扇区/磁道(360KB)

FFH　5.25″,双面,8 扇区/磁道(320KB)

FAT 的第二个入口项如果为 FFFFH,表示软盘的 FAT 入口项是 12 位的;如果为 FFFFFFH,表示硬盘的 FAT 入口项是 16 位的。

FAT 的第三个入口项是指针项,对应于数据区的每一簇。目录项中的 1AH ～ 1BH 字节保存的是文件起始簇的地址,文件的下一个簇地址及以后的簇地址由 FAT 的指针项给出。实际上,指针项是文件的地址(簇号)链表。

下面的例子有助于了解 FAT 的结构。假设软盘上只有一个名为 CUSTOMER.FIL

的文件,它存储于簇 2、簇 3 和簇 4 中。这个文件的目录项含有文件名 CUSTOMER 和扩展名 FIL、文件属性、日期、文件第一个簇的地址以及文件的大小。12 位的 FAT 入口项如下所示:

| FAT 入口项: | F0F | FFF | 003 | 004 | FFF | --- | --- | --- | --- | --- |

FAT 的头两项,F0 表示双面、9 扇区(1.44MB)的软盘,紧接着的 FFFFH 表示 FAT 入口项是 12 位的。当系统在目录项中搜索到文件名 CUSTOMER 和扩展名 FIL 后,将文件的起始簇号 002 取出,系统即把簇 002 扇区中的数据传送到内存。

文件所在的下一个簇,由 FAT 的指针项给出簇号,第三、四项中的 003 和 004 表明文件的后续部分在 003 簇和 004 簇。系统根据这两个簇地址分别从扇区中读出数据送到内存。

FAT 第五项中的 FFFH 表明文件结束,没有簇再分配给文件了。

由于 FAT 指针项的链表作用,文件在磁盘中的簇号不一定是连续的,但系统一般按递增序号来分配簇。例如,一个文件可能存储在磁盘的 8、9、10、14、17 和 18 簇。

11.2　文件代号式磁盘存取

在 DOS 2.0 以上的版本中,为了支持层次结构,引用了树形结构目录,因此相应增加了一个新的存取方式即文件代号式存取方式(file handles access)。这种方式将有关文件的各种信息都包括在 DOS 中,对用户程序是透明的。在处理指定文件时,必须使用一个完整的路径名(path name),一旦文件的路径名被送入操作系统,就被赋予一个简单的文件代号(file handle),这个文件代号是一个 16 位的数。以后对该文件进行读写操作时,就用这个文件代号去查找相应的文件。对于每一个打开的文件,DOS 还为其管理一个读写指针(read/write pointer),读写指针总是指向下一次要存取的文件中的字节,这个读写指针可以移动到文件的任意位置,从而能满足随机存取的要求。

文件代号式存取方式对各种错误采取了更统一的处理方法。在操作过程中,AX 中回送错误代码,这些错误代码对所有代号式存取功能都是相同的,这为用户进行分析提供了方便。

DOS 2.0 提供的有关代号式文件管理功能都包括在扩充 DOS 功能调用中。表 11.3 列出了其中常用的部分功能调用。

表 11.3　代号式文件管理功能调用

AH	功　能	调用参数	返　回　参　数
3CH	建文件	DS=ASCIZ 串的段地址 DX=ASCIZ 串的偏移地址 CX=文件属性	CF=0 操作成功 AX=文件代号 CF=1 操作出错 AX=错误代码

续表

AH	功能	调用参数	返回参数
3DH	打开文件	DS=ASCIZ 串的段地址 DX=ASCIZ 串的偏移地址 AL=存取代码	CF=0 操作成功 AX=文件代号 CF=1 操作出错 AX=错误代码
3EH	关闭文件	BX=文件代号	CF=0 操作成功 CF=1 出现错误 AX=错误代码
3FH	读文件或设备	DS=数据缓冲区段地址 DX=数据缓冲区偏移地址 BX=文件代号 CX=读取的字节数	CF=0 读成功 AX=实际读入的字节数 AX=0 文件结束 CF=1 读出错 AX=错误代码
40H	写文件或设备	DS=数据缓冲区的段地址 DX=数据缓冲区的偏移地址 BX=文件代号 CX=写入的字节数	CF=0 读成功 AX=实际读入的字节数 CF=1 出现错误 AX=错误代码
42H	移动文件指针	CX=所需字节的偏移地址(高位) DX=所需字节的偏移地址(低位) AL=方式码 BX=文件代号	CF=0 操作成功 DX:AX=新指针位置 CF=1 操作失败 AX=错误代码
43H	检验或改变 文件属性	AL=0 检验文件属性 AL=1 置文件属性 CX=新属性 DS=ASCIZ 串的段地址 DX=ASCIZ 串的偏移地址	CF=0 操作成功 AL=0 CX=属性 CF=1 操作失败 AX=错误代码

11.2.1 路径名和 ASCIZ 串

当使用扩展功能处理磁盘文件时,首先要告诉 DOS 一个 ASCIZ 串的地址。这个 ASCIZ 串包括文件的路径名和一个全 0 的字节。路径名说明文件的位置,包括磁盘驱动器、目录路径和文件名。下面是两个 ASCIZ 串:

```
PATHNM1    DB    'B:\TEST.ASM',00
PATHNM2    DB    'C:\UTILITY\NU.EXE',00
```

串中的后斜线,也可能是前斜线,起分割各项的作用。上面两个字符串都用一个 0 字节作为结束,所以叫做 ASCIZ 串(ASCII_ZERO)。路径名的最大长度允许 63 个字节,对于请求 ASCIZ 串的中断调用,要求把 ASCIZ 串的地址装在 DX 寄存器中,例如,可用指令把串的地址送 DX 寄存器:

```
LEA   DX,PATHNM1
```

11.2.2 文件代号和错误返回代码

存取文件要借助于文件代号,文件代号是由打开文件功能(3DH)和建立文件功能(3CH)传送到 AX 的一个 16 位数。对标准设备不必打开就可直接使用它们的文件代号,因为 DOS 已经预定义了它们的文件代号:

0 = 标准输入设备
1 = 标准输出设备
2 = 标准错误输出设备
3 = 标准辅助设备
4 = 标准打印设备

对建立或打开的文件,其代号从 6 开始顺序排列,在任一时刻最多只能同时打开 5 个文件,可见 16 位长的文件代号实际上并没有全部用上。打开文件的操作,使系统通过它的目录和 FAT 入口项寻找磁盘文件并修改这些入口项。当程序执行时,调用的每一个文件都必须分配一个惟一的文件代号。

对于存取磁盘文件,首先用一个 ASCIZ 串指定文件并调用 DOS 功能 3CH 建立或打开文件。如果成功,操作置 CF 为 0,并把文件代号传送到 AX 中,这时文件和代号建立了对应关系,所以要注意保存这个代号。例如,把它保存在程序中的一个字数据项中,否则丢失了代号相当于丢失了文件。如果操作不成功,CF 被置成 1,AX 中包含的是错误代码,这些错误代码都取自一个统一的错误信息表,表 11.4 列出了 01~36 的错误代码,其他的代码与网络有关。

表 11.4 错误返回代码

01	非法功能号		19	磁盘写保护
02	文件未找到		20	未知单元
03	路径未找到		21	驱动器没有准备好
04	同时打开的文件太多		22	未知命令
05	拒绝存取		23	CRC 数据错
06	非法文件代号		24	请求指令长度错
07	内存控制块被破坏		25	搜索错
08	内存不够		26	未知的介质类型
09	非法存储块地址		27	扇区未发现
10	非法环境		28	打印机纸出界
11	非法格式		29	写故障
12	非法存取代码		30	读故障
13	非法数据		31	一般性失败
14	(未用)		32	共享违例
15	非法指定设备		33	锁违例
16	试图删除目录		34	非法磁盘更换
17	设备不一致		35	FCB 无效
18	已没有文件		36	共享缓冲区溢出

11.2.3 文件属性

文件属性是一个说明文件特性的字节,它保存在文件目录项中的0BH字节。比如一个文件被赋予的属性是"只读",则这个文件只能由用户读取,而不能往文件中写入任何内容,这实际上是保护文件的一种方法,文件属性字节的各位含义如图11.2所示。

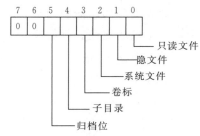

01——只读文件,该文件不能为写而打开。
02——隐文件,用 DIR 查不到该文件。
03——系统文件,用 DIR 查不到该文件。
08——软盘的卷标号。
10——子目录。
20——已写入并关闭了文件(硬盘用)。

图 11.2 文件属性字节

属性字节的 6 和 7 位是保留位,应当总是 0。每当修改了文件,归档位就被设置为 1。如果入口项中有子目录名,子目录位就被设置为 1。如果磁盘用 / V 选项格式化就生成一个卷名,卷标位则是磁盘卷名的标示。系统文件位表明该文件是操作系统的一部分。隐文件位使文件隐藏起来,使它不能被常用的DOS命令搜索到,它的文件名也不会在目录中列出。只读位可防止文件被删除或被修改。

一个文件可以同时具有几种属性,也就是说可以把字节中的几位同时置1。例如,IBMBIO.COM 和 IBMDOS.COM 文件既是只读文件,又是隐文件和系统文件。

对一般的程序,最有用的是只读位和隐文件位,因为有时文件不希望别人改动也不希望别人知道。然而大多数文件都不使用这些特征位,所以属性字节通常为 00,属性字节只用了 6 位,但规定用一个 16 位的寄存器 CX 存放这个字节,这可能是为了以后扩充方便而这样设置的。

使用改变文件属性功能(43H)可以改变现有文件的属性。

```
MOV     AH,43H              ; check or change attribute
MOV     AL,01               ; set file attribute
MOV     CX,01               ; read only
MOV     DX,SEG FNAME        ; addr of ASCIZ string
MOV     DS,DX
MOV     DX,OFFSET FNAME
INT     21H                 ; call DOS
```

INT 21H 的功能 43H 可以检验或改变目录中文件的属性,DX 中放 ASCIZ 串的地址,为了检验文件属性,AL 置为 00,操作后当前的属性值返回到 CX 寄存器中。改变文件属性时,置 AL 为 01,CX 放入新的属性,操作把这个新属性置入目录项。操作失败,AX 中返回的错误码为 01、02、03 或 05。在网络环境下,除了归档位外,要改变任何文件属性都必须具有文件访问权。改变归档位没有任何约束。

11.2.4 写磁盘文件

写一个新文件或用同一个文件名重写一个旧文件时,首先要建立文件并赋给它一个属性,如果 DOS 发现要建立的文件已经存在,那么原来的文件就被破坏。

建立文件的功能调用是 3CH,调用该功能时,在 DX 中装入 ASCIZ 串的地址,在 CX 中装入文件的属性。例如建立一个有正常属性文件的指令序列如下:

```
PATHNM1    DB    'E:\ACCOUNTS.FIL',00H
HANDLE1    DW    ?                    ; file handle
...        ...
           MOV   AH,3CH               ; request create
           MOV   CX,00                ; normal file
           LEA   DX,PATHNM1           ; ASCIZ string
           INT   21H                  ; call DOS
           JC    ERROR                ; exit if error
           MOV   HANDLE1,AX           ; save handle in DW
```

对于一次成功的操作,DOS 用给定的属性建立目录项,清除进位位,并把文件代号回送到 AX 寄存器,以后对文件的所有操作都使用这个代号。在文件打开的同时,文件指针设置为 0。如果文件已经存在,操作将文件长度置为 0,新文件就对老文件进行重写。如果操作把 CF 置为 1,则说明建文件有错误,错误代码回送到 AX 寄存器,可能出现的错误为 03,04 或 05。错误码 05(拒绝存取)说明了目录已经满了或所请求的文件是一个只读文件。

写磁盘文件是利用功能调用 40H,在 BX 中装入文件代号,要写入的字节数放在 CX 中,输入缓冲区的地址放在 DX 中。下面的指令是把 OUTREC 数据区中的 256 个字节写入磁盘文件。

```
HANDLE1    DW    ?
OUTREC     DB    256 DUP(?)
...        ...
           MOV   AH,40H               ; request write
           MOV   BX,HANDLE1           ; handle
           MOV   CX,256               ; number of bytes
           LEA   DX,OUTREC            ; addr of output area
           INT   21H                  ; call interrupt service
           JC    ERROR2               ; special action if error
           CMP   AX,256               ; all bytes written?
           JNE   ERROR3               ; if not,error
```

如文件从内存写入磁盘的操作成功,则 CF 清 0,并把实际写入的字节数送入 AX 寄存器。如果磁盘空间满,实际写入的字节数可能会和要求写入的字节数不同,但是系统没有把这种情况报告为一个错误,程序员只能测试返回到 AX 中的值。非法操作的标志也是 CF=1,并在 AX 中回送错误码 05(拒绝存取)或 06(非法文件代号)。

当写入文件操作完成后,必须用 DOS 功能调用 3EH 来关闭文件,以确保操作系统将文件记录在磁盘上,这个操作只要求在 BX 中放入文件代号。关闭文件时,DOS 把内存缓冲区(等待输入字符满 512 个字节以填满一个扇区)中的数据写入磁盘,并用日期和文件大小修改目录和 FAT(文件分配表)。

```
MOV     AH,3EH              ; request close
MOV     BX, HANDLE1         ; file handle
INT     21H                 ; call DOS
```

关闭文件操作在 AX 中返回的错误码只可能是 06(非法文件代号)。

例 11.1 用文件代号建立文件。

该程序接收一个由姓名组成的文件。它包括以下几个子程序:

```
CREATH   利用功能调用 3CH 来建立文件,并把文件代号保存在 HANDLE 数据项中。
PROCH    键盘接收输入并把输入缓冲区中其余的单元用空(blank)填入。
WRITH    利用功能调用 40H 写文件。
CLSE     利用功能调用 3EH 来关闭文件以建立相应的目录项。
         TITLE     HANCREAT.EXE ————Creat disk file of names
                   .model  small
                   .stack  64
;------------------------------------------------------------------
                   .data
namepar label byte                   ;name parameter list
maxlen   db 30
namelen  db ?
namerec  db 30 dup(' '),0dh,0ah      ;entered name
                                     ;CR/LF for writing
clrf     db 13,10,'$'
errcde   db 0
handle   dw ?
pathnam  db 'D:\NAME.DAT',0
prompt   db 'name ? '
row      db 01
opnmsg   db '* * * open error * * *',0dh,0ah,'$'
wrtmsg   db '* * * write error * * *',0dh,0ah,'$'
;------------------------------------------------------------------
                   .code
begin    proc far
         mov       ax,@data          ;initialize DS
         mov       ds,ax
         mov       es,ax
```

```
                mov     ax,0600h
                call    scren               ;clear screen
                call    curs                ;set cursor
                call    creath              ;create file
                cmp     errcde,0            ;create error?
                jz      contin              ;yes,continue
                jmp     back                ;no,exit
contin:
                call    proch
                cmp     namelen,0           ;end of input
                jne     contin              ;no,continue
                call    clseh               ;yes ,close
back:           mov     ax,4c00h            ;return to DOS
                int     21h
begin   endp
;----------------------------------------------------------------
; create disk file
creath  proc    near
                mov     ah,3ch              ;request create
                mov     cx,0                ;normal attribute
                lea     dx,pathnam
                int     21h
                jc      a1                  ;error?
                mov     handle,ax           ;no,save handle
                ret
a1:             lea     dx,opnmsg           ;error message
                call    errm
                ret
creath  endp
;----------------------------------------------------------------
; accept input
proch   proc    near
                mov     ah,40h              ;request display
                mov     bx,01               ;01 = ouput device
                mov     cx,06               ;length of prompt
                lea     dx,prompt           ;display prompt
                int     21h

                mov     ah,0ah              ;request input
                lea     dx,namepar          ;accept name
                int     21h
                cmp     namelen,0           ;is there a name?
                jne     b1                  ;yes
```

```
                ret                         ;no,exit
b1:
                mov     al,20h              ;blank for storing
                sub     ch,ch
                mov     cl,namelen          ;length
                lea     di,namerec
                add     di,cx               ;address + length
                neg     cx                  ;calculate remaining
                add     cx,30               ;length
                cld
                rep     stosb               ;set to blank
                call    writh               ;write disk record
                call    scrl                ;check for scroll
                ret
proch   endp
;------------------------------------------------------------------
; check for scroll
scrl    proc    near
                cmp     row,18h             ;bottom of screen?
                jae     c1                  ;yes ,bypass
                mov     ah,09               ;no,to next row
                lea     dx,clrf
                int     21h
                inc     row                 ;no,add to row
                ret
c1:             mov     ax,0601h            ;scroll one row
                call    scren
                call    curs                ;reset cursor
                ret
scrl    endp
;------------------------------------------------------------------
; write disk record
writh   proc    near
                mov     ah,40h              ;request write
                mov     bx,handle
                mov     cx,32               ;for name and CR/LF
                lea     dx,namerec
                int     21h
                jnc     d1                  ;valid write?
                lea     dx,wrtmsg           ;no
                call    errm                ;call error routine
                mov     namelen,0
d1:             ret
```

```
writh     endp
```
;--
; close disk file
```
clseh     proc    near
          mov     namerec,1ah        ;set EOF mark
          call    writh
          mov     ah,3eh             ;request close
          mov     bx,handle
          int     21h
          ret
clseh     endp
```
;--
; scroll screen
```
scren     proc    near               ;AX set on entry
          mov     bh,1eh             ;set yellow on blue
          mov     cx,0
          mov     dx,184fh
          int     10h                ;scroll
          ret
scren     endp
```
;--
; set cursor
```
curs      proc    near
          mov     ah,02
          mov     bh,0
          mov     dh,row             ;set cursor
          mov     dl,0
          int     10h
          ret
curs      endp
```
;--
; disk error routine
```
errm      proc    near               ;DX contains address
          mov     ah,40h             ; of message
          mov     bx,01              ;handle
          mov     cx,21              ;length
          int     21h
          mov     errcde,01          ;set error code
          ret
errm      endp
```
;--
```
          end     begin
```

输入缓冲区(NAMEREC)有30个字节,包括回车换行符共32个字节。此程序把32个字节作为一个固定长度的记录写入。在程序开始显示提示符和程序末尾显示错误信息都采用的是代号式I/O,显示器预定义的文件代号是01,所以可以直接使用此文件代号向设备写入信息。在11.3节将专门介绍"字符设备的文件代号式I/O"。

11.2.5 读磁盘文件

调用读文件或设备功能(3FH),要求把文件打开取得文件代号,然后按照指定的字节数从磁盘中把文件读出,送入内存中预先定义好的数据缓冲区。如果读入的字节数大于缓冲区空间,那么,这些多余的数据将送到程序所占空间之上的存储器中。

打开文件操作(3DH)要检查文件名是否合法,文件是否有效。文件名是一个ASCIZ串,其地址装入DX寄存器,并在AL中设置存取代码。

表11.5中的存取代码告诉操作系统打开文件的目的是什么,例如打开一个属性是只读的文件,而目的是为了向文件中写入内容(AL中存取代码01),则操作将回送一个错误码05(拒绝访问)。因此可以看出,文件属性和存取代码结合起来能防止非法读写文件。

表 11.5 文件存取代码

位	存取代码	位	存取代码
0~2	000=为读而打开文件	3	1=保留
	001=为写而打开文件	4~6	共享方式
	010=为读和写打开文件	7	继承标志

下面是一个为读而打开文件的例子:

```
HANDLE2   DW    ?                 ; file handle
     ...        ...
     MOV        AH,3DH            ; request open
     MOV        AL,00             ; read only
     LEA        DX,PATHNM1        ; ASCIZ string
     INT        21H               ; call DOS
     JC         ERROR4            ; exit if error
     MOV        HANDLE2,AX        ; save handle in DW
```

如果指定的文件存在,打开文件操作将把记录长度置为1,确定它的属性,设置文件指针为0(指向文件的开始),清除CF标志位,并把文件代号放入AX。

如果指定的文件不存在,操作置位CF,并在AX中返回一个错误代码:02,04,05或12。所以在打开文件之后一定要检查CF位。

读文件调用DOS功能3FH,首先在BX中设置文件代号,在CX中装入要读取的字节数,在DX中放入输入数据缓冲区的地址,然后调用DOS。

下面的指令读取一个512字节的记录:

```
HANDLE2   DW    ?
```

```
INPREC   DB    512   DUP(' ')
         ...   ...
         MOV   AH,3FH              ; request read
         MOV   BX,HANDLE2          ; handle
         MOV   CX,512              ; record length
         LEA   DX,INPREC           ; addr of input area
         INT   21H                 ; call DOS
         JC    ERROR5              ; test for error
         CMP   AX,0                ; zero bytes read?
         JE    ENDFILE             ; yes,end of file
```

如果读操作成功地把记录读入存储器,则清除 CF 位,并把实际读入的字节数放入 AX。假如 AX=0,表明试图从文件尾开始读,这是一个警告信息,而不是错误信息。

如果读操作不成功,把 CF 位置 1 并在 AX 中返回错误码 05(拒绝存取)或 06(非法文件代号)。

因为 DOS 限制一次打开的文件数,所以在读取文件之后必须关闭这些文件。

一个文件分几次读取,取决于文件的大小和输入缓冲区的空间。如果文件比较小,又有足够的输入缓冲区,则用一次调用就能读入整个文件的内容。如果文件很大,程序中不能设置如此大的缓冲区,则必须分几次重复调用读功能,直到整个文件结束。

例 11.2 利用文件代号顺序读取一个文件

读取的文件是由例 11.1 建立的 NAME.DAT 文件。主要的过程如下:

BEGIN   调用子程序 OPENH,READH,DISPH 并关文件。
OPENH   用 INT 21H 的功能 3DH 打开文件。
READH   用 INT 21H 的功能 3FH 来读取记录,并把文件代号保存在 HANDLE 中。
DISPH   显示记录并卷屏,因为建立文件时,每个记录后面都跟着成对的回车/换行符,所以本程序在显示记录时不必移动光标。

```
         TITLE    HANREAD.EXE
         ;  Read disk records created by hancreat
         ;----------------------------------------------------------------
                  .model   small
                  .stack   64
                  .data
endcde   db    0                        ; end process indicator
handle   dw    ?
ioarea   db    32 dup(' ')
pathnam  db    'D:\NAME.DAT',0
openmsg  db    '* * * open error * * *',0dh,0ah
readmsg  db    '* * * read error * * *',0dh,0ah
row      db    0
         dseg ends
         ;----------------------------------------------------------------
```

```
            .code
begin       proc    far
            mov     ax,@data            ;initialize DS and ES
            mov     ds,ax
            mov     es,ax
            mov     ax,0600h
            call    scren               ;clear screen
            call    curs                ;set cursor
            call    openh               ;open file, set DTA
            cmp     endcde,0            ;valid open?
            jnz     a1                  ;no, terminate
contin:
            call    readh               ;read disk record
            cmp     endcde,0            ;normal read?
            jnz     a1
            call    disph               ;yes,display name
            jmp     contin              ;continue
a1:
            mov     ax,4c00h            ;terminate
            int     21h
begin       endp
;-----------------------------------------------------------------
; open file
openh       proc    near
            mov     ah,3dh              ;request open
            mov     al,0                ;normal file
            lea     dx,pathnam
            int     21h
            jc      b1                  ;error ?
            mov     handle,ax           ;no, save handle
            ret
b1:
            mov     endcde,01           ;yes
            lea     dx,openmsg          ;display
            call    errm                ; error message
            ret
openh       endp
;-----------------------------------------------------------------
; read disk record
readh       proc    near
            mov     ah,3fh              ;request read
            mov     bx,handle
```

```
            mov     cx,32             ;for name and CR/LF
            lea     dx,ioarea
            int     21h
            jc      c1                ;error on read?
            cmp     ax,0              ;end of file?
            je      c2
            cmp     ioarea,1ah        ;EOF marker?
            je      c2                ;yes,exit
            ret
c1:         lea     dx,readmsg        ;no,invalid read
            call    errm
c2:         mov     endcde,01         ;force_ end
            ret
readh       endp
;------------------------------------------------------------
; display name
disph       proc    near
            mov     ah,40h            ;request display
            mov     bx,01             ;set handle
            mov     cx,32             ;and length
            lea     dx,ioarea
            int     21h
            cmp     row,24            ;bottom of screen?
            jae     d1                ;yes ,bypass
            inc     row
            ret
d1:
            mov     ax,0601h
            call    scren             ;scroll
            call    curs              ;set cursor
            ret
disph       endp
;------------------------------------------------------------
; scroll screen
scren       proc    near              ;AX set on entry
            mov     bh,1eh            ;set color
            mov     cx,0
            mov     dx,184fh          ;request scroll
            int     10h
            ret
scren       endp
;------------------------------------------------------------
```

```
; set cursor
curs        proc    near
            mov     ah,2            ;request set cursor
            mov     bh,0
            mov     dh,row          ;row
            mov     dl,0            ;column
            int     10h
            ret
curs        endp
;------------------------------------------------------------
; disk error routine
errm        proc    near
            mov     ah,40h          ;DX contains address
            mov     bx,01           ;handle
            mov     cx,20           ;length of message
            int     21h
            ret
errm        endp
;------------------------------------------------------------
            end     begin
```

11.2.6 移动读写指针

利用文件代号存取文件是以字节为存取单位的,一个文件被看作由许多字节组成,每次读写的字节数可任意指定,但一般还是为输入输出缓冲区的大小所限制。所以一个比较大的文件总是要分几次读写,每次读写的字节称为记录。另外像名单、零件表、账目等,每项具有同类型数据和同样格式的文件,按记录来进行处理更为方便。那么,在读写文件时,每次读写的记录是如何"拼接"起来的呢?原来是操作系统为文件保存了一个称为读写指针(read/write pointer)的变量,由它指示应从文件的什么地方读出,或应往文件的什么地方写入。

为了存取文件中间某一特定的记录,首先要使读写指针指向这个记录。DOS 提供了移动读写指针功能 42H,该功能要求在 BX 中指定文件代号,由 AL 中的代码确定改变指针的三种方式。在每种方式中,由 CX 和 DX 指定一个双字长的偏移值,低位字在 DX 中,高位字在 CX 中,这个偏移值是一个带符号的整数,它可以是正数,也可以是负数。

1. AL=00 绝对移动方式

偏移值从文件首开始计算。例如,偏移值是 182,则读写指针指向文件的第 182 字节。为了使指针指向文件首,可以在 CX,DX,AL 中都送入 0,那么以后的读写操作就从文件首开始。下例是从文件首开始,移动指针 1024 字节:

```
        MOV     AH,42H          ; request move pointer
```

```
            MOV    AL,00                ; to start of file
            MOV    BX,HANDLE1           ; set file handle
            MOV    CX,00                ; high_order word of offset
            MOV    DX,1024              ; low_order word of offset
            INT    21H                  ; call DOS
            JC     ERROR                ; Check error
```

2. AL=01　相对移动方式

当前的指针值加上偏移值作为新的指针值,也就是说,偏移值指出了从当前的读写位置起移动的字节数。根据偏移值的正负可正向或反向移动指针。

例如,想从当前的指针位置向前或向后移 N 个字节,则可用方式 1。假定偏移值 N 的范围在 -32768～32767 之间。

```
            MOV    BX,HANDLE            ; file handle
            MOV    CX,0                 ; high_order word
            MOV    DX,N                 ; low_order word
            CMP    DX,0                 ; if N>0
            JGE    POINT                ; yes
            NOT    CX                   ; extending the sign
    POINT:  MOV    AL,1                 ; method 1
            MOV    AH,42H               ; request move pointer
            INT    21H                  ; call DOS
            JC     ERROR                ; Check error
```

如果希望得到指针的当前值,可以使用方式 1 和偏移值 0。

3. AL=2　绝对倒移方式

新的指针位置通过把偏移值和文件尾的位置相加而确定。如果文件的记录长度是 32 字节,那么在 AL 中送入 2,DX 中送入 -32,CX 中送入 0FFFFH(负号扩展到高位字),则读写指针指向文件的第一个记录。如果 CX 和 DX 为 0,AL 为 2,则移动后的指针将指向文件尾,实际上这个新指针值就是文件的长度,用这种方式可以检测文件的大小。

如果要在一个已存在的文件后面添加记录,则利用方式 2 在写之前把指针指向文件尾:

```
            MOV    AH,42H               ; request move pointer
            MOV    AL,2                 ; method 2
            MOV    BX,HANDLE            ; handle
            MOV    CX,0                 ; offset
            MOV    DX,0
            INT    21H                  ; call DOS
            JC     ERROR                ; check error
            ...
```

移动读写指针功能可能出现的错误码是 01 和 06,错误码 01 说明 AL 中的方式值是不合法的,错误码 06 说明 BX 中的文件代号不合法。

如果指针移动成功,AX 和 DX 将是移动后的指针值,AX 中是低位字,DX 是高位字(调用之前,DX 是偏移值的低位字,CX 是偏移值的高位字)。

移动指针的功能使用起来很简单,一般在程序中打开文件之后,使用这个功能的某种方式将读写指针移到文件中需要的位置,以后的读写就从文件的这个地方开始,从而提供了在文件中随机存取的能力。

当程序要求随机地处理一个记录时,读操作在目录中找到这个记录所在的扇区,并把整个扇区的数据从磁盘读到缓冲区,然后把所要求的记录传送给程序。例如,4 个 128 字节长度的记录存放在一个扇区中,当请求读取 21♯记录时,同一扇区中的 4 个记录都被读入缓冲区:

记录 20♯	记录 21♯	记录 22♯	记录 23♯

当程序再次随机地请求读取一个记录,比如是 23♯记录,读操作首先检查缓冲区。如果所要求的记录已经在缓冲区了,则直接将这个记录(23♯)传送给程序;如果所要求的记录不在缓冲区中,则读操作又再次利用目录确定此记录的位置,并把它所在扇区的数据全部读入缓冲区,接着把所要求的记录传送给程序。可以看出,在随机存取的方式下,如果请求存取的记录在文件中是连续的,那么真正从磁盘存取记录的次数不会很多。

例 11.3 利用文件代号随机读取磁盘文件。

程序随机读取例 11.1 建立的文件 NAME.DAT。若键入文件中的相对记录号,屏幕上则显示出所要求的记录。如果文件有 24 个记录,那么合法的记录号是 01~24。从键盘上输入的记录号是 ASCII 码的形式,所以记录号应当是 1 位数字或 2 位数字,文件记录的长度是 16 字节。

```
OPENRAN    打开文件并得到文件代号。
RECNUM     从键盘接收记录号并检验它的长度,有三种情况:
00            处理请求结束
01            1 位记录号,保存在 AL 中
02            2 位记录号,保存在 AX 中
```

这个子程序还要把输入的记录号转换为文件中的字节数。例如,从键盘输入数字 12,AX 中得到 3132,AND 指令将它变为 0102,AAD 指令又将它转换为 000CH(12),SHL 使它再左移 4 位(乘 16)变成 00C0H(192)。结果,从文件 192 开始的 16 个字节(一个记录)的内容被显示出来。

```
READRAN    利用移动文件指针功能 42H 和相对字节地址来设置文件指针,用
           3FH 功能读取所要求的记录。
DISPREC    显示读取的记录。
TITLE      READR.EXE－－－－ Read disk records randomly
           .model    small
           .stack    64
```

```
;------------------------------------------------------------------
            .data
handle      dw      ?                           ; file handle
recindx     dw      ?                           ; record index
errcde      db      00                          ; read error indicator
prompt      db      'RECORD NUMBER?'
ioarea      db      32 dup(' ')                 ;disk record area
pathnam     db      'D:\NAME.DAT',0
openmsg     db      '* * *   open error    * * *',0dh,0ah
readmsg     db      '* * *   read error    * * *',0dh,0ah
row         db      00
col         db      00

recdpar     label   byte                        ; input parameter list:
maxlen      db      3                           ; maximum length
actlen      db      ?                           ; actual length
recdno      db      3 dup(' ')                  ; record number
;------------------------------------------------------------------
            .code
            .386
main        proc    far
            mov     ax,@data                    ; Initialize
            mov     ds,ax                       ;   sehment
            mov     es,ax                       ;   registers
            mov     ax,0600h
            call    scrn                        ; clear screen
            call    curs                        ; set cursor
            call    openran                     ; open file
            cmp     errcde,0                    ; valid open?
            jnz     exit                        ; no, exit
callsub:
            call    recnum                      ; requst record #
            cmp     actlen,0                    ; any more requests?
            je      exit                        ; no, exit
            call    readran                     ; read disk record
            cmp     errcde,0                    ; normal read?
            jnz     next                        ;   no, bypass
            call    disp                        ;   yes, display name
next:       jmp     callsub                     ; continue
exit:       mov     ax,4c00h                    ; end processing
            int     21h
main        endp
```

```
;--------------------------------------------------------------------
;   open file
openran     proc    near
            mov     ah,3dh              ; request open
            mov     al,0                ; normal file
            lea     dx,pathnam
            int     21h
            jc      errm1               ; error?
            mov     handle,ax           ; no, save handle
            ret
errm1:      mov     errcde,01           ; yes
            lea     dx,openmsg          ; display
            call    errorp              ; error message
            ret
openran     endp
;--------------------------------------------------------------------
; get record number
recnum      proc    near
            mov     ah,40h              ; requst display promp
            mov     bx,01               ; file handle
            mov     cx,15               ; 15 characters
            lea     dx,prompt
            int     21h
            mov     ah,0ah              ; requst input
            lea     dx,recdpar          ;   of record number
            int     21h
            cmp     actlen,01           ; check length
            jb      endd                ; length 0,terminate
            ja      twodgt
            xor     ah,ah               ; length 1
            mov     al,recdno
            jmp     conv
twodgt:
            mov     ah,recdno           ; length 2
            mov     al,recdno+1
conv:
            and     ax,0f0fh            ; clear ASCII 3s
            aad                         ; convert to binary
            dec     ax                  ; adjust (1st record is 0)
            mov     cl,04
            shl     ax,cl               ; multiply by 16
            mov     recindx,ax          ; save index
```

```
endd:       mov     col,20
            call    curs
            ret
recnum      endp
;----------------------------------------------------------------
; read disk record randomly
readran     proc    near
            mov     ah,42h              ; request set file pointer
            mov     al,0                ;   to start of file
            mov     bx,handle           ; file handle
            mov     cx,0                ; upper portion of offset
            mov     dx,recindx          ; lower portion of offset
            int     21h
            jc      errm2               ; error condition?
                                        ;   yes, bypass
            mov     ah,3fh              ; request read
            mov     bx,handle
            mov     cx,32
            lea     dx,ioarea
            int     21h
            jc      errm2
            cmp     ioarea,1ah          ; EOF marker?
            je      exit2               ; yes, exit
            jmp     back
errm2:      lea     dx,readmsg          ; invalid read
            call    errorp              ; display message
exit2:      mov     errcde,01           ; force end
back:       ret
readran     endp
;----------------------------------------------------------------
; display name
disp        proc    near
            mov     ah,40h              ; request display
            mov     bx,01               ; set handle
            mov     cx,32               ;   and length
            lea     dx,ioarea
            int     21h
            mov     col,0               ; clear column
            cmp     row,20              ; bottom of screen
            jae     scrol               ;   yes, bypass
            inc     row                 ; no, increment row
            jmp     return
```

```
scrol：
          mov     ax,0601h
          call    scrn                    ;scroll
          call    curs                    ;set cursor
return：   ret
disp      endp
;------------------------------------------------------------------
;scroll screen
scrn      proc    near                    ;ax set on entry
          mov     bh,1eh                  ;set color
          mov     cx,0
          mov     dx,184fh                ;request scroll
          int     10h
          ret
scrn      endp
;------------------------------------------------------------------
;set cursor
curs      proc    near
          mov     ah,02                   ;request set
          mov     bh,0                    ;cursor
          mov     dh,row                  ;row
          mov     dl,col                  ;colum
          int     10h
          ret
curs      endp
;------------------------------------------------------------------
;display disk error message
errorp    proc    near
          mov     ah,40h                  ;DX contains address
          mov     bx,01                   ;handle
          mov     cx,20                   ;length
          int     21h                     ;of message
          inc     row
          ret
errorp    endp
;------------------------------------------------------------------
          end     main
```

11.3 字符设备的文件代号式 I/O

例 11.1～例 11.3 程序实例中,已使用了键盘、显示器等设备的文件代号式 I/O 功能,这是 DOS 2.0 为用户提供的另外一组独立于硬件的 DOS 功能。常用字符设备的文件代号都是由 DOS 预先定义好的。当一个用户程序得到控制权后,它就得到了五个已打开的文件代号,这五个文件代号是:

  0000 标准输入设备,通常是键盘
  0001 标准输出设备,通常是显示器
  0002 错误输出设备,总是显示器
  0003 标准辅助设备,一般为通信端口
  0004 标准打印机(♯0 打印机)

  设备和文件代号建立了对应关系,用户就可将这些设备视为文件。前两种设备的 I/O 功能允许改向操作,比如,允许用户程序可以从键盘文件输入也可以从别的文件输入,可以向显示器文件输出也可以向别的文件输出,而且不必打开或关闭这些文件,这些设备文件也没有读写指针。键盘输入和显示器、打印机输出等设备使用代号式 I/O 是非常简单的。例如,从键盘输入一行字符的指令序列为:

```
BUFFER   DB    80   DUP(?)
...      ...
MOV      AH,3FH              ; request read func
MOV      BX,0                ; handle 0=keyboard
MOV      CX,80               ; number of bytes
MOV      DX,SEG BUFFER       ; addr of input area
MOV      DS,DX
MOV      DX,OFFSET BUFFER
INT      21H                 ; call DOS
JC       ERROR
```

从键盘实际读入的字符数返回到 AX 寄存器。

  代号 0001 和 0002 可用于传送字符串并完成显示。例如,利用功能 40H(写文件或设备)将"hello"写到屏幕上,程序段如下:

```
BUFFER   DB   'hello'
...      ...
MOV      AH,40H              ;request write func
MOV      BX,01               ;handle 1= CRT
MOV      CX,5                ;length of string
MOV      DX,SEG BUFFER       ;DS:DX= addr of string
MOV      DS,DX
MOV      DX,OFFSET BUFFER
INT      21H                 ;call DOS
```

  该功能调用返回时,AX 中包含实际写的字符数。除去输出被改向为写磁盘文件而磁盘溢出的情况外,实际所写的字节数应与要求写的字节数相等。

  显示输出还可用错误输出设备(文件代号 0002),该文件代号总是指向控制台设备,

并且它是不可改向的,所以使用0002文件代号,一定是输出到了显示器,不可能输出到其他设备文件。

使用代号式I/O来完成打印输出比其他控制方式更为简单。打印机预先定义的文件代号为0004,用下列指令可将字符串'hello'输出到打印机:

```
BUFFER  DB    'hello'
        ...    ...
        MOV   AH,40H            ; request write func
        MOV   BX,04             ; handle 04= printer
        MOV   CX,5              ; leng of string
        MOV   DX, SEG BUFFER    ; addr of output area
        MOV   DS, DX
        MOV   DX, OFFSET BUFFER
        INT   21H               ; call DOS
        JC    ERROR             ; CF=1 if error
```

打印机输出功能返回时,AX中包含实际写到打印机的字符数,正常情况下,AX中的值应该与所需写的字符长度一样,而且CF标志位为0。但若输出数据中有文件结束标志(Ctrl-2)时,输出结束,返回的CF位置1,表明程序有致命错误或是破坏了操作系统。

采用代号式I/O,可以分别写几个列表设备(如LPT1,LPT2)。方法是由INT 21H的3DH功能打开指定的设备,然后根据返回的文件代号调用写功能40H存取某一指定的打印机。

和其他字符设备一样,串行口通信也可以通过文件代号式读写功能实现。作为辅助设备的串行接口,其预定义的代号为0003H。下面的程序可将'hello'写到串行口:

```
BUFFER  DB    'hello'
        ...    ...
        MOV   AH,40H            ; request write
        MOV   BX,03             ; 03= communication
        MOV   CX,5              ; length of string
        MOV   DX,SEG BUFFER     ; addr of output area
        MOV   DS,DX
        MOV   DX,OFFSET BUFFER
        INT   21H               ; call DOS
        JC    ERROR             ; CF=1 if error
```

如果系统中有多个串行口或辅助设备,仍需要通过打开文件功能(3DH)使设备和文件代号建立关系,然后根据返回的文件代号去完成读写功能。

从以上的介绍可以看出,使用文件代号式I/O,其优点是不用考虑不同设备之间差异很大的硬件接口特性,只要指出不同的文件代号;在输入设备上读文件,在输出设备上写文件,其程序基本都是相同的,而且对错误的判断也有统一的标志(CF=1),所以使用起来相当简便。如果用户不愿意使用预定义的标准设备,可先关闭这些文件代号(调用

INT 21H 的 3EH 功能),这样就为另一些文件或设备释放了文件代号。

为了进一步加深理解,下面再编写一个完整的程序例子。

例 12.4 利用文件代号式 I/O,从键盘上输入文件并从打印机输出。

```
        TITLE   TYPER.EXE -- Using file handle I/O
        .model  small
        .stack  64
keyboard        equ     0
crt             equ     1
printer         equ     4
;------------------------------------------------------------
        .data
typebuff        db      130 dup(' ')
errmsg          db      '* * * error * * *',0dh,0ah
;------------------------------------------------------------
        .code
main    proc    far
        mov     ax,@data         ; initialize DS
        mov     ds,ax
        mov     es,ax
        sti
        cld
        mov     ah,1             ; initializing printer
        mov     dx,0             ; printer #
        int     17h              ; call BIOS
input:
        mov     bx,keyboard      ; handle 0
        mov     ah,3fh           ; request read func.
        mov     cx,130           ; number of bytes
        lea     dx,typebuff      ; addr of input area
        int     21h              ; call DOS
        jc      error
        cmp     typebuff,1ah     ; type CTRL+Z to exit
        je      exit             ; end of kbd input
output:
        mov     cx,ax            ; number of char.
        mov     bx,printer       ; handle 4
        mov     ah,40h           ; request write func
        int     21h              ; call DOS
        jc      error
        jmp     input
error:
```

```
            mov     bx,crt              ; handle 1
            mov     ah,40h              ; output function
            mov     cx,15               ; length of string
            lea     dx,errmsg           ; addr of string
            int     21h                 ; call DOS
exit:
            mov     ax,4c00h            ; return to DOS
            int     21h
main        endp
;------------------------------------------------------------------------
            end     main
```

11.4 BIOS 磁盘存取功能

可以直接在 BIOS 一级上编写磁盘处理程序，但 BIOS 不能像 DOS 功能那样自动地支持目录处理、文件结束操作和记录分块。BIOS 磁盘操作 INT 13H 处理的记录都是一个扇区的大小，都是以实际的磁道号和扇区号寻址的。

读、写和检验磁盘文件之前，先把下列寄存器初始化：

AH 要执行的操作：读、写、检验或格式化等
AL 扇区数
CH 柱面/磁道号（0 为起始号）
CL 起始的扇区号（1 为起始号）
DH 磁头/盘面号，对软盘是 0 或 1
DL 驱动器号。软盘：0＝驱动器 A，1＝驱动器 B，…
       硬盘：80H＝驱动器 1，81H＝驱动器 2，…
ES:BX 数据区中 I/O 缓冲区的地址（除检验操作外）

11.4.1 BIOS 磁盘操作

BIOS INT 13H 要求在 AH 寄存器中指定操作。

AH＝00 复位磁盘系统

这个操作执行对磁盘控制器的硬件复位。如果在其他磁盘操作之后调用这个功能，则返回一系列错误。

AH＝01 读取磁盘状态

该操作在 AL 中返回最后一次磁盘 I/O 操作之后的状态。

AH＝02 读磁盘

该操作把同一磁道上的若干个扇区中的数据读取到内存。BX 中存放缓冲区的内存

地址,但是注意 BX 中的偏移地址在附加段(ES),这样输入缓冲区的地址应是 ES:BX。下面的例子把一个扇区的内容读入缓冲区 INSECT,这个缓冲区应足够大以接收一个扇区的所有内容。

```
INSECT  DB   512    DUP(?)        ;Area for input
        ...         ...
        MOV         AH,02         ;request read
        MOV         AL,01         ;one sector
        LEA         BX,INSECT     ;input buffer at ES:BX
        MOV         CH,05         ;track 05
        MOV         CL,03         ;sector 03
        MOV         DH,00         ;head 00
        MOV         DL,01         ;drive 01(B)
        INT         13H           ;call BIOS
```

调用返回时,AL 中是实际读取的扇区数,DS,BX,CX 和 DX 寄存器的内容不变。

在大多数情况下,程序只指定读一个扇区或一个磁道上的全部扇区。读操作只是顺序读取 CH 和 CL 指定的扇区内容,并递增 CH 和 CL 中的磁道号和扇区号。如果扇区号超过了磁道的最大扇区号,必须把扇区号重新置为 01,并把磁道号增 1,或者把双面盘的 0 面变为 1 面。

AH=03 写磁盘

写磁盘操作把指定内存区中的数据写到一个扇区或几个扇区,这个内存区域一般为一个或几个 512 字节。除了 AH=03 外,其他寄存器的设置和读磁盘一样,调用返回时,AL 中是实际写入的扇区数,DS,BX,CX 和 DX 寄存器不变。

AH=04 检验磁盘扇区

这个操作只简单地检测指定的扇区是否能找到,并且执行奇偶校验。有时为了更可靠的输出,可以在写操作之后,花费一些 I/O 时间来检验扇区的地址。这个操作因为没有数据传送,所以不必设置 ES:BX。返回时,AL 中是实际检测的扇区数,DX,DS,BX 和 CX 寄存器不变。例如,判断在驱动器 A 中是否有格式化的盘时,用下面的指令序列:

```
        MOV    AH,04        ;request verify
        MOV    DL,00        ;drive 0
        MOV    DH,00        ;head 0
        MOV    CH,00        ;track 0
        MOV    CL,01        ;sector1
        MOV    AL,01        ;one sector
        INT    13H          ;call BIOS
        JC     ERROR        ;bad disk
        ...    ...
```

AH=5 格式化盘磁道

用这个操作可以对一个或几个磁道进行格式化，IBM PC 对磁道格式化的标准大小是 512。读写操作都要求装入一个指定扇区的格式化信息。格式化操作要求 ES:BX 寄存器存放格式化参数表首址,对软盘磁道上的每一个扇区,必须有一个格式为 T/H/S/B 的四字节的表项,这里

T＝磁道号
H＝磁头号
S＝扇区号
B＝每区的字节数(00＝128 字节,01＝256 字节,02＝512 字节,03＝1024 字节)

例如,把磁道 3,磁头 0 格式化为每扇区 512 字节,那么 1 扇区的第一项内容就为 03000102。

对硬盘上的每个扇区,有一个二字节的表项：

字节 0 —— 标志字节(00H＝好,01H＝坏)
字节 1 —— 扇区号

11.4.2 状态字节

对上面介绍的 BIOS 磁盘操作(AH＝02,03,04,05)如果操作成功,则 CF 和 AH 置为 0;如果操作失败,CF 置为 1。AH 中返回表示出错原因的状态代码,见表 11.6。

表 11.6　BIOS 磁盘操作错误返回码

AH	状　态
01	给磁盘 I/O 传送了非法命令,控制器不能识别
02	磁盘上没有发现地址标记
03	试图往写保护磁盘上写
04	非法的磁道号/扇区号
05	复位操作失败
06	驱动器参数错
08	DMA 超限运行(数据传输太快)
09	DMA 超过 64K 的限制
10	读盘数据错(CRC)
20	磁盘控制器出错
40	搜索操作失败(硬件错)
80	软驱门开或无盘；硬盘超时
AAH	驱动器没有准备好
BBH	未定义错
CCH	写失败

如果磁盘操作返回一个错误码,通常就把磁盘复位(AH＝00),并且把这个操作重复执行三次。如果仍然得到一个错误码,则显示出错信息,以给用户掉换磁盘的机会。

11.4.3 BIOS 磁盘操作举例

现在利用 BIOS 指令来编写一个读磁盘的程序 BIOREAD。这个程序的特点是：

(1) 程序要计算每一个磁盘地址,每次读操作之后,扇区号加 1,当扇区号加到 10,则重新置扇区号为 01;如果盘面是 1,则增加磁道号,然后改变盘面,或由 0 改变为 1,或由 1 改变为 0。

(2) 数据项 CURADR 包含起始的磁道/扇面,ENDADR 包含结束的磁道/扇面地址。

```
        TITLE    BIOREAD.COM ———— Read disk sectors via BIOS
                 .model    small
                 .stack    64
;------------------------------------------------------------------
                 .data
recdin   db      512 dup(' ')            ;input area
endcde   db      0
curadr   dw      0304h                   ;beginning track/sector
endadr   dw      0501h                   ;ending track/sector
readmsg  db      '* * * read error * * *'
side     db      0
user     db      1,1 dup(?)              ;user input
;------------------------------------------------------------------
                 .code
main     proc    near
         mov     ax,@data                ;Initialize
         mov     ds,ax                   ;  segment reg.
         mov     es,ax
         mov     ax,0600h                ;request scroll
rept1:
         call    scren                   ;clear screen
         call    curs                    ;set cursor
         call    addrs                   ;calculate disk addr.
         mov     cx,curadr
         mov     dx,endadr
         cmp     cx,dx                   ;at ending sector?
         je      exit                    ;yes,exit
         call    reads                   ;read disk record
         cmp     endcde,0                ;normal read?
         jnz     exit                    ;no,exit
         call    disps                   ;display sector
         jmp     rept1                   ;repeat
```

```
exit:
        mov     ax,4c00h                ;terminate
        int     21h
main    endp
;----------------------------------------------------------------
; calculate next disk address
addrs   proc    near
        mov     cx,curadr               ;get track/sector
        cmp     cl,10                   ;past last sector
        jne     return                  ;no,exit
        cmp     side,0                  ;bypass if side 0
        je      chs
        inc     ch                      ;increment track
chs:
        xor     side,01                 ;change side
        mov     cl,01                   ;set sector to 1
        mov     curadr,cx
return:
        ret
addrs   endp
;----------------------------------------------------------------
; read disk sector
reads   proc    near
        mov     ah,02                   ;request read
        mov     al,01                   ;No,of sector
        lea     bx,recdin               ;addr of buffer
        mov     cx,curadr               ;track/sector
        mov     dh,side                 ;side
        mov     dl,00                   ;drive A
        int     13h
        cmp     ah,0                    ;normal read?
        jz      incrt                   ;yes,exit
        mov     endcde,01               ; no,
        call    errm                    ;   invalid read
incrt:
        inc     curadr                  ;increment sector
        ret
reads   endp
;----------------------------------------------------------------
; display sector
disps   proc    near
        mov     ah,40h                  ;request display
        mov     bx,01                   ;handle
```

```
                mov     cx,512              ;length
                lea     dx,recdin           ;address of input
                int     21h
wait1:
                mov     ah,0ah              ;user input
                lea     dx,user
                int     21h
                cmp     user+1,0dh
                je      wait1               ;wait until user input Return
                ret
disps           endp
;------------------------------------------------------------
; clear screen
scren           proc    near
                mov     ax,0600h            ;full screen
                mov     bh,1eh              ;set color
                mov     cx,0                ;request scroll
                mov     dx,184fh
                int     10h
                ret
scren           endp
;------------------------------------------------------------
; set cursor
curs            proc    near
                mov     ah,02               ;request set cursor
                mov     bh,0
                mov     dx,0
                int     10h
                ret
curs            endp
;------------------------------------------------------------
; disk error routine
errm            proc    near
                mov     ah,40h              ;request display
                mov     bx,01               ;handle
                mov     cx,18h              ;length of message
                lea     dx,readmsg
                int     21h
                ret
errm            endp
;------------------------------------------------------------
                end     begin
```

建议在 DEBUG 下运行这个程序,用 G 命令执行这个程序并观察输入缓冲区的内容。

注意:如果用 DOS 建立了一个文件,那么这个文件可能插在几个扇区中,它们在磁盘上的位置不一定是相邻的,所以不能用 BIOS INT 13H 把这个文件顺序地读出来。

习 题

11.1 写出文件代号式磁盘存取操作的错误代码:
(1) 非法文件代号 (2) 路径未发现 (3) 写保护磁盘

11.2 使用 3CH 功能建一文件,而该文件已经存在,这时会发生什么情况?

11.3 从缓冲区写信息到一个文件,如果没有关文件,可能会出现什么问题?

11.4 下面的 ASCIZ 串有什么错误?
PATH_NAME   DB   'C:\PROGRAMS\TEST.DAT'

11.5 下面为保存文件代号定义的变量有什么错误?
FILE_HNDL   DB   ?

11.6 在 ASCPATH 字节变量中为驱动器 D 的文件 PATIENT.LST,请定义 ASCIZ 串。

11.7 对 11.6 题中的文件,它的每个记录包含:

病历号(patient number):	5 字符,	姓名(name):	20 字符,
城市(city):	20 字符,	街道(street address):	20 字符,
出生年月(mmddyy):	6 字符,	性别(M/Fcode):	1 字符,
病房号(room number):	2 字符,	床号(bed number):	2 字符,

(1) 定义病人记录的各个域 (2) 定义保存文件代号的变量 FHANDLE
(3) 建文件 (4) 把 PATNTOUT 中的记录写入
(5) 关文件 (6) 以上文件操作包括测试错误

11.8 对 11.7 题的文件,用文件代号式编写一个完整的读文件程序,读出的每个记录存入 PATNTIN 并在屏幕上显示。

11.9 编写建立并写入磁盘文件的程序,允许用户从键盘键入零件号(3 字符),零(配)件说明(12 字符),单价(一个字)。程序使用文件代号式建立含有这些信息的文件。注意要把单价从 ASCII 码转换为二进制数。下面是输入数据的例子:

Part#	Description	Price	Part#	Description	Price
023	Assembler	00315	122	Lifters	10520
024	Linkages	00430	124	Processors	21335
027	Compilers	00525	127	Labtlers	00960
049	Compressors	00920	232	Bailers	05635
114	Extractors	11250	237	Grinders	08250
117	Haulers	00630	999		000

11.10 编写一个程序使用文件代号式读出并显示 11.9 题建立的文件。注意,要把二进制表示的单价转换为 ASCIIZ 码。

11.11 对 11.9 题建立的文件按下面的要求编写程序:

(1) 把所有的记录读入内存的数据缓冲区 TABLE;

(2) 显示字符串提示用户输入零(配)件号及其数量;

(3) 按零件号搜索 TABLE;

(4) 如果发现所要求的零件,用它的单价计算出总价(单价×数量);

(5) 显示零(配)件说明及总价值。

11.12 用随机处理记录的方式编写程序,将用户需要的零(配)件记录读取到 TABLE,并根据键入的数量,计算出总价值,然后显示出零(配)件说明及总价值。

附 录

附录 1 80x86 指令系统一览

助记符	汇编语言格式	功　能	操作数	时钟周期数	字节数	标志位 O D I T S Z A P C	备注
AAA	AAA	(AL)←把 AL 中的和调整到非压缩的 BCD 格式 (AH)←(AH)＋调整产生的进位值		3	1	u - - - u u x u x	
AAD	AAD	(AL)←10 * (AH)＋(AL) (AH)←0		10	2	u - - - x x x u x	
AAM	AAM	实现除法的非压缩的 BCD 调整 (AX)←把 AH 中的积调整到非压缩的 BCD 格式		18	2	u - - - x x u x u	
AAS	AAS	(AL)←(AH)－调整到非压缩的 BCD 格式 (AH)←(AH)－调整产生的借位值		3	1	u - - - u u x u x	
ADC	ADC dst,src	(dst)←(src)＋(dst)＋CF	reg,reg	1	2	x - - - x x x x x	
			reg,mem	2	2~7		
			mem,reg	3	2~7		
			reg,imm	1	3~6		
			ac,imm	1	2~5		
			mem,imm	3	3~11		
ADD	ADD dst,src	(dst)←(src)＋(dst)	reg,reg	1	2	x - - - x x x x x	
			reg,mem	2	2~7		

续表

助记符	汇编语言格式	功　能	操作数	时钟周期数	字节数	标志位 O D I T S Z A P C	备注
AND	AND　dst,src	(dst)←(src)∧(dst)	mem,reg reg,imm ac,imm mem,imm reg,reg reg,mem mem,reg reg,imm ac,imm mem,imm	3 1 1 3 1 2 3 1 1 3	2~7 3~6 2~5 3~11 2 2~7 2~7 3~6 2~5 3~11	0 - - - x x u x 0	
ARPL	ARPL　dst,src	调整选择器的 RPL 字段		7	2~7	- - - - - x - - - -	自 286 起有 系统指令
BOUND	BOUND　rsg,mem	测数组下标(reg)是否在指定的上下界(mem)之内， 在内，则往下执行； 不在内，产生 INT 5		8 INT+32	2~5	- - - - - - - - - -	自 286 起有
BSF	BSF　reg,src	自右向左扫描(src)，遇第一个为 1 的位，则 ZF←0,该位位置装入 reg；如 (src)=0,则 ZF←1	reg16,reg16 reg32,reg32 reg16,mem16 reg32,mem32	6~34 6~42 6~35 6~43	3 3~8	u - - - u x u u u	自 386 起有
BSR	BSR　reg,src	自左向右扫描(src)，遇第一个为 1 的位，则 ZF←0,该位位置装入 reg；如 (src)=0,则 ZF←1	reg16,reg16 reg32,reg32 reg16,mem16 reg32,mem32	7~39 7~71 7~40 7~72	3 3~8	u - - - u x u u u	自 386 起有
BSWAP	BSWAP　r32	(r32)字节次序变反		1	2	- - - - - - - - - -	自 486 起有

续表

助记符	汇编语言格式	功 能	时钟周期数	字节数	标志位 O D I T S Z A P C	备 注
BT	BT dst,src	把由(src)指定的(dst)中的位内容送 CF	4	3	u - - - u u u u x	自 386 起有
			9	3~8		
			4	4		
			4	4~9		
BTC	BTC dst,src	把由(src)指定的(dst)中的位内容送 CF, 并把该位变反	7	3	u - - - u u u u x	自 386 起有
			13	3~8		
			7	4		
			8	4~9		
BTR	BTR dst,src	把由(src)指定的(dst)中的位内容送 CF, 并把该位置 0	7	3	u - - - u u u u x	自 386 起有
			13	3~8		
			7	4		
			8	4~9		
BTS	BTS dst,src	把由(src)指定的(dst)中的位内容送 CF, 并把该位置 1	7	3	u - - - u u u u x	自 386 起有
			13	3~8		
			7	4		
			8	4~9		
CALL	CALL dst				- - - - - - - - -	
		段内直接: push(IP 或 EIP) (IP)←(IP)+D16 或 (EIP)←(EIP)+D32	1	1~5		
		段内间接: push(IP 或 EIP) (IP 或 EIP)←(EA)	2	2~7	reg	
			2		mem	
		段间直接: push(CS) push(IP 或 EIP) (IP 或 EIP)←dst 指定的偏移地址 (CS)←dst 指定的段地址	4*	2~6		

· 437 ·

续表

助记符	汇编语言格式	功能	时钟周期数	操作数	字节数	O	D	I	T	S	Z	A	P	C	备注
		段间间接：push(CS) push(IP 或 EIP) (IP 或 EIP)←(EA) (CS)←(EA+2 或 4)	5*		2~7										
CBW	CBW	(AL)符号扩展到(AH)	3		1	-	-	-	-	-	-	-	-	-	
CWDE	CWDE	(AX)符号扩展到(EAX)	3		1	-	-	-	-	-	-	-	-	-	自 386 起有
CLC	CLC	进位置 0	2		1	-	-	-	-	-	-	-	-	0	
CLD	CLD	方向标志置 0	2		1	-	0	-	-	-	-	-	-	-	
CLI	CLI	中断标志置 0	7		1	-	-	0	-	-	-	-	-	-	
CLTS	CLTS	清除 CR0 中的任务切换标志	10		2	-	-	-	-	-	-	-	-	-	自 386 起有 系统指令
CMC	CMC	进位变反	2		1	-	-	-	-	-	-	-	-	x	
CMP	CMP opr1,opr2	(opr1)-(opr2)	1 2 2 1 2	reg,reg reg,mem mem,reg reg,imm ac,imm mem,imm	2 2~7 2~7 3~6 2~4 3~6	x	-	-	-	x	x	x	x	x	
CMPS	CMPSB CMPSW CMPSD	((SI 或 ESI))-((DI 或 EDI)) (SI 或 ESI)←(SI 或 ESI)±1 或 2 或 4 (DI 或 EDI)←(DI 或 EDI)±1 或 2 或 4	5		1	x	-	-	-	x	x	x	x	x	
CMPXCHG	CMPXCHG dst,reg	(ac)-(dst) 相等：ZF←1，(dst)←(reg) 不相等：ZF←0，(ac)←(dst)	5 6	reg,reg mem,reg	3 4~8	x	-	-	-	x	x	x	x	x	自 486 起有

续表

助记符	汇编语言格式	功能	操作数	时钟周期数	字节数	标志位 O D I T S Z A P C	备注
CMPXCHG8B	CMPXCHG8B dst	(EDX,EAX)←(dst) 相等:ZF←1, (dst)←(ECX,EBX) 不相等:ZF←0, (EDX,EAX)←(dst)		10	3~8	- - - - - - x - -	自586起有
CPUID	CPUID	(EAX)←CPU 识别信息		14	1	- - - - - - - - -	自586起有
CWD	CWD	(AX)符号扩展到(DX)		2	1	- - - - - - - - -	
CDQ	CDQ	(EAX)符号扩展到(EDX)		2	1	- - - - - - - - -	自386起有
DAA	DAA	(AL)←把 AL 中的和调整到压缩的BCD格式		3	1	u - - - x x x x x	
DAS	DAS	(AL)←把 AL 中的差调整到压缩的BCD格式		3	1	u - - - x x x x x	
DEC	DEC opr	(opr)←(opr)-1	reg mem	1 3	1或2 2~7	x - - - x x x x -	
DIV	DIV src	(AL)←(AX)/(src)的商 (AH)←(AX)/(src)的余数 (AX)←(DX,AX)/(src)的商 (DX)←(DX,AX)/(src)的余数 (EAX)←(EDX,EAX)/(src)的商 (EDX)←(EDX,EAX)/(src)的余数	reg8 reg16 reg32 mem8 mem16 mem32	17 25 41 17 25 41	2 2~7	u - - - u u u u u	
ENTER	ENTER imm16,imm8	建立堆栈帧 imm16 为堆栈帧的字节数 imm8 为堆栈帧的层数 L	L=0 L=1 L>1	11 15 15+2L	4	- - - - - - - - -	自386起有
HLT	HLT	停机		1	1	- - - - - - - - -	系统指令

续表

助记符	汇编语言格式	功 能	操作数	时钟周期数	字节数	标志位 O D I T S Z A P C	备 注
IDIV	IDIV src	(AL)←(AX)/(src)的商 (AH)←(AX)/(src)的余数 (AX)←(DX,AX)/(src)的商 (DX)←(DX,AX)/(src)的余数 (EAX)←(EDX,EAX)/(src)的商 (EDX)←(EDX,EAX)/(src)的余数	reg8 reg16 reg32 mem8 mem16 mem32	22 30 46 22 30 46	2 2~7 	u - - u u u u u	
IMUL	IMUL src	(AX)←(AL) * (src) (DX,AX)←(AX) * (src) (EDX,EAX)←(EAX) * (src) src 为:	reg8 reg16 reg32 mem8 mem16 mem32	11 11 10 11 11 10	2 2~7 	x - - u u u u x	
IMUL	IMUL reg,src	(reg16)←(reg16) * (src) (reg32)←(reg32) * (src) src 为:	reg8 reg16 reg32 mem8 mem16 mem32	10 10 10 10 10 10	3 3~8 		自286起有
IMUL	IMUL reg,src,imm	(reg16)←(src) * imm (reg32)←(src) * imm	reg8 reg16 reg32 mem8 mem16	10 10 10 10 10	3~6 3~11 		自286起有

续表

助记符	汇编语言格式	功能	操作数	时钟周期数	字节数	O	D	I	T	S	Z	A	P	C	备注
IN	ac,PORT	(ac)←(PORT)		10	2	-	-	-	-	-	-	-	-	-	
IN	ac,DX	(ac)←((DX))		7*	1	-	-	-	-	-	-	-	-	-	
INC	opr	(opr)←(opr)+1	reg	1	1或2	x	-	-	-	x	x	x	x	-	
			mem	3	2~7										
INSB		((DI 或 EDI))←((DX))		9*	1	-	-	-	-	-	-	-	-	-	自286起有
INSW		(DI 或 EDI)←(DI 或 EDI)±1 或 2 或 4													
INSD															
INT	type	push(FLAGS)	type≠3	INT+6	2	-	-	0	0	-	-	-	-	-	
INT	(当 type=3 时)	push(CS)	type=3	INT+5	1										
		push(IP)													
		(IP)←(type * 4)													
		(CS)←(type * 4 + 2)													
INTO		若 OF=1,则		4(OF=0)	1	-	-	0	0	-	-	-	-	-	
		push(FLAGS)		INT+5 (OF=1)											
		push(CS)													
		push(IP)													
		(IP)←(10H)													
		(CS)←(12H)													
INVD		使高速缓存无效		15	2	-	-	-	-	-	-	-	-	-	自486起有 系统指令
INVLPG	opr	使TLB入口无效		29	3~8	-	-	-	-	-	-	-	-	-	自486起有 系统指令
IRET		(IP)←POP( )		7*	1	r	r	r	r	r	r	r	r	r	
		(CS)←POP( )													
		(FLAGS)←POP( )													
IRETD		(EIP)←POP( )		7*	1	r	r	r	r	r	r	r	r	r	自386起有
		(CS)←POP( )													
		(EFLAGS)←POP( )													

续表

助记符	汇编语言格式	功能	操作数	时钟周期数	字节数	标志位 O D I T S Z A P C	备注
Jcc		满足条件则转移	8位位移量	1	2	- - - - - - - - -	
			16/32位位移量	1	4~6	- - - - - - - - -	自386起有
	JZ/JE opr	ZF=1则转移					
	JNZ/JNE opr	ZF=0则转移					
	JS opr	SF=1则转移					
	JNS opr	SF=0则转移					
	JO opr	OF=1则转移					
	JNO opr	OF=0则转移					
	JP/JPE opr	PF=1则转移					
	JNP/JPO opr	PF=0则转移					
	JC/JB/JNAE opr	CF=1则转移					
	JNC/JNB/JAE opr	CF=0则转移					
	JBE/JNA opr	CF∨ZF=1则转移					
	JNBE/JA opr	CF∨ZF=0则转移					
	JL/JNGE opr	SF∀OF=1则转移					
	JNL/JGE opr	SF∀OF=0则转移					
	JLE/JNG opr	(SF∀OF)∨ZF=1则转移					
	JNLE/JG opr	(SF∀OF)∨ZF=0则转移					
JCXZ	opr	(CX)=0则转移		6/5	2	- - - - - - - - -	
JECXZ	opr	(ECX)=0则转移		6/5	2	- - - - - - - - -	自386起有
JMP	opr	无条件转移				- - - - - - - - -	
		段内直接短 (IP或EIP)←(IP或EIP)+D8		1	2		
		段内直接近 (IP)←(IP)+D16 或 (EIP)←(EIP)+D32		1	3~5		
		段内间接 (IP或EIP)←(EA)	reg	2	2		
			mem	2	2~7		

续表

助记符	汇编语言格式	功能	操作数	时钟周期数	字节数	标志位 O D I T S Z A P C	备注
		段间直接					
		(IP 或 EIP)←opr 指定的偏移地址		3*	2~6	- - - - - - - - -	
		(CS)←opr 指定的段地址					
		段间间接					
		(IP 或 EIP)←(EA)		4*	2~7	- - - - - - - - -	
		(CS)←(EA+2 或 4)					
LAHF	LAHF	(AH)←(FLAGS 的低字节)		2	1	- - - - - - - - -	
LAR	LAR reg,src	取访问权字节	reg,reg	8	3	- - - - - x - - -	自 286 起有
			reg,mem	8	3~8		系统指令
LDS	LDS reg,src	(reg)←(src)		4~13	2~7	- - - - - - - - -	
		(DS)←(src+2 或 4)					
LEA	LEA reg,src	(reg)←src		1	2~7	- - - - - - - - -	
LEAVE	LEAVE	释放堆栈帧		3	1	- - - - - - - - -	
LES	LES reg,src	(reg)←(src)		4~13	2~7	- - - - - - - - -	
		(ES)←(src+2 或 4)					
LFS	LFS reg,src	(reg)←(src)		4~13	3~8	- - - - - - - - -	自 386 起有
		(FS)←(src+2 或 4)					
LGDT	LGDT mem	装入全局描述符表寄存器		6	3~8	- - - - - - - - -	自 286 起有
		(GDTR)←(mem)					系统指令
LGS	LGS reg,src	(reg)←(src)		4~13	3~8	- - - - - - - - -	自 386 起有
		(GS)←(src+2 或 4)					
LIDT	LIDT mem	装入中断描述符表寄存器		6	3~8	- - - - - - - - -	自 286 起有
		(IDTR)←(mem)					系统指令
LLDT	LLDT src	装入局部描述符表寄存器	reg	8	3	- - - - - - - - -	自 286 起有
		(LDTR)←(src)	mem	8	3~8		系统指令
LMSW	LMSW src	装入机器状态字(在 CR0 寄存器中)	reg	8	3	- - - - - - - - -	自 286 起有
		(MSW)←(src)	mem	8	3~8		系统指令
LOCK	LOCK	插入 LOCK # 信号前缀		1	1	- - - - - - - - -	
LODS	LODSB	(ac)←((SI 或 ESI))		2	1	- - - - - - - - -	
	LODSW	(SI 或 ESI)←(SI 或 ESI)±1 或 2 或 4					系统指令

续表

助记符	汇编语言格式		功　能	操作数	时钟周期数	字节数	标志位 O D I T S Z A P C	备　注
LODSD								
LOOP	LOOP	opr	(CX 或 ECX)≠0 则循环		5/6	2	- - - - - - - - -	
LOOPZ/LOOPE	LOOPZ/LOOPE	opr	ZF=1 且(CX 或 ECX)≠0 则循环		7/8	2	- - - - - - - - -	
LOOPNZ/LOOPNE	LOOPNZ/LOOPNE	opr	ZF=0 且(CX 或 ECX)≠0 则循环		7/8	2	- - - - - - - - -	
LSL	LSL	reg,src	取段界限	reg,reg	8	3	- - - - x - - - -	自 286 起有 系统指令
				reg,mem	8	3~8		
LSS	LSS	reg,src	(reg)←(src) (SS)←(src+2 或 4)		4~13*	3~8	- - - - - - - - -	自 386 起有
LTR	LTR	src	装入任务寄存器	reg	10	3	- - - - - - - - -	自 286 起有 系统指令
				mem	10	3~8		
MOV	MOV	dst,src	(dst)←(src)	reg,reg	1	2	- - - - - - - - -	
				reg,mem	1	2~7		
				mem,reg	1	2~7		
				reg,imm	1	2~6		
				mem,imm	1	3~11		
				ac,mem	1	2~5		
				mem,ac	1	2~5		
MOV	MOV	reg,CR0-4(控制寄存器)	(reg)←(CR0-4)	CR0,reg	4	3	u - - - u u u u u	自 386 起有 系统指令
	MOV	CR0-4,reg	(CR0-4)←(reg)	CR2,reg	22	3		
				CR3,reg	12	3		
				CR4,reg	21	3		
					14	3		
MOV	MOV	reg,DR(调试寄存器)	(reg)←(DR)	reg,DR0-3	11	3	u - - - u u u u u	自 386 起有 系统指令
				reg,DR4-5	12	3		
				reg,DR6-7	2	3		
MOV	MOV	DR,reg	(DR)←(reg)	DR0-3,reg	11	3		
				DR4-5,reg	12	3		
				DR6-7,reg	11	3		

续表

助记符	汇编语言格式	功能	操作数	时钟周期数	字节数	标志位 O D I T S Z A P C	备注
MOV							
	MOV dst,SR(段寄存器)	(dst)←(SR)	reg	1	2	- - - - - - - - -	
			mem	1	2~7		
	MOV SR,src	(SR)←(src)	reg	2~11	2	- - - - - - - - -	如段寄存器为 SS
			mem	2~11*	2~7		
				3			
MOVS	MOVSB	((DI 或 EDI))←((SI 或 ESI))		3~12*	2~7	- - - - - - - - -	如段寄存器为 SS
	MOVSW	(SI 或 ESI)←(SI 或 ESI)±1 或 2 或 4		4	1		
	MOVSD	(DI 或 EDI)←(DI 或 EDI)±1 或 2 或 4					
MOVSX	MOVSX dst,src	(dst)←符号扩展(src)	reg,reg	3	3	- - - - - - - - -	自 386 起有
			reg,mem	3	3~8		
MOVZX	MOVZX dst,src	(dst)←零扩展(src)	reg,reg	3	3	- - - - - - - - -	自 386 起有
			reg,mem	3	3~8		
MUL	MUL src	(AX)←(AL)*(src)	src 为 reg8	11	2	x - - - u u u u x	
		(DX,AX)←(AX)*(src)	reg16	11			
		(EDX,EAX)←(EAX)*(src)	reg32	10			
			mem8	11			
			mem16	11			
			mem32	10			
NEG	NEG opr	(opr)←−(opr)	reg	1	2	x - - - x x x x x	
			mem	3	2~7		
NOP	NOP	无操作		1	1	- - - - - - - - -	
NOT	NOT opr	(opr)←(opr)	reg	1	2	- - - - - - - - -	
			mem	3	2~7		
OR	OR dst,src	(dst)←(dst) V (src)	reg,reg	1	2	0 - - - x x x u x 0	
			reg,mem	2	2~7		
			mem,reg	3	2~7		
			reg,imm	1	3~6		
			ac,imm	1	2~5		
			mem,imm	3	3~11		

· 445 ·

续表

助记符	汇编语言格式		功　能	操作数	时钟周期数	字节数	标志位 O D I T S Z A P C	备注
OUT	OUT	port,ac	(port)←(ac)		12*	2	- - - - - - - - -	
	OUT	DX,ac	((DX))←(ac)		12*	1	- - - - - - - - -	
OUTS	OUTSB		((DX))←((SI 或 ESI))		13*	1	- - - - - - - - -	
	OUTSW		(SI 或 ESI)←(SI 或 ESI)±1 或 2 或 4					
	OUTSD							
POP	POP	dst	(dst)←((SP 或 ESP))	reg	1	1 或 2	- - - - - - - - -	
			(SP 或 ESP)←(SP 或 ESP)+2 或 4	mem	3	2～7	- - - - - - - - -	
				SR	3～12	1	- - - - - - - - -	
			如段寄存器为 SS		3～12*			
			如段寄存器为 FS,GS		3～12	2	- - - - - - - - -	
POPA/POPAD	POPA		出栈送 16 位通用寄存器		5	1	- - - - - - - - -	自 286 起有
	POPAD		出栈送 32 位通用寄存器		5	1	- - - - - - - - -	自 386 起有
POPF/POPFD	POPF		出栈送 FLAGS		4*	1	r r r r r r r r r	
	POPFD		出栈送 EFLAGS		4*	1	r r r r r r r r r	自 386 起有
PUSH	PUSH	src	(SP 或 ESP)←(SP 或 ESP)−2 或 4	reg	1	1 或 2	- - - - - - - - -	
			((SP 或 ESP))←(src)	mem	2	2～7	- - - - - - - - -	
				imm	1	1～5	- - - - - - - - -	
				SR	1	1～2	- - - - - - - - -	
PUSHA/PUSHAD	PUSHA		16 位通用寄存器进栈		5	1	- - - - - - - - -	自 286 起有
	PUSHAD		32 位通用寄存器进栈		5	1	- - - - - - - - -	自 386 起有
PUSHF/PUSHFD	PUSHF		FLAGS 进栈		3*	1	- - - - - - - - -	自 286 起有
	PUSHFD		EFLAGS 进栈		3*	1	- - - - - - - - -	自 386 起有
RCL	RCL	opr,cnt	带进位循环左移	reg,1	1	2	x - - - - - - - x	
				mem,1	3	2～7		
				reg,CL	7～24	2		
				mem,CL	9～26	2～7	u - - - - - - - x	
				reg,imm8	8～25	3		
				mem,imm8	10～27	3～8		
RCR	RCR	opr,cnt	带进位循环右移	reg,1	1	2	x - - - - - - - x	自 286 起有

续表

助记符	汇编语言格式	功能	操作数	时钟周期数	字节数	标志位 O D I T S Z A P C	备注
			mem,1	3	2~7	- - - - - - - - -	
			reg,CL	7~24	2	- - - - - - u - x	
			mem,CL	9~26	2~7		
			reg,imm8	8~25	3		自286起有
			mem,imm8	10~27	3~8		自286起有
RDMSR	RDMSR	读模型专用寄存器 (EDX,EAX)←MSR[ECX]		20~24	2	- - - - - - - - -	自586起有 系统指令
REP	REP string primitive	当(CX 或 ECX)=0,退出重复;否则, (CX 或 ECX)←(CX 或 ECX)-1, 执行其后的串指令					
	REP INS		C=0	11+3C	2	- - - - - - - - -	
	REP LODS		C>0	7	2	- - - - - - - - -	
				7+3C			
	REP MOVS		C=0	6	2	- - - - - - - - -	
			C=1	13			
			C>1	13+C			
	REP OUTS		C=0	13+4C	2	- - - - - - - - -	
	REP STOS		C>0	6	2	- - - - - - - - -	
				9+C			
REPE/REPZ	REPE/REPZ string primitive	当(CX 或 ECX)=0 或 ZF=0,退出 重复;否则, (CX 或 ECX)←(CX 或 ECX)-1, 执行其后的串指令					
	REPE CMPS		C=0	7	2	x - - x x x x x x	
			C>0	8+4C			
	REPE SCAS		C=0	7	2	x - - x x x x x x	
			C>0	8+4C			
REPNE/REPNZ	REPNE/REPNZ string primitive	当(CX 或 ECX)=0 或 ZF=1 退出重复;					

· 447 ·

续表

助记符	汇编语言格式	功能	操作数	时钟周期数	字节数	O D I T S Z A P C	备注
	REPNE CMPS	否则,(CX 或 ECX)←(CX 或 ECX)-1,执行其后的串指令	C=0	7	2	x - - - x x x x x	
			C>0	9+4C	2		
	REPNZ SCAS		C=0	7	2	x - - - x x x x x	
			C>0	8+4C	2		
RET	RET	段内:(IP)←POP( )		2	1	- - - - - - - - -	
		段间:(IP)←POP( )		4*	1	- - - - - - - - -	
		(CS)←POP( )					
		段内:(IP)←POP( )		3	3	- - - - - - - - -	
		(SP 或 ESP)←(SP 或 ESP)+D16					
	RET exp	段间:(IP)←POP( )		4*	3	- - - - - - - - -	
		(CS)←POP( )					
		(SP 或 ESP)←(SP 或 ESP)+D16					
ROL	ROL opr,cnt	循环左移	reg,1	1	2	x - - - - - - - x	
			mem,1	3	2~7		
			reg,CL	4	2		
			mem,CL	4	2~7		
			reg,imm8	1	3		自 286 起有
			mem,imm8	3	3~8		自 286 起有
ROR	ROR opr,cnt	循环右移	reg,1	1	2	x - - - - - - - x	
			mem,1	3	2~7		
			reg,CL	4	2		
			mem,CL	4	2~7		
			reg,imm8	1	3		自 286 起有
			mem,imm8	3	3~8		自 286 起有
RSM	RSM	从系统管理方式恢复		2	2	x x x x x x x x x	系统指令
SAHF	SAHF	(FLAGS 的低字节)←(AH)		2	1	- - - - r r r r r	

· 448 ·

续表

助记符	汇编语言格式		功　能	操作数	时钟周期数	字节数	标志位 O D I T S Z A P C	备　注
SAL	SAL	opr,cnt	算术左移	reg,1	1	2	x - - - x x u x x	
				mem,1	3	2~7		
				reg,CL	4	2		
				mem,CL	4	2~7		
				reg,imm8	1	3		自286起有
				mem,imm8	3	3~8		自286起有
SAR	SAR	opr,cnt	算术右移	reg,1	1	2	x - - - x x u x x	
				mem,1	3	2~7		
				reg,CL	4	2		
				mem,CL	4	2~7		
				reg,imm8	1	3		自286起有
				mem,imm8	3	3~8		自286起有
SBB	SBB	dst,src	(dst)←(dst)−(src)−CF	reg,reg	1	2	x - - - x x x x x	
				reg,mem	2	2~7		
				mem,reg	3	2~7		
				reg,imm	1	3~6		
				ac,imm	1	2~5		
				mem,imm	3	3~11		
SCAS	SCASB		(ac)−((DI 或 EDI))		4	1	x - - - x x x x x	
	SCASW		(DI 或 EDI)←(DI 或 EDI)±1 或 2 或 4					
	SCASD							
SET	SETcc	dst	条件设置	reg	1	3	- - - - - - - - -	自386起有
				mem	2	3~8		
SGDT	SGDT	mem	从全局描述符表寄存器取 (mem)←(GDTR)		4	3~8	- - - - - - - - -	自286起有 系统指令
SHL	SHL	opr,cnt	逻辑左移	与 SAL 相同				
SHLD	SHLD	dst,reg,cnt	双精度左移	reg,reg,imm8	4	4	u - - - x x u x x	自386起有
				mem,reg,imm8	4	4~9		
				reg,reg,CL	4	3		

续表

助记符	汇编语言格式	功能	操作数	时钟周期数	字节数	标志位 O D I T S Z A P C	备注
SHR	SHR opr,cnt	逻辑右移	mem,reg,CL	5	3～8	- - - - - - - - -	
			reg,1	1	2	x - - - x x u x x	
			mem,1	3	2～7		
			reg,CL	4	2		
			mem,CL	4	2～7		
			reg,imm8	1	3		自286起有
			mem,imm8	3	3～8		自286起有
SHRD	SHRD dst,reg,cnt	双精度右移	reg,reg,imm8	4	4	u - - - x x u x x	自386起有
			mem,reg,imm8	4	4～9		
			reg,reg,CL	4	3		
			mem,reg,CL	5	3～8		
SIDT	SIDT mem	从中断描述符表取 (mem)←(IDTR)	mem	4	3～8	- - - - - - - - -	自286起有 系统指令
SLDT	SLDT dst	从局部描述符表取 (dst)←(LDTR)	reg	2	3	- - - - - - - - -	自286起有 系统指令
			mem	2	3～8		
SMSW	SMSW dst	从机器状态字取 (dst)←(MSW)	reg	4	3	- - - - - - - - -	自286起有 系统指令
			mem	4	3～8		
STC		进位置1		2	1	- - - - - - - - 1	
STD		方向标志置1		2	1	- 1 - - - - - - -	
STI		中断标志置1		7	1	- - 1 - - - - - -	
STOS	STOSB STOSW STOSD	((DI或EDI))←(ac) (DI或EDI)±1或2或4		3	1	- - - - - - - - -	
STR	STR dst	从任务寄存器取 (dst)←(TR)	reg	2	3	- - - - - - - - -	自286起有 系统指令
			mem	2	3～8		
SUB	SUB dst,src	(dst)←(dst)−(src)	reg,reg	1	2	x - - - x x x x x	
			reg,mem	2	2～7		
			mem,reg	3	2～7		
			reg,imm	1	3～6		

续表

助记符	汇编语言格式	功　能	操作数	时钟周期数	字节数	O D I T S Z A P C	备　注
TEST	TEST opr1,opr2	(opr1)∧(opr2)	ac,imm	1	2~5		
			mem,imm	3	3~11		
			reg,reg	2	2	0 - - x x x u x 0	
			reg,mem	1	2~7		
			reg,imm	1	3~6		
			ac,imm	1	2~5		
			mem,imm	2	3~11		
VERR	VERR opr	检验 opr 中的选择器所表示的段是否可读	reg	7	3	- - - - - - - x - -	自286起有系统指令
			mem	7	3~8		
VERW	VERW opr	检验 opr 中的选择器所表示的段是否可写	reg	7	3	- - - - - - - x - -	自286起有系统指令
			mem	7	3~8		
WAIT	WAIT	等待		1	1	- - - - - - - - - -	
WBINVD	WBINVD	写回并使高速缓存无效		2000+	2	- - - - - - - - - -	自486起有系统指令
WRMSR	WRMSR	写入模型专用寄存器 MSR(ECX)←(EDX,EAX)		30~45	2	- - - - - - - - - -	自586起有系统指令
XADD	XADD dst,src	TEMP←(src)+(dst) (src)←(dst) (dst)←TEMP	reg,reg	3	3	x - - - x x x x x x	自486起有系统指令
			mem,reg	4	3~8		
XCHG	XCHG opr1,opr2	(opr1)↔(opr2)	reg,reg	3	2	- - - - - - - - - -	
			ac,reg	2	1		
			mem,reg	3	2~7		
XLAT	XLAT	(AL)←((BX 或 EBX)+(AL))		4	1	- - - - - - - - - -	
XOR	XOR dst,src	(dst)←(dst)∀(src)	reg,reg	1	2	0 - - - x x x u x 0	
			reg,mem	2	2~7		
			mem,reg	3	2~7		
			reg,imm	1	3~6		
			ac,imm	1	2~5		
			mem,imm	3	3~11		

续表

助记符	汇编语言格式	功能	操作数	时钟周期数	字节数	标志位 O D I T S Z A P C	备注
前缀字节							
	地址长度前缀			1	1		
	操作数长度前缀			1	1		
	段前缀			1	1		
	LOCK			1	1		
外部中断							
NMI	非屏蔽中断			INT+14			
页故障				INT+6			
虚86模式异常				INT+40			
CLI				INT+9			
STI				INT+9			
INT n				INT+9			
PUSHF				INT+9			
POPF				INT+9			
IRET				INT+9			
IN				INT+34			
OUT				INT+34			
INS				INT+34			
OUTS				INT+34			
REP INS				INT+34			
REP OUTS				INT+34			

说明：

(1) 附录一的指令表来源于 Intel Pentium 的用户手册，其中提供的有关数据只适用于 Pentium。

(2) 本表列出了 80x86 中面向应用程序设计的指令和面向系统程序设计的指令，其中后者在备注栏中以"系统指令"注明。本表未收入 80x86 的浮点指令，如有需要读者可从 Intel 80x86 的用户手册中查到。

(3) 表中所用符号说明如下：

① 操作类型中的 ac——累加器，reg——通用寄存器，mem——存储单元，imm——立即数。如其后跟以 8,16,32，则表示其长度，如 imm8 表示 8 位立即数，reg16 表示 16 位通用寄存器，余类推。

② 时钟周期数中的 *——实模式下的时钟周期数。在保护模式下，由于情况比较复杂，在本表中未提供。
L/NL——表示循环指令中的循环/不循环时钟周期数。如 6/5 表示如执行循环则时钟周期数为 6，如不执行循环则时钟周期数为 5。
INT——表示一次中断所用的时钟周期数。一般实模式下与其所用门的类型和所在特权级均有关，这里不提供。

③ 标志位所用符号：0——置 0,1——置 1,x——根据结果设置，-——不影响，u——无定义，r——恢复原先保存的值。

附录 2 伪操作与操作符

附表 2.1 伪操作

类 型	伪操作名	格 式	说 明	适用版本
处理机选择	.8086	.8086	允许使用 8086 和 8088 指令系统及 8087 专用指令	MASM 1–6
	.8087	.8087	允许使用 8087 指令	MASM 4–6
	.286	.286	在 8086 指令基础上,允许使用 80286 实模式指令和 80287 指令	MASM 3–6
	.286P	.286P	在 8086 指令基础上,允许使用包括保护模式在内的 80286 指令系统和 80287 指令	MASM 3–6
	.287	.287	在 8087 指令基础上,允许使用 80287 指令	MASM 3–6
	.386	.386	在 8086,80286 指令基础上,允许使用 80386 实模式指令和 80387 指令	MASM 5–6
	.386P	.386P	在 8086,80286 指令基础上,允许使用包括保护模式在内的 80386 指令系统和 80387 指令	MASM 5–6
	.387	.387	在 8087,80287 指令基础上,允许使用 80387 指令	MASM 5–6
	.486	.486	在 8086,80286,80386 指令基础上,允许使用 80486 实模式指令和 80387,80487 指令	MASM 6
	.486P	.486P	在 8086,80286,80386 指令基础上,允许使用包括保护模式在内的 80486 指令系统和 80387,80487 指令	MASM 6
	NO87	NO87	不允许使用协处理器指令	
段定义	SEGMENT ENDS	segname SEGMENT [align][combine][use]['class'] : segname ENDS	定义段 align 说明段起始地址的边界值。它们可以是 PARA,BYTE,WORD,DWORD 或 PAGE	MASM 1–6

续表

类型	伪操作名	格式	说明	适用版本
	ASSUME	ASSUME segreg:segname[,…] ASSUME datareg:qualified type[,…]	combine 说明连接时的段合并方式。它们可以是 PRIVATE, PUBLIC, COMMON, AT expression, MEMORY 或 STACK use 指出段大小。它们可以是 USE16 或 USE32。 'class' 指定类别 规定段所属的段寄存器 指定寄存器所指向数据的类型 如:ASSUME BX:PTR WORD 表示 BX 指向一个字数组,其下的指令中如有[BX]就不必再加类型说明	MASM 1—6
	ASSUME	reg:ERROR[,…]	用来限制使用某些寄存器 如:ASSUME SI:ERROR 则其后程序不允许再用 SI 寄存器	
	ASSUME	reg:NOTHING	用来取消前面已指定的连接关系。如: ASSUME BX:NOTHING 或 ASSUME SI:NOTHING 均可取消前面指定的限制。也可用 ASSUME ES:NOTHING 表示段寄存器 ES 并未和任一段相关	
存储模型及 简化段定义	.MODEL	.MODEL memory-model[,model options]	存储模型选择,用于所有简化段定义之前。memory-model 指定所用存储模型,它们可以是 TINY, SMALL, MEDIUM, COMPACT, LARGE, HUGE 或 FLAT。model options 可以指定三种高级语言接口,可以是 C, BASIC, FORTRAN, PASCAL, SYSCALL 或 STDCALL。操作系统,可以是 OS—DOS 或 OS—OS2。堆栈距离,可以是 NEARSTACK 或 FARSTACK	MASM 5—6
	.CODE	.CODE [name]	定义代码段。对于一个代码段的模型,.name 为可选项;对于多个代码段的模型,则应为每个代码段指定段名	MASM 5—6
	.DATA	.DATA	定义初始化数据段	MASM 5—6
	.DATA?	.DATA?	定义未初始化数据段	MASM 5—6
	.FARDATA	.FARDATA [name]	定义远初始化数据段。可指定段名	MASM 5—6
	.FARDATA?	.FARDATA? [name]	定义远未初始化数据段。可指定段名	MASM 5—6

续表

类 型	伪操作名	格　　式	说　　　明	适用版本
	.CONST	.CONST	定义常数数据段	MASM 5—6
	.STACK	.STACK [size]	定义堆栈段。可指定堆栈段大小（以字节为单位）；如不指定，默认值为1KB	MASM 5—6
段组定义	GROUP	grpname GROUP segname [,segname...]	允许用户把多个定义文件段归于一个段组中	MASM 1—6
	NAME	NAME module_name	可用来指定目标文件模块名。如不使用，则汇编程序自动用源文件名作为模块名	MASM 1—6
程序的开始和结束	END	END [label]	表示源文件结束。label指定程序开始执行的起始地址；在多个模块相连接时，只有主模块需要指定label，其他模块则不必指定	MASM 1—6
	.STARTUP	.STARTUP	定义程序的入口点，并产生设置DS、SS和SP的代码。在使用.STARTUP时，END后的label将不必指定	MASM 6
	.EXIT	.EXIT [return_value]	可产生退出程序并返回操作系统的代码。return_value为返回给操作系统的代码	MASM 6
段排列	.SEQ	.SEQ	指示MASM按段文件中的次序写入目标文件在默认情况下，段排列与有.SEQ时相同	MASM 5—6
	.ALPHA	.ALPHA	指示MASM按段名的字母序次写入目标文件	MASM 5—6
	DOSSEG	DOSSEG	指示MASM用DOS所规定的方式排列段，即代码段在低地址区，然后是数据段，最后是堆栈段	MASM 5—6
数据定义及存储器分配	DB	[variable] DB operand[,...] 重复从句 repeat_count DUP(operand[,...])	定义字节变量	MASM 1—6
	BYTE	[variable] BYTE operand[,...]	定义字节变量	MASM 6
	SBYTE	[variable] SBYTE operand[,...]	定义带符号字节变量	MASM 6
	DW	[variable] DW operand[,...]	定义字变量	MASM 1—6
	WORD	[variable] WORD operand[,...]	定义字变量	MASM 6
	SWORD	[variable] SWORD operand[,...]	定义带符号字变量	MASM 6
	DD	[variable] DD operand[,...]	定义双字变量（允许单精度浮点数）	MASM 1—6

续表

类型	伪操作名	格式	说明	适用版本
类型	DWORD	[variable] DWORD operand[,…]	定义双字变量(不允许浮点数)	MASM 6
	SDWORD	[variable] SDWORD operand[,…]	定义带符号双字变量	MASM 6
	DF	[variable] DF operand[,…]	定义6字节变量，一般用来存放远指针	MASM 5-6
	FWORD	[variable] FWORD operand[,…]	定义6字节变量，一般用来存放远指针	MASM 6
	DQ	[variable] DQ operand[,…]	定义4字节变量	MASM 1-6
	QWORD	[variable] QWORD operand[,…]	定义4字节变量	MASM 6
	DT	[variable] DT operand[,…]	定义10字节变量	MASM 1-6
	TBYTE	[variable] TBYTE operand[,…]	定义10字节变量	MASM 6
	REAL4	[variable] REAL4 operand[,…]	定义4字节浮点数	MASM 6
	REAL8	[variable] REAL8 operand[,…]	定义8字节浮点数	MASM 6
	REAL10	[variable] REAL10 operand[,…]	定义10字节浮点数	MASM 6
	LABEL	name LABEL type	定义name的类型；如name为变量，则type可以是BYTE,WORD,DWORD等；如name为标号，则type可以是NEAR,FAR或PROC	MASM 1-6
	TYPEDEF	typename TYPEDEF [distance] PTR qualified_type	建立指针类型；distance可以是NEAR,NEAR16,FAR,FAR32或空。对16位段：NEAR是2字节,FAR是4字节；对32位段：NEAR是4字节,FAR是6字节。默认时，由存储模型控制；qualified_type说明类型为typename的指针所指向目标的类型，可以是BYTE,WORD,DWORD等	MASM 6
赋值	EQU	name EQU expression	赋值	MASM 1-6
	=	name = expression	赋值	MASM 1-6
	TEXTEQU	name TEXTEQU ⟨string⟩ name TEXTEQU tmname name TEXTEQU %(x+y)	赋值。与EQU等价，但EQU可用于数字表达式，TEXTEQU可用于文本串	MASM 6
对准	ORG	ORG constant_expression	地址计数器(可用$表示)设成constant_expression的值。	MASM 1-6
	EVEN	EVEN	使地址计数器成为偶数	

续表

类型	伪操作名	格 式	说 明	适用版本
基数控制	ALIGN	ALIGN boundary	使地址计数器成为 boundary 的整数倍，boundary 必须是 2 的幂	MASM 5—6
	.RADIX	.RADIX expression	改变当前基数为 expression 的值（用十进制数 2~16 表示）	MASM 1—6
文本串处理	CATSTR	newstring CATSTR string1,string2	合并串。连接 string1 和 string2 生成 newstring	MASM 5.1,6
	INSTR	pos INSTR start,string,substring	获取子串位置。获取 substring 在 string 中的位置 pos,start 为搜索的起始位置,在 pos 中以此点为 1 计	MASM 5.1,6
	SUBSTR	part SUBSTR string,startpos,length	抽取子串。从 string 中抽取起始位置为 startpos,长度为 length 的子串 part	MASM 5.1,6
结构、联合和记录	SIZESTR	strsize SIZESTR string	判定串长度。strsize 为 string 的长度	MASM 5.1,6
	STRUC 或 STRUCT	structure_name STRUC [alignment,NONUNIQUE] ⋮ structure_name ENDS 结构预置语句的格式为: variable structure_name ⟨preassignment specifications⟩	定义结构。结构中各域顺序分配不同的内存位置 alignment 可以是 1,2 或 4 NONUNIQUE 强制定子结构中的所有域名	MASM 1—6
	UNIOU	union_name UNIOU [alignment,NONUNIQUE] ⋮ union_name ENDS	定义联合。结构中所有域均共享同一内存位置	MASM 6
	RECORD	record_name RECORD fieldname:width[,…] 位感值语句: variable record_name ⟨initial_values⟩	定义记录。在字或字节内定义位模式 fieldname 为字段名,width 为该字段的位宽度	MASM 1—6
模块化程序设计 过程	PORC	porcname PORC [NEAR 或 FAR] procname ENDP	过程定义	MASM 1—6
	PROC	procname PROC [attributes field][USES registerlist][,parameter field]	过程定义	MASM 5.1,6

续表

类型	伪操作名	格式	说　明	适用版本
	:	LOCAL　vardef[,vardef]	attributes field 由以下各项组成： distance language type visibility prologue 其中 distance 可用 NEAR 或 FAR； language type 可用 C,PASCAL,BASIC,FORTRAN 或 STDCALL； visibility 可用 PRIVATE 或 PUBLIC； prologue 控制与过程的入口和出口有关代码的宏名 USES 字段允许用户指定所需保存和恢复的寄存器 parameter field 允许指定过程所用参数模型，格式为： identifier:type[,identifier:type] 其中 identifier 为参数的符号名，type 为参数模型 LOCAL 可以为局部变量申请空间，格式为： label,label;type 或 label[count]:type 第一种未指定类型，按 word 分配空间； 第二种可指定类型，如 byte,word,dword 等； 第三种可用来申请数组空间,label 为数组名,count 为元素数,type 为类型	
	procname　ENDP			
INVOKE	INVOKE　procname[,arguments]		调用过程。完成类型检查、转换参数、参数入栈、调用过程的工作，并在过程返回时清除堆栈 arguments 可以是地址表达式、立即数、寄存器对或是由 ADDR 引导的一个列表（传递地址时应在地址前加前缀 ADDR） INVOKE 要求在其前所调用过程已经由 PROC 定义或由 PROTO 建立该过程原型 INVOKE 使用 AX,EAX,DX,EDX	MASM 6
PROTO	label　PROTO　[distance][language type][parameters]		建立过程原型	MASM 6

续表

类型	伪操作名	格式	说明	适用版本
	INCLUDE	INCLUDE filename	label 为过程名，它是外部的或公用的符号。其他参数说明与 PROC 相同	MASM 1—6
			把名为 filename 的文件插入到当前 INCLUDE 语句所在位置。filename 也可以是完整的路径名	MASM 1—6
	INCLUDELIB	INCLUDELIB libname	指定目标程序要与名为 libname 的库文件相连接。libname 只能是文件名，不允许使用完整的路径名	MASM 1—6
	EXTRN 或 EXTERN	EXTRN name:type[,…]	说明在本模块中使用的外部符号 name，如为变量，则 type 可为 BYTE,WORD,DWORD 等；如为标号，则 type 可为 NEAR,FAR 或 PROC	MASM 1—6
		EXTRN [language] name:type[,…]	允许用户指定调用语言，language 可以是 C 或 PASCAL	MASM 5,1
	PUBLIC	PUBLIC symbol[,…]	说明在本模块中定义的外部符号	MASM 1—6
		PUBLIC [language] symbol[,…]	允许用户指定调用语言，language 可以是 C 或 PASCAL	MASM 5,1
	COMM	COMM [NEAR 或 FAR] var:size[;number]	定义公共变量。该变量是一个未初始化的全局变量。NEAR 或 FAR 说明对该变量的访问是偏移地址还是段地址；偏移地址；size 为变量类型，可用 BYTE,WORD,DWORD 等；number 为变量个数（默认值为 1）。COMM 必须放在数据段中	MASM 5—6
	EXTRNDEF	EXTRNDEF [language-type] name:type[,…]	说明公共和外部符号。既可以与 EXTRN 等同，又可与 PUBLIC 等同。language-type 可以是 C, PASCAL, BASIC, FORTRAN, SYSCALL 或 STDCALL。其他符号各义与 EXTRN 相同	MASM 6
宏	MACRO ENDM	macro_name MACRO [dummylist] ⋮ ENDM 宏调用：macro_name [paramlist]	宏定义	MASM 1—6
	LOCAL	LOCAL symbol[,…]	说明宏中的局部符号。MASM 将对其指定的每个 symbol 建立从 0000～0FFFFH 的符号	MASM 1—6

续表

类型	伪操作名	格式	说明	适用版本
			LOCAL 必须是宏定义中的第一个语句。有关 LOCAL 在过程定义中的作用见 PROC	
	PURGE	PURGE macro_name [,…]	删除指定的宏定义	MASM 1-6
	EXITM	EXITM	从宏(包括条件块及重复块)中退出	MASM 1-6
	EXITM	EXITM ⟨return_value⟩	从宏函数退出,并返回字符串值	MASM 6
	GOTO	GOTO label	在宏定义体中,用来跳转到 label 处目标符号的格式是 :label	MASM 6

条件	IFxx	IFxx argument
		statements_1
		ELSE
		statements_2
		ENDIF
			argument 为真,则汇编 statements_1,否则汇编 statements_2	MASM 1-6

	ELSE			
	ENDIF			
	IF	IF expression	表达式不为零则为真	
	IFE	IFE expression	表达式为零则为真	
	IF1	IF1	汇编的第一遍扫视为真(MASM 6 不支持)	
	IF2	IF2	汇编的第二遍扫视为真(MASM 6 不支持)	
	IFDEF	IFDEF symbol	符号已定义为真	
	IFNDEF	IFNDEF symbol	符号未定义为真	
	IFB	IFB ⟨argument⟩	自变量为空则为真	
	IFNB	IFNB ⟨argument⟩	自变量不空则为真	
	IFIDN	IFIDN ⟨arg_1⟩⟨arg_2⟩	arg_1 和 arg_2 相同时为真	
	IFIDNI	或 IFIDNI ⟨arg_1⟩ ⟨arg_2⟩	arg_1 和 arg_2 相同时为真,但参数比较与大小写相关	
	IFDIF	IFDIF ⟨arg_1⟩ ⟨arg_2⟩	arg_1 和 arg_2 不相同时为真	
	IFDIFI	或 IFDIFI ⟨arg_1⟩⟨arg_2⟩	arg_1 和 arg_2 不相同时为真,但参数比较与大小写无关	
	ELSExx	IFxx expression_1 … ELSEIFxx expression_2	允许用 ELSEIFxx 编写嵌套做条件汇编	MASM 5.1,6

续表

类型	伪操作名	格式	说明	适用版本
		ELSEIFxx　expression_3 ⋮ ENDIF	条件与相对应的 IFxx 相同	
	ELSEIF	ELSEIF		
	ELSEIFE	ELSEIFE		
	ELSEF1	ELSEF1	(MASM 6 不支持)	
	ELSEF2	ELSEF2	(MASM 6 不支持)	
	ELSEDEF	ELSEDEF		
	ELSENDEF	ELSENDEF		
	ELSEIFB	ELSEIFB		
	ELSEIFNB	ELSEIFNB		
	ELSEIFIDN	ELSEIFIDN		
	ELSEIFDIF	ELSEIFDIF		
重复	REPT	REPT　expression ⋮ ENDM	REPT 和 ENDM 之间的语句重复由表达式的值所指定的次数	MASM 1-6
	REPEAT	REPEAT　count ⋮ ENDM	与 REPT 相同	MASM 6
	IRP	IRP　dummy,⟨arg1,arg2,…⟩ ⋮ ENDM	重复 IRP 和 ENDM 之间的语句,每次重复用自变量表中的一项取代语句中的哑元	MASM 1-6
	FOR	FOR　dummy,⟨arg1,arg2,…⟩ ⋮ ENDM	与 IRP 相同	MASM 6
	IRPC	IRPC　dummy,string	重复 IRPC 和 ENDM 之间的语句,每次重复用字符	MASM 1-6

续表

类型	伪操作名	格　式	说　明	适用版本
		... ENDM	串中的下一个字符取代语句中的哑元	
	FORC	FORC dummy,string ... ENDM	与 IRPC 相同	MASM 6
高级语言宏	.IF .ELSEIF .ELSE .ENDIF	.IF expression_1 statements_1 .ELSEIF expression_2 statements_2 .ELSEIF expression_3 statements_3 ... .ELSE statements_n .ENDIF	生成相当于高级语言 if,then,else,endif 的语句	MASM 6
	.WHILE .ENDW	.WHILE expression statements .ENDW	生成相当于高级语言中建立 while 循环的语句	MASM 6
	.REPEAT .UNTIL	.REPEAT statements .UNTIL expression	生成相当于高级语言中建立 until 循环的语句	MASM 6
	.REPEAT .UNTILCXZ	.REPEAT statements .UNTILCXZ [expression]	与.REPEAT/.UNTIL 类似,但其不用 expression 时可用 CX 存放循环计数值;使用 expression 时,可以再增加退出循环的条件	MASM 6
	.BREAK	.BREAK .BREAK .IF expression	可提前退出.WHILE 或.REPEAT 循环。前一种不带参数的格式表示无条件退出;后一种带参数的格式给出退出循环的条件	MASM 6

续表

类型	伪操作名	格式	说明	适用版本
	.CONTINUE	.CONTINUE	控制直接跳转到.WHILE或.REPEAT循环的测试条件。第一种不带参数的格式表示无条件跳转；后一种带参数的格式给出跳转的条件	MASM 6
		.CONTINUE .IF expression		
列表格式	PAGE	PAGE lines_per_page,chars_per_line	设置列表文件每页的行数(10—255)(默认为50)和每行的字符数(60—132)(默认为80)	MASM 1—6
		PAGE	开始一个新页	MASM 1—6
		PAGE+	开始一个新行	MASM 1—6
	TITLE	TITLE text_string	指示文本串(不超过60字符)作为标题。该标题打印在列表文件的每一行上	MASM 1—6
	SUBTITLE 或 SUBTTL	SUBTITLE text_string	指示文本串(不超过60字符)作为子标题。该子标题打印在列表文件每一页的标题下面	MASM 1—6
有关源程序	.LIST	.LIST	在列表文件中开始列出包括源语句	MASM 1—6
	.XLIST	.XLIST	在列表文件中停止列出包括源语句	MASM 1—6
	.NOLIST	.NOLIST	含义与.XLIST相同	MASM 6
	.LISTALL	.LISTALL	在列表文件中列出程序的所有语句	MASM 6
有关宏	.LALL	.LALL	在列表文件中列出宏展开的所有语句	MASM 1—6
	.LISTMACROALL	.LISTMACROALL	含义与.LALL相同	MASM 6
	.XALL	.XALL	在列表文件中只列出宏展开产生代码或数据的语句(默认)	MASM 1—6
	.LISTMACRO	.LISTMACRO	含义与.XALL相同	MASM 6
	.SALL	.SALL	在列表文件中不列出宏展开的所有语句(只列出宏调用)	MASM 1—6
	.NOLISTMACRO	.NOLISTMACRO	含义与.SALL相同	MASM 6
有关条件汇编	.LFCOND	.LFCOND	在列表文件中列出条件块中的所有语句，包括测试条件为假而未被汇编的条件块	MASM 3—6
	.LISTIF	.LISTIF	含义与.LFCOND相同	MASM 6

续表

类型	伪操作名	格式	说明	适用版本
	.SFCOND	.SFCOND	在列表文件中不列出测试条件为假而未被汇编的条件块（默认）	MASM 1—6
	.NOLISTIF	.NOLISTIF	含义与.SFCOND 相同	MASM 6
	.TFCOND	.TFCOND	用来切换列出测试条件为假而未被汇编的条件块的状态。即如已设置.LISTIF 则.TFCOND 可把它转换为.NOLISTIF；反之亦然	MASM 1—6
有关交叉引用信息	.CREF	.CREF	使在交叉引用文件中出现有关其后符号的信息	MASM 1—6
	.XCREF	.XCREF	使在交叉引用文件中不出现有关其后符号的信息	MASM 1—6
	.NOCREF	.NOCREF	含义与.XCREF 相同	MASM 6
其他	%OUT	%OUT text	在汇编过程中使标准输出设备显示一行文本	MASM 1—6
	ECHO	ECHO text string	含义与%OUT 相同	MASM 6
	COMMENT	COMMENT comment-delimiter	标记成块注释（适用于单行注释）	MASM 1—6
	PUSHCONTEXT	PUSHCONTEXT context	保存 MASM 状态。context 可以是 ASSUMES,RADIX,LISTIGN,CPU 或 ALL	MASM 6
	POPCONTEXT	POPCONTEXT context	恢复 MASM 状态	MASM 6
条件出错	.ERR	.ERR	用来测试给定条件，如出错则退出汇编，并产生相应的出错信息	MASM 4—6
	.ERRE	.ERRE expression	如表达式的值为假（为零）则产生出错，出错编号为 90，出错信息为："Forced error—expression true(0)"	MASM 4—6
	.ERRNZ	.ERRNZ expression	如表达式的值为真（不为零）则产生出错，出错编号为 91，出错信息为："Forced error—expression false(not 0)"	MASM 4—6
	.ERR1	.ERR1	在第一遍扫视中出错	MASM 1—5
	.ERR2	.ERR2	在第二遍扫视中出错	MASM 1—5
	.ERRDEF	.ERRDEF symbol	如符号已定义则产生出错，出错编号为 93，出错信息为："Forced error—symbol defined"	MASM 4—6

续表

类型	伪操作名	格式	说明	适用版本
	.ERRNDEF	.ERRNDEF symbol	如符号未定义则产生错误，出错编号为92，出错信息为"Forced error—symbol not defined"	MASM 4-6
	.ERRB	.ERRB arg	如自变量为空则产生错误，出错编号为94，出错信息为"Forced error—string blank"	MASM 4-6
	.ERRNB	.ERRNB arg	如自变量不空则产生错误，出错编号为95，出错信息为"Forced error—string not blank"	MASM 4-6
	.ERRIDN 或 .ERRIDNI	.ERRIDN arg1,arg2	如arg1和arg2相同则产生错误，出错编号为96，出错信息为"Forced error—string identical"。如用.ERRIDNI参数比较与大小写无关	MASM 4-6
	.ERRDIF 或 .ERRDIFI	.ERRDIF arg1,arg2	如arg1和arg2不同则产生错误，出错编号为97，出错信息为"Forced error—string defferent"。如用.ERRDIFI参数比较与大小写相关	MASM 4-6

附表 2.2 操 作 符

类型	操作符名	格式	说明	适用版本
算术	+	expression1 + expression2	相加	MASM1-6
	-	expression1 - expression2	相减	MASM1-6
	*	expression1 * expression2	相乘	MASM1-6
	/	expression1 / expression2	相除	MASM1-6
	MOD	expression1 MOD expression2	表达式1除以表达式2所得余数	MASM1-6
	.	offset notation . fieldname	访问结构数据中的变量。offset notation为结构数据的首地址，fieldname为字段名	MASM1-6
	[ ]	expression1 [expression2]	回送expression1加上expression2的偏移地址之和	MASM1-6
逻辑和移位	AND	expression1 AND expression2	两个表达式按位与	MASM1-6
	OR	expression1 OR expression2	两个表达式按位或	MASM1-6
	XOR	expression1 XOR expression2	两个表达式按位异或	MASM1-6
	NOT	NOT expression	将表达式中的值按位求反	MASM1-6
	SHL	expression SHL numshift	将表达式左移numshift位，如numshift大于15，则结果为0	MASM1-6

续表

类型	操作符名	格式	说明	适用版本
关系	SHR	expression SHR numshift	将表达式右移 numshift 位,如 numshift 大于 15,则结果为 0	MASM1-6
	EQ	expression1 EQ expression2	如两个表达式相等则回送真(0FFFFH),否则回送假(0)	MASM1-6
	NE	expression1 NE expression2	如两个表达式不相等则回送真,否则回送假	MASM1-6
	LT	expression1 LT expression2	如 expression1 小于 expression2 则回送真,否则回送假	MASM1-6
	GT	expression1 GT expression2	如 expression1 大于 expression2 则回送真,否则回送假	MASM1-6
	LE	expression1 LE expression2	如 expression1 小于或等于 expression2 则回送真,否则回送假	MASM1-6
	GE	expression1 GE expression2	如 expression1 大于或等于 expression2 则回送真,否则回送假	MASM1-6
数值回送	TYPE	TYPE expression	回送表达式类型。如为变量,则回送其每个元素的字节数;如为标号,则回送代表其类型的数值:NEAR 为 −1,FAR 为 −2	MASM1-6
	LENGTH	LENGTH variable	回送变量所定义数据项个数。对于使用 DUP 的情况,回送 DUP 的计数值;如未使用 DUP 则回送 1	MASM1-6
	LENGTHOF	LENGTHOF variable	回送变量所定义的数据项个数	MASM1-6
	SIZE	SIZE variable	回送变量分配给变量的字节数。它是 LENGTH 值与 TYPE 值的乘积	MASM1-6
	SIZEOF	SIZEOF variable	回送变量分配给变量的字节总数	MASM1-6
	OFFSET	OFFSET variable 或 label	回送变量或标号的偏移地址值	MASM1-6
	SEG	SEG variable 或 label	回送变量或标号的段地址值	MASM1-6
	MASK	MASK fieldname	回送记录定义中表示指定字段名所占有的位置的值:其所占位为 1,其他位为 0	MASM1-6
	WIDTH	WIDTH fieldname	回送记录定义中指定字段名的位的宽度	MASM1-6
属性	PTR	type PTR expression	建立表达式的类型。expression 可以是标号、变量或指令中用各种寻址方式表示的存储单元;type 可以是 NEAR、FAR 或 PROC,也可以是 BYTE、WORD、DWORD 等	MASM1-6
	段操作符	segname 或 segreg:expression	用段名(包括段组名)或段寄存器来表示一个标号或地址表达式的段属性	MASM1-6
	SHORT	SHORT label	表示 JMP 指令中转向地址 label 的属性,指出 label 是在 JMP 的下一条指令的±127 字节的范围内	MASM1-6
	THIS	THIS type	指定与当前地址计数器相等的一个地址单元的类型。type 可以是 BYTE、WORD、DWORD 等或 PROC;也可以是 NEAR、FAR 等	MASM1-6

续表

类型	操作符名	格式	说明	适用版本
	HIGH	HIGH expression	回送字表达式的高位字节	MASM1-6
	LOW	LOW expression	回送字表达式的低位字节	MASM1-6
	HIGHWORD	HIGHWORD expression	回送双字表达式的高位字	MASM1-6
	LOWWORD	LOWWORD expression	回送双字表达式的低位字	MASM1-6
	OPATTR	OPATTR expression	以位标志返回表达式的属性。位 0 表示标号；位 1 表示变量；位 2 表示常量；位 3 表示方法直接使用返回表达式的存储器寻址方式；位 4 表示寄存器；位 5 表示使用未定义符号且未出错；位 6 表示为 SS 相关的存储器表达式；位 7 表示外部符号；位 8-11 为语言类型：其中 001 为 C；010 为 SYSCALL；011 为 STDCALL；100 为 PASCAL；101 为 FORTRAN；110 为 BASIC	MASM1-6
宏	&	&parameter	在宏定义文体中作为哑元的前缀，展开时可把 & 前后两个符号合并而形成一个符号	MASM1-6
	%	% expression	把表达式的值转换成当基数下的数，展开时用此数代哑元	MASM1-6
	!	! char	取消该字符的特殊功能	MASM1-6
	;;	;;	注释开始符。这种注释在宏展开时不出现	MASM1-6

附录 3　中断向量地址一览

附表 3.1　80x86 中断向量

I/O 地址	中断类型	功　能
0～3	0	除法溢出中断
4～7	1	单步(用于 DEBUG)
8～B	2	非屏蔽中断(NMI)
C～F	3	断点中断(用于 DEBUG)
10～13	4	溢出中断
14～17	5	打印屏幕
18～1F	6、7	保留

附表 3.2　8259 中断向量

I/O 地址	中断类型	功　能
20～23	8	定时器(IRQ0)
24～27	9	键盘(IRQ1)
28～2B	A	彩色/图形(IRQ2)
2C～2F	B	串行通信 COM2(IRQ3)
30～33	C	串行通信 COM1(IRQ4)
34～37	D	LPT2 控制器中断(IRQ5)
38～3B	E	磁盘控制器中断(IRQ6)
3C～3F	F	LPT1 控制器中断(IRQ7)

附表 3.3　BIOS 中断

I/O 地址	中断类型	功　能
40～43	10	视频显示 I/O
44～47	11	设备检验
48～4B	12	测定存储器容量
4C～4F	13	磁盘 I/O
50～53	14	RS-232 串行口 I/O
54～57	15	系统描述表指针
58～5B	16	键盘 I/O
5C～5F	17	打印机 I/O
60～63	18	ROM BASIC 入口代码
64～67	19	引导装入程序
68～6B	1A	日时钟

附表 3.4 提供给用户的中断

I/O 地址	中断类型	功 能
6C~6F	1B	Ctrl-Break 控制的软中断
70~73	1C	定时器控制的软中断

附表 3.5 参数表指针

I/O 地址	中断类型	功 能
74~77	1D	视频参数块
78~7B	1E	软盘参数块
7C~7F	1F	图形字符扩展码

附表 3.6 DOS 中断

I/O 地址	中断类型	功 能
80~83	20	DOS 中断返回
84~87	21	DOS 系统功能调用
88~8B	22	程序终止时 DOS 返回地址(用户不能直接调用)
8C~8F	23	Ctrl-Break 处理地址(用户不能直接调用)
90~93	24	严重错误处理(用户不能直接调用)
94~97	25	绝对磁盘读功能
98~9B	26	绝对磁盘写功能
9C~9F	27	终止并驻留程序
A0~A3	28	DOS 安全使用
A4~A7	29	快速写字符
A8~AB	2A	Microsoft 网络接口
B8~BB	2E	基本 SHELL 程序装入
BC~BF	2F	多路服务中断
CC~CF	33	鼠标中断
104~107	41	硬盘参数块
118~11B	46	第二硬盘参数表
11C~3FF	47~FF	BASIC 中断

附录 4 DOS 系统功能调用(INT 21H)

附表 4.1

AH[①]	功　能	调用参数	返回参数
00	程序终止(同 INT 21H)	CS=程序段前缀 PSP	
01	键盘输入并回显		AL=输入字符
02	显示输出	DL=输出字符	
03	辅助设备(COM1)输入		AL=输入数据
04	辅助设备(COM1)输出	DL=输出字符	
05	打印机输出	DL=输出字符	
06	直接控制台 I/O	DL=FF(输入) DL=字符(输出)	AL=输入字符
07	键盘输入(无回显)		AL=输入字符
08	键盘输入(无回显) 检测 Ctrl-Break 或 Ctrl-C		AL=输入字符
09	显示字符串	DS:DX=串地址 字符串以'$'结尾	
0A	键盘输入到缓冲区	DS:DX=缓冲区首址 (DS:DX)=缓冲区最大字符数	(DS:DX)=实际输入的字符数
0B	检验键盘状态		AL=00 有输入 AL0FF 无输入
0C	清除缓冲区并 请求指定的输入功能	AL=输入功能号(1,6,7,8)	
0D	磁盘复位		清除文件缓冲区
0E	指定当前默认的磁盘驱动器	DL=驱动器号 (0=A,1=B,…)	AL=系统中驱动器数
0F	打开文件 (FCB)	DS:DX=FCB 首地址	AL=00 文件找到 AL=FF 文件未找到
10	关闭文件(FCB)	DS:DX=FCB 首地址	AL=00 目录修改成功 AL=FF 目录中未找到文件
11	查找第一个目录项(FCB)	DS:DX=FCB 首地址	AL=00 找到匹配的目录项 AL=FF 未找到匹配的目录项
12	查找下一个目录项(FCB)	DS:DX=FCB 首地址 使用通配符进行目录项查找	AL=00 找到匹配的目录项 AL=FF 未找到匹配的目录项
13	删除文件(FCB)	DS:DX=FCB 首地址	AL=00 删除成功 AL=FF 文件未删除

① AH=0~2E 适用 DOSV1.0 以上；AH=2F~57 适用 DOSV2.0 以上；AH=58~62 适用 DOSV3.0 以上；AH=63~6C 适用 DOSV4.0 以上

续表

AH	功　能	调用参数	返回参数
14	顺序读文件(FCB)	DS:DX=FCB 首地址	AL＝00 读成功 　＝01 文件结束,未读到数据 　＝02 DTA 边界错误 　＝03 文件结束,记录不完整
15	顺序写文件(FCB)	DS:DX=FCB 首地址	AL＝00 写成功 　＝01 磁盘满或是只读文件 　＝02 DTA 边界错误
16	建文件 (FCB)	DS:DX=FCB 首地址	AL＝00 建文件成功 　＝FF 磁盘操作有错
17	文件改名(FCB)	DS:DX=FCB 首地址	AL＝00 文件被改名 　＝FF 文件未改名
19	取当前默认磁盘驱动器		AL＝00 默认的驱动器号 0＝A,1＝B,2＝C,…
1A	设置 DTA 地址	DS:DX=DTA 地址	
1B	取默认驱动器 FAT 信息		AL＝每簇的扇区数 DS:BX＝指向介质说明的指针 CX＝物理扇区的字节数 DX＝每磁盘簇数
1C	取指定驱动器 FAT 信息		同上
1F	取默认磁盘参数块		AL＝00 无错 　＝FF 出错 DS:BX＝磁盘参数块地址
21	随机读文件(FCB)	DS:DX=FCB 首地址	AL＝00 读成功 　＝01 文件结束 　＝02 DTA 边界错误 　＝03 读部分记录
22	随机写文件(FCB)	DS:DX=FCB 首地址	AL＝00 写成功 　＝01 磁盘满或是只读文件 　＝02 DTA 边界错误
23	测定文件大小(FCB)	DS:DX=FCB 首地址	AL＝00 成功,记录数填入 FCB 　＝FF 未找到匹配的文件
24	设置随机记录号	DS:DX=FCB 首地址	
25	设置中断向量	DS:DX=中断向量 AL=中断类型号	
26	建立程序段前缀 PSP	DX=新 PSP 段地址	
27	随机分块读(FCB)	DS:DX=FCB 首地址 CX=记录数	AL＝00 读成功 　＝01 文件结束 　＝02 DTA 边界错误 　＝03 读部分记录 CX＝读取的记录数

续表

AH	功 能	调 用 参 数	返 回 参 数
28	随机分块写(FCB)	DS:DX=FCB首地址 CX=记录数	AL=00 写成功 ＝01 磁盘满或是只读文件 ＝02 DTA 边界错误
29	分析文件名字符串(FCB)	ES:DI=FCB首址 DS:SI=ASCIZ 串 AL=分析控制标志	AL=00 标准文件 ＝01 多义文件 ＝FF 驱动器说明无效
2A	取系统日期		CX=年（1980—2099） DH=月（1～12） DL=日（1～31） AL=星期（0～6）
2B	置系统日期	CX=年（1980—2099） DH=月（1～12） DL=日（1～31）	AL＝00 成功 ＝FF 无效
2C	取系统时间		CH:CL=时:分 DH:DL=秒:1/100 秒
2D	置系统时间	CH:CL=时:分 DH:DL=秒:1/100 秒	AL＝00 成功 ＝FF 无效
2E	设置磁盘检验标志	AL＝00 关闭检验 ＝FF 打开检验	
2F	取 DTA 地址		ES:BX=DTA 首地址
30	取 DOS 版本号		AL=版本号 AH=发行号 BH=DOS 版本标志 BL:CX=序号(24 位)
31	结束并驻留	AL=返回码 DX=驻留区大小	
32	取驱动器参数块	DL=驱动器号	AL=FF 驱动器无效 DS:BX=驱动器参数块地址
33	Ctrl-Break 检测	AL＝00 取标志状态	DL＝00 关闭 Ctrl-Break 检测 ＝01 打开 Ctrl-Break 检测
35	取中断向量	AL=中断类型	ES:BX=中断向量
36	取空闲磁盘空间	DL=驱动器号 0=默认,1=A,2=B,…	成功:AX=每簇扇区数 BX=可用簇数 CX=每扇区字节数 DX=磁盘总簇数
38	置/取国别信息	AL＝00 或取当前国别信息 ＝FF 国别代码放在 BX 中 DS:DX=信息区首地址 DX=FFFF 设置国别代码	BX=国别代码 (国际电话前缀码) DS:DX=返回的信息区首址 AX=错误代码
39	建立子目录	DS:DX=ASCIZ 串地址	AX=错误码
3A	删除子目录	DS:DX=ASCIZ 串地址	AX=错误码
3B	设置目录	DS:DX=ASCIZ 串地址	AX=错误码

续表

AH	功能	调用参数	返回参数
3C	建立文件(handle)	DS:DX=ASCIZ 串地址 CX=文件属性	成功:AX=文件代号 失败:AX=错误码
3D	打开文件(handle)	DS:DX=ASCIZ 串地址 AL=访问和文件共享方式 0=读,1=写,2=读/写	成功:AX=文件代号 失败:AX=错误码
3E	关闭文件(handle)	BX=文件代号	失败:AX=错误码
3F	读文件或设备(handle)	DS:DX=数据缓冲区地址 BX=文件代号 CX=读取的字节数	成功:AX=实际读入的字节数 AX=0 已到文件尾 失败:AX=错误码
40	写文件或设备(handle)	DS:DX=数据缓冲区地址 BX=文件代号 CX=写入的字节数	成功:AX=实际写入的字节数 失败:AX=错误码
41	删除文件	DS:DX=ASCIZ 串地址	成功:AX=00 失败:AX=错误码
42	移动文件指针	BX=文件代号 CX:DX=位移量 AL=移动方式	成功:DX:AX=新指针位置 失败:AX=错误码
43	置/取文件属性	DS:DX=ASCIZ 串地址 AL=00 取文件属性 AL=01 置文件属性 CX=文件属性	成功:CX=文件属性 失败:AX=错误码

44	设备驱动程序控制	BX=文件代号
		AL=设备子功能代码(0～11H)
		  0=取设备息,
		  1=置设备信息
		  2=读字符设备
		  3=写字符设备
		  4=读块设备
		  5=写块设备
		  6=取输入状态
		  7=取输出状态,…
		BL=驱动器代码
		CX=读/写的字节数
			成功:DX=设备信息
			    AX=传送的字节数
			失败:AX=错误码

45	复制文件代号	BX=文件代号1	成功:AX=文件代号2 失败:AX=错误码
46	强行复制文件代号	BX=文件代号1 CX=文件代号2	失败:AX=错误码
47	取当前目录路径名	DL=驱动器号 DS:SI=ASCIZ 串地址 (从根目录开始的路径名)	成功: DS:SI=当前 ASCIZ 串地址 失败:AX=错误码

续表

AH	功 能	调 用 参 数	返 回 参 数
48	分配内存空间	BX＝申请内存字节数	成功： AX＝分配内存的初始段地址 失败：AX＝错误码 BX＝最大可用空间
49	释放已分配内存	ES＝内存起始段地址	失败：AX＝错误码
4A	修改内存分配	ES＝原内存起始段地址 BX＝新申请内存字节数	失败：AX＝错误码 BX＝最大可用空间
4B	装入/执行程序	DS：DX＝ASCIZ 串地址 ES：BX＝参数区首地址 AL＝00 装入并执行程序 　　＝01 装入程序，但不执行	失败：AX＝错误码
4C	带返回码终止	AL＝返回码	
4D	取返回代码		AL＝子出口代码 AH＝返回代码 　　00＝正常终止 　　01＝用 Ctrl-c 终止 　　02＝严重设备错误终止 　　03＝用功能调用 31H 终止
4E	查找第一个匹配文件	DS：DX＝ASCIZ 串地址 CX＝属性	失败：AX＝错误码
4F	查找下一个匹配文件	DTA 保留 4EH 的原始信息	失败：AX＝错误码
50	置 PSP 段地址	BX＝新 PSP 段地址	
51	取 PSP 段地址		BX＝当前运行进程的 PSP
52	取磁盘参数块		ES：BX＝参数块链表指针
53	把 BIOS 参数块（BPB）转换为 DOS 的驱动器参数块（DPB）	DS：SI＝BPB 的指针 ES：BP＝DPB 的指针	
54	取写盘后读盘的检验标志		AL＝00 检验关闭 　　＝01 检验打开
55	建立 PSP	DX＝建立 PSP 的段地址	
56	文件改名	DS：DX＝当前 ASCIZ 串地址 ES：DI＝新 ASCIZ 串地址	失败：AX＝错误码
57	置/取文件日期和时间	BX＝文件代号 AL＝00 读取日期和时间 AL＝01 设置日期和时间 （DX：CX）＝日期：时间	失败：AX＝错误码
58	取/置内存分配策略	AL＝00 取策略代码 AL＝01 置策略代码 BX＝策略代码	成功：AX＝策略代码 失败：AX＝错误码
59	取扩充错误码	BX＝00	AX＝扩充错误码 BH＝错误类型 BL＝建议的操作 CH＝出错设备代码

续表

AH	功　能	调　用　参　数	返　回　参　数
5A	建立临时文件	CX=文件属性 DS:DX=ASCIZ 串（以\结束）地址	成功:AX=文件代号 DS:DX=ASCIZ 串地址 失败:AX=错误代码
5B	建立新文件	CX=文件属性 DS:DX=ASCIZ 串地址	成功:AX=文件代号 失败:AX=错误代码
5C	锁定文件存取	AL=00 锁定文件指定的区域 　=01 开锁 BX=文件代号 CX:DX=文件区域偏移值 SI:DI=文件区域的长度	失败:AX=错误代码
5D	取/置严重错误标志的地址	AL=06 取严重错误标志地址 AL=0A 置 ERROR 结构指针	DS:SI=严重错误标志的地址
60	扩展为全路径名	DS:SI=ASCIZ 串的地址 ES:DI=工作缓冲区地址	失败:AX=错误代码
62	取程序段前缀地址		BX=PSP 地址
68	刷新缓冲区数据到磁盘	AL=文件代号	失败:AX=错误代码
6C	扩充的文件打开/建立	AL=访问权限 BX=打开方式 CX=文件属性 DS:SI=ASCIZ 串地址	成功:AX=文件代号 CX=采取的动作 失败:AX=错误代码

附录 5 BIOS 功能调用

附表 5.1

INT	AH	功能	调用参数	返回参数
10	0	设置显示方式	AL＝00 40×25 黑白文本,16 级灰度 　＝01 40×25 16 色文本 　＝02 80×25 黑白文本,16 级灰度 　＝03 80×25 16 色文本 　＝04 320×200 4 色图形 　＝05 320×200 黑白图形,4 级灰度 　＝06 640×200 黑白图形 　＝07 80×25 黑白文本 　＝08 160×200 16 色图形（MCGA） 　＝09 320×200 16 色图形（MCGA） 　＝0A 640×200 4 色图形（MCGA） 　＝0D 320×200 16 色图形（EGA/VGA） 　＝0E 640×200 16 色图形（EGA/VGA） 　＝0F 640×350 单色图形（EGA/VGA） 　＝10 640×350 16 色图形（EGA/VGA） 　＝11 640×480 黑白图形（VGA） 　＝12 640×480 16 色图形（VGA） 　＝13 320×200 256 色图形（VGA）	
10	1	置光标类型	$(CH)_{0-3}$＝光标起始行 $(CL)_{0-3}$＝光标结束行	
10	2	置光标位置	BH＝页号 DH/DL＝行/列	
10	3	读光标位置	BH＝页号	CH＝光标起始行 CL＝光标结束行 DH/DL＝行/列
10	4	读光笔位置		AX＝0 光笔未触发 　＝1 光笔触发 CH/BX＝像素行/列 DH/DL＝字符行/列
10	5	置当前显示页	AL＝页号	
10	6	屏幕初始化 或上卷	AL＝0 初始化窗口 AL＝上卷行数 BH＝卷入行属性 CH/CL＝左上角行/列号 DH/DL＝右上角行/列	

续表

INT	AH	功能	调用参数	返回参数
10	7	屏幕初始化或下卷	AL=0 初始化窗口 AL=下卷行数 BH=卷入行属性 CH/CL=左上角行/列号 DH/DL=右上角行/列	
10	8	读光标位置的字符和属性	BH=显示页	AH/AL=字符/属性
10	9	在光标位置显示字符和属性	BH=显示页 AL/BL=字符/属性 CX=字符重复次数	
10	A	在光标位置显示字符	BH=显示页 AL=字符 CX=字符重复次数	
10	B	置彩色调色板	BH=彩色调色板 ID BL=和 ID 配套使用的颜色	
10	C	写像素	AL=颜色值 BH=页号 DX/CX=像素行/列	
10	D	读像素	BH=页号 DX/CX=像素行/列	AL=像素的颜色值
10	E	显示字符 （光标前移）	AL=字符 BH=页号 BL=前景色	
10	F	取当前显示方式		BH=页号 AH=字符列数 AL=显示方式
10	10	置调色板寄存器（EGA/VGA）	AL=0,BL=调色板号,BH=颜色值	
10	11	装入字符发生器（EGA/VGA）	AL=0~4 全部或部分装入字符点阵集 AL=20~24 置图形方式显示字符集 AL=30 读当前字符集信息	ES:BP=字符集位置
10	12	返回当前适配器设置的信息（EGA/VGA）	BL=10H（子功能）	BH=0 单色方式 =1 彩色方式 BL=VRAM 容量 （0=64K,1=128K,…） CH=特征位设置 CL=EGA 的开关设置
10	13	显示字符串	ES:BP=字符串地址 AL=写方式(0~3) CX=字符串长度 DH/DL=起始行/列 BH/BL=页号/属性	

续表

INT	AH	功 能	调 用 参 数	返 回 参 数
11		取设备信息		AX＝返回值（位映像） 0＝设备未安装 1＝设备未安装
12		取内存容量		AX＝字节数（KB）
13	0	磁盘复位	DL＝驱动器号 （00,01为软盘，80h,81h,…为硬盘）	失败：AH＝错误码
13	1	读磁盘驱动器状态		AH＝状态字节
13	2	读磁盘扇区	AL＝扇区数 $(CL)_{6,7}(CH)_{0\sim7}$＝磁道号 $(CL)_{0\sim5}$＝扇区号 DH/DL＝磁头号/驱动器号 ES:BX＝数据缓冲区地址	读成功： AH＝0 AL＝读取的扇区数 读失败： AH＝错误码
13	3	写磁盘扇区	同上	写成功： AH＝0 AL＝写入的扇区数 写失败： AH＝错误码
13	4	检验磁盘扇区	AL＝扇区数 $(CL)_{6,7}(CH)_{0\sim7}$＝磁道号 $(CL)_{0\sim5}$＝扇区号 DH/DL＝磁头号/驱动器号	成功：AH＝0 　　　AL＝检验的扇区数 失败：AH＝错误码
13	5	格式化盘磁道	AL＝扇区数 $(CL)_{6,7}(CH)_{0\sim7}$＝磁道号 $(CL)_{0\sim5}$＝扇区号 DH/DL＝磁头号/驱动器号 ES:BX＝格式化参数表指针	成功：AH＝0 失败：AH＝错误码
14	0	初始化串行口	AL＝初始化参数 DX＝串行口号	AH＝通信口状态 AL＝调制解调器状态
14	1	向通信口写字符	AL＝字符 DX＝通信口号	写成功：$(AH)_7$＝0 写失败：$(AH)_7$＝1 $(AH)_{0\sim6}$＝通信口状态
14	2	从通信口读字符	DX＝通信口号	读成功：$(AH)_7$＝0 　　（AL）＝字符 读失败：$(AH)_7$＝1
14	3	取通信口状态	DX＝通信口号	AH＝通信口状态 AL＝调制解调器状态
14	4	初始化扩展COM		
14	5	扩展COM控制		
15	0	启动盒式磁带机		
15	1	停止盒式磁带机		

续表

INT	AH	功能	调用参数	返回参数
15	2	磁带分块读	ES:BX=数据传输区地址 CX=字节数	AH＝状态字节 ＝00 读成功 ＝01 余检验错 ＝02 无数据传输 ＝04 无引导 ＝80 非法命令
15	3	磁带分块读	DS:BX=数据传输区地址 CX=字节数	AH＝状态字节 （同上）
16	0	从键盘读字符		AL＝字符码 AH＝扫描码
16	1	取键盘缓冲区 状态		ZF＝0 AL＝字符码 　　　AH＝扫描码 ZF＝1 缓冲区无按键 　　　等待
16	2	取键盘标志字节		AL＝键盘标志字节
17	0	打印字符 回送状态字节	AL＝字符 DX＝打印机号	AH＝打印机状态字节
17	1	初始化打印机 回送状态字节	DX＝打印机号	AH＝打印机状态字节
17	2	取打印机状态	DX＝打印机号	AH＝打印机状态字节
18		ROM BASIC 语言		
19		引导装入程序		
1A	0	读时钟		CH:CL＝时:分 DH:DL＝秒:1/100 秒
1A	1	置时钟	CH:CL＝时:分 DH:DL＝秒:1/100 秒	
1A	6	置报警时间	CH:CL＝时:分（BCD） DH:DL＝秒:1/100 秒（BCD）	
1A	7	清除报警		
33	00	鼠标复位	AL＝00	BX＝鼠标的键数
33	00	显示鼠标光标	AL＝01	显示鼠标光标
33	00	隐藏鼠标光标	AL＝02	隐藏鼠标光标
33	00	读鼠标状态	AL＝03	BX＝键状态 CX/DX＝鼠标水平/垂直位置
33	00	设置鼠标位置	AL＝04 CX/DX＝鼠标水平/垂直位置	
33	00	设置图形光标	AL＝09 BX/CX＝鼠标水平/垂直中心 ES:DX＝16×16 光标映像地址	安装了新的图形光标

续表

INT	AH	功　能	调　用　参　数	返　回　参　数
33	00	设置文本光标	AL=0A BX=光标类型 CX=像素位掩码或起始的扫描线 DX=光标掩码或结束的扫描线	设置的文本光标
33	00	读移动计数器	AL=0B	CX/DX=鼠标水平/垂直距离
33	00	设置中断子程序	AL=0C　CX=中断掩码 ES:DX=中断服务程序的地址	

参 考 文 献

［1］ Barry B. Brey：“8086/8088，80286，80386，AND 80486 assembly Language Programming" Macmillan Publishing Compang USA. 1994.
［2］ Muhammad Ali Mazidi and Janice Grillispie Mazidi："The 80x86 IBM PC and Compatible Computer" Second edition, Prentice Hall Inc. USA 1998.
［3］ Peter Abel："IBM PC Assembly Language and Programming" Fourth edition, Prentice Hall, Inc. USA. 1998.
［4］ Yu-cheng Liu Glenn A. Gibson："Microcomputer System：The 8086/8088 Family—Architecture, Programming, and Design" Prentice Hall, Inc. ,1984.
［5］ Leo J. Scanlon："Assembly Language Programming with the IBM PC AT" Brady CommunicaYion Company, Inc., 1986.
［6］ David C. Willen and Jeffrey I. Krantz："8088 Assembler Language Programming：The IBM PC" Howard W. Sams & Co., Inc., 1983.
［7］ Donna N. Tabler："IBM PC Assembly Language" John Wiley & Sons, Inc., 1985.
［8］ Thomas P. Skinner："An Introduction to Assembly Language Programming for the 8086 Family", John Wiley & Sons, Inc., 1985.
［9］ KIP R. IRVINE, Assembly Language for the IBM-PC(Second Edition), Prentice Hall, Inc, 1993.
［10］ Julio Sanchez and Maria P. Canton, Programming Solutions Handbook for IBM Microcomputers, McHraw-Hill, Inc, 1991.
［11］ Barry B. Brey, Programming the 80286, 80386, 80486, and Pentium-Based Personal Computer, Prentice Hall, Inc, 1996.
［12］ Pentium™ Family User's Manual , Intel Corporation 1994.
［13］ Intel 486™ DX2 Microprocessor Data Book, Intel Corporation 1992.
［14］ 东阳生等编著："宏汇编语言 MASM6 实用大全"，科学出版社,1993。
［15］ Ray Duncan 著,贺志强、李昌译："DOS 磁盘操作系统—高级程序员指南"，中国科学院计算所新技术发展公司,1988.
［16］ 田学锋,周豫滨译："80386/80486 编程指南"，电子工业出版社,1994.
［17］ 求伯君主编："新编深入 DOS 编程"，学苑出版社,1994.